D1207108

THE ASTRONOMICAL SCRAPBOOK

Skywatchers, Pioneers, and Seekers in Astronomy

The Astronomical Scrapbook

Skywatchers, Pioneers, and Seekers in Astronomy

by

JOSEPH ASHBROOK

Edited by

LEIF J. ROBINSON

Introduction *by*

OWEN GINGERICH

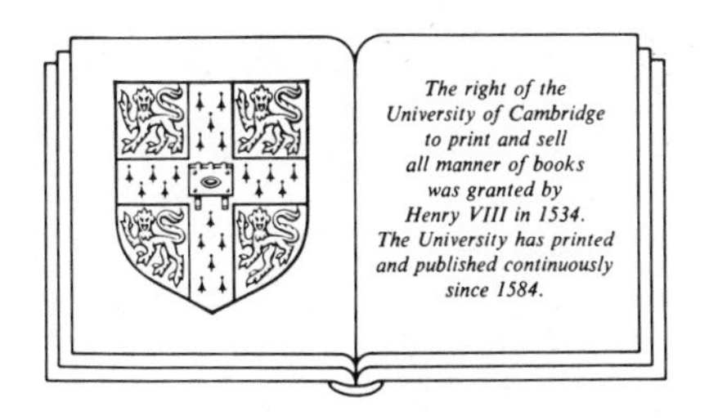

CAMBRIDGE UNIVERSITY PRESS

Cambridge

London New York New Rochelle

Melbourne Sydney

&

SKY PUBLISHING CORPORATION

Cambridge, Massachusetts

Published by the Press Syndicate of the University of Cambridge
The Pitt Building, Trumpington Street, Cambridge CB2 1RP
32 East 57th Street, New York, NY 10022, USA
10 Stamford Road, Oakleigh, Melbourne 3166, Australia

And by Sky Publishing Corporation
49 Bay State Road
Cambridge, Massachusetts 02238

First published 1984

Printed in the United States of America

Library of Congress Cataloging in Publication Data
Ashbrook, Joseph, d. 1980.
The astronomical scrapbook.
Bibliography: p.
Includes index.
1. Astronomy—Addresses, essays, lectures.
I. Robinson, Leif J. II. Title.
QB51.A77 1984 520 84-12036
ISBN 0-521-30045-2
ISBN 0-933346-24-7 (Sky Pub. Corp.)

Int[roduction]

JOSEPH ASHBROOK'S NAME first came to my attention through a vividly memorable incident in the early 1950's. One of Annie Jump Cannon's monumental but lesser known tasks at Harvard had been the compilation of an enormous card catalogue of variable star observations, but in a general observatory cleanup it was decided that the cards were no longer of use. "Let's send it to Yale for Joe Ashbrook, since he knows the entire variable star catalogue by heart." This remark was all the more impressive, coming as it did from Cecilia Payne-Gaposchkin, herself a veritable walking encyclopedia of star lore.

But Joe Ashbrook did not stay long at Yale. Quick of mind but slow of tongue, he found the editorial responsibilities of *Sky & Telescope* more challenging and congenial than classroom teaching. For the small *Sky & Telescope* operation, the hiring of an astronomy PhD was a bold step forward. Joe Ashbrook brought his encyclopedic expertise and his stabilizing authority to Cambridge, where he acted as anchor man for the accelerating growth of the magazine.

His profound interest in the history of astronomy was not so much from the perspective of an historian as that of an astronomer who simply had read all the old journals and remembered most of them. Hence he constituted a one-man resource for historical anecdotes, biographical details, and unusual illustrations. From this wellspring of lore flowed his long "Astronomical Scrapbook" series in *Sky & Telescope*. With astonishing regularity this column appeared in the magazine every other month, rarely missing a beat from August, 1954, until his untimely death in August, 1980.

Connoisseurs of the "Astronomical Scrapbook" (myself included) could never quite guess what Joe Ashbrook would come up with next. To be sure, there were periodic columns on selenography, since the ongoing saga of Moon-mapping was one of his favorite topics. From time to time he would rescue from oblivion some obscure amateur astronomer who had once risen to fleeting fame through some unexpected and perhaps everlasting contribution to the science. Especially to be savored were those accounts by Joseph

Ashbrook, detective, in which some half-forgotten mystery of the last century would be exhumed and tidily solved.

At their best, Ashbrook's columns were carefully sculpted cameos, exquisite trifles whose slight scale concealed their craftsmanship. Even before Joe's series came to an end, his admirers coveted the chance to have them brought together in an anthology. Fortunately, this desire has not been thwarted. At last the book is in hand, a collection of his best and well worth waiting for. *"Si monumentum requiris:" ecce!*

Cambridge, March, 1984 — OWEN GINGERICH

Editor's Preface

SEVERAL MONTHS BEFORE HIS SUDDEN DEATH in August, 1980, Joseph Ashbrook and I began to prepare an anthology of his "Astronomical Scrapbook" articles from *Sky & Telescope* magazine. Joe had published these often little-known asides to the history of astronomy nearly every other month for 26 years. Their preparation was a special activity for Joe, who loved to ferret out astronomical curiosities among the crumbling and forgotten literature at Harvard Observatory's library.

Work on this anthology stopped when Joe's duties as editor of *Sky & Telescope* became mine. But early in 1984 the project was reborn. With critical advice from astronomical historian Owen Gingerich and Alan MacRobert, I combined into this volume's 83 chapters 91 of the 159 Scrapbooks Joe prepared. In some cases I have added new information that became available after the original article was published. Also, with the invaluable assistance of Stephen J. O'Meara, I've tried to augment the text with unusual and interesting illustrations.

The most difficult chore was to decide which Scrapbooks should be omitted. Some were left out because they didn't quite measure up to Joe's unusually high standards. Others became anachronisms, such as his "Review of a Yet-Unwritten Moon Book" (December, 1966). In a few cases his historical questions have been answered. "Hertzsprung's Enigmatic Object" (December, 1967), which he thought from the evidence then available to be a probable comet, was shown in 1983 to be a plate defect. And, in a couple of instances, Joe was just plain wrong. In his zeal to remind contemporary technocrats of historical observations and the care with which they were made, Joe "picked up the story by the wrong end," as he would say. "What Is the True Shape of the Sun?" (October, 1967) is a case in point.

Nevertheless, every one of Joe's articles resounds with Ashbrookian curiosity, integrity, and scholarship. Joe was a most eclectic person – he remembered almost everything he ever read and took any opportunity to calculate something. He also had a particular fondness for misfits, ill-conceived projects, and far-away places. Such preoccupations punctuate this collection.

Joe did not leave a dedication. I'll guess that it would have been to his wife, Martha, and his daughters Cathy, Betsy, Jane, and Susan.

Cambridge, March, 1984 LEIF J. ROBINSON

Contents

Phenomena of the Earth, Moon, and planets

Studies and students of the Moon

Planets and other Solar System objects

Stars and stellar systems

Star atlases and other publications

STAR-CROSSED LIVES

1. Tycho Brahe's nose

LARGER THAN LIFE SIZE accurately describes Tycho Brahe, both as to his present-day scientific reputation and as to the physical appearance he presented to his contemporaries.

In the 16th century, a profound change was beginning in western ideas about celestial bodies and their relation to the world. Within a few generations the traditional concept of a small, Earth-centered cosmos was to give way to the realization that our globe is one of several planets swinging in elliptical orbits around the Sun, in a vast space thinly sprinkled with distant stars.

Among the many thinkers who brought about this change, Tycho Brahe will always be remembered for three outstanding observational contributions. His thorough studies of the brilliant new star of 1572 refuted the orthodox doctrine of the immutability and perfection of the celestial spheres. His work on the great comet of 1577 demonstrated that such objects move through interplanetary space, instead of being mere atmospheric phenomena as Aristotle had taught. And most important of all, his thousands of accurate positional measurements of heavenly bodies were the material from which Kepler was to deduce the laws of planetary motions.

But the big red-bearded Danish nobleman preferred to think of himself as an aristocrat rather than an astronomer. His father had a considerable estate in Scania, the southern province of present-day Sweden that was Danish until 1658. There is a tradition that Tycho always wore court costume while observing. The most obvious explanation is not that Tycho had some unusual reverence for astronomy, but that he sought to escape being branded with the low social status of scientific practitioners in 16th-century Europe.

This proud and quarrelsome man was the feudal lord of the island of Hven, assigned to him in 1576 by King Frederick II of Denmark. Here Tycho lived in state for two decades amid his retainers and scientific aides, in a combined manor house and observatory which also had a private jail. He was an oppressive landlord to the farmers on his island, who finally rose against him. To many of his contemporaries he must have seemed more a hot-tempered country squire than one of the greatest scientists of his time.

His temper gave rise to a celebrated mishap that was described in P. Gassendi's biography of 1654 and repeated in J. L. E. Dreyer's *Tycho Brahe* (1890). As a young student at Rostock, Germany, in 1566, Tycho fell out with another Danish nobleman, named Manderup Parsbjerg. They first quarreled at a dance on December 10th (four days before Tycho's 20th birthday), renewed their dispute at a Christmas party, and two nights later fought with swords. In the duel Tycho lost part of his nose. To hide the disfigurement he replaced the lost piece by a device of gold and silver. Gassendi was told by a pupil of the Danish astronomer that the latter always carried in his pocket a box of ointment, which he frequently rubbed on his nose. Dreyer adds: "The various portraits which we possess of Tycho show distinctly that there was something strange about the appearance of his nose, but one cannot see with certainty whether it was the tip or the bridge that was injured, though it seems to be the latter."

The story that Tycho lost his nose in a duel and thereafter wore an artificial one became widely spread and acquired various apocryphal details. One little indication of its diffusion in the early years of the 19th century is a stanza in the famous Astronomer's Drinking Song, which Augustus de Morgan quoted in his *Budget of Paradoxes:*

The noble Tycho placed the stars,
Each in its due location;
He lost his nose by spite of Mars,
But that was no privation:
Had he but lost his mouth, I grant
He would have felt dismay, sir,
Bless you! he *knew what he should want*
To drink his bottle a day, sir!

When Dreyer published his standard life of Tycho in 1890, the episode seemed hardly better founded than George Washington's cherry tree or Isaac Newton's apple. But belatedly fresh evidence appeared. On June 24, 1901, 300 years after Tycho Brahe's death, the city officials of Prague opened his marble tomb in the Tyn church, in the presence of Prof. A. Schrutz and Dr. H. Matiegka of the university medical school. There were two skeletons inside, one of a man clad in a red silk robe, the other of a woman with crossed hands.

The appearance of Tycho's nose seems a bit strange in this little-known portrait that resides at Gavno Castle, Naestved, Denmark. It is one of two depicting this famous Danish astronomer discovered in the castle's collection by Fletcher G. Watson during a visit in 1976. According to the inscription, the painting was done when Tycho was 50, presumably in 1596. This portrait is recorded as a copy, and the name of the original artist is unknown. Courtesy Baron Otto Reedtz-Thott.

A thorough scientific examination of the former verified that it was indeed the remains of Tycho. The measurements of height, the massive bones, and the great mustaches and beard agreed with descriptions and portraits. There was a narrow, curved mark on the skull at the upper end of the nasal opening, and this opening was rimmed by a bright green stain of copper, except for its lowest third.

Dr. Matiegka's account identifies the curved mark as unquestionably the reaction to a wound and shows that the green stain came from some object that remained in postmortem contact with bone for a long time. "This object that contained copper was," says Dr. Matiegka, "in all likelihood the oft-mentioned prosthetic device, the artificial nose. Although it may have been described as silver or gold, it certainly had a high copper content. Tycho Brahe, who was expert in casting metal, would naturally spend much time and effort in obtaining a composition that would be both durable and nearly the color of his skin, so it is understandable that he chose a copper alloy."

2. The cosmic vision of Robert Burton

GALILEO BUILT HIS FIRST TELESCOPES IN 1609, and the following year announced to the world his epoch-making observations of the satellites of Jupiter, the phases of Venus, the rugged lunar surface, and the resolution of the Milky Way into hosts of faint stars. Some of the earliest reports in English literature of these revolutionary discoveries can be found in Robert Burton's oddly titled *Anatomy of Melancholy,* together with much other astronomical material old and new.

Burton, who was born in 1577 and died in 1640, two years before Galileo, was primarily not an astronomer but an Oxford don of the widest intellectual interests. He spent most of his "silent, sedentary, solitary, private life" in omnivorous reading within one of Europe's best libraries. The distillate of this reading was the *Anatomy of Melancholy*. In this strange but fascinating book, says a modern author, Burton "explored the human mind with the aid of all the learning of the classical world. He is a freebooting scholar, who finds his prizes all equally worthwhile, and all equally relevant to the great purpose he has in mind. He examines the disease of melancholy, Hamlet's disease. . . . Few volumes in English are so full of curiosities, and this eccentric writer has given pleasure to discerning minds in all the centuries since his death."

The astronomical allusions in the *Anatomy* occur chiefly in a long section wherein Burton enthusiastically proposes study of the heavens and Earth as a powerful remedy for care-ridden minds. This passage, entitled "Digression of Air," depicts an imaginary intended journey through space, giving its author the opportunity to list all manner of marvels and scientific puzzles.

On the outward leg of his cosmic voyage, Burton proposes to fly over the arctic regions to discover the cause of terrestrial magnetism. "I will first see . . . whether there be . . . a great rock of Loadstones, which may cause the needle in the Compass still to bend that

Galileo Galilei (1564–1642). Courtesy American Museum of Natural History.

way, and what should be the true cause of the variation of the Compass. Is it a magnetical rock, or the Pole-star, as Cardan will; or some other star in the Bear, as Marsilius Ficinus; or a magnetical meridian, as Maurolicus; or situated in a vein of the earth, as Agricola; or the nearness of the next Continent, as Cabeus will; or some other cause . . . ; why at the Azores it looks directly North, otherwise not?"

Other geographical problems that Burton lists for investigation are whether Hudson's new-found Bay in fact exists, and whether the latest maps are correct in informing us that "California is not a Cape, but an Island." There are also controversies to settle about the Earth's interior: "Franciscus Ribera will have Hell a material & local fire in the Center of the earth, 200 Italian miles in diameter. . . . But Lessius will have this local Hell far less, one Dutch mile in Diameter, all filled with fire and brimstone; because, as he there demonstrates, that space cubically multiplied will make a Sphere able to hold eight hundred thousand millions of damned bodies (allowing each body six foot square) which will abundantly suffice."

In Burton's lifetime Copernican ideas were still fighting for acceptance, and it is therefore natural that the *Anatomy* contains at great length the arguments for an Earth-centered solar system as well as the newer heliocentric system. Burton is up to date enough to dismiss as "absurd and ridiculous" the notion that the vast machinery of concentric spheres, epicycles, and eccentrics has a material existence. At best he regarded them as concepts arbitrarily invented to help describe mathematically the motions of the Moon, Sun, and planets.

The Medicean Stars, as Burton calls the four bright satellites of Jupiter, are referred to repeatedly. The Oxford scholar had seen them himself, "by the help of a glass eight feet long" he tells us in a footnote. But he also mentions some very early erroneous discoveries of satellites to other planets. "Those several Planets have their several Moons about them, as the earth hath her's, as Galileo hath already evinced by his glasses: four about Jupiter, two about Saturn (though Sitius the Florentine, Fortunius Licetus, and Julius Caesar la Galla cavil at it): yet Kepler, the Emperor's Mathematician, confirms out of his experience that he saw as much by the same help, & more about Mars, Venus; and the rest they hope to find out, peradventure even amongst the fixed stars, which Brunus & Brutius have already averred."

The "two [Moons] about Saturn" is an allusion to Galileo's first telescopic observations of the planet in 1610, which led him to announce Saturn as being triple. Not until nearly two generations later was Saturn's ring recognized as such; Galileo's moons were merely imperfect views of it. The first bona-fide Saturnian satellite was Titan, discovered by Christiaan Huygens in 1655. The early suggestion of planetary systems of other stars is worth notice.

Selenography is mentioned in this book published only 12 years after Galileo took his first look at the Moon through a telescope. Burton notes that Galileo, Kepler, and others "find by their glasses those spots on the face of the moon, the brighter parts are Earth, the dusky Sea, which Thales, Plutarch, and Pythagoras formerly taught: and manifestly discern Hills and Dales, and such like concavities. . . ."

Sunspots raised much controversy among the earliest telescopic observers, and some even regarded them as inferior planets (the Bourbonian Stars) seen in transit upon the solar disk. Burton tells us about this: "One saith the Sun stands, another he moves; a third comes in, taking them all at rebound, and, lest there should any paradox be wanting, he finds certain spots and clouds in the Sun by the help of glasses, which multiply (saith Kepler) a thing seen a thousand times bigger in plane, and makes it come 32 times nearer to the eye of the beholder: but see the demonstration of this glass in Tarde, by means of which the Sun must turn round upon his Center, or they about the Sun. Fabricius puts only three, and those in the Sun: Apelles 15, and those without the Sun, floating like the Cyanean Isles in the Euxine Sea. Tarde the Frenchman hath observed 33, and those neither spots nor clouds, as Galileo supposeth, but Planets concentrick with the Sun, and not far from him, with regular motions. Christopher Scheiner . . . divides them into Spots and Torchlets, and will have them fixed on the Sun's surface, and to absolve their periodical and regular motion in 27 or 28 days, holding withal the rotation of the Sun upon his Center: and are all so confident, that they have made schemes and tables of their motions. The Hollander [Fabricius], in his controversy with Apelles, censures all: and thus they disagree amongst themselves, old and new, irreconcileable in their opinions. . . ."

Burton clearly recognizes that if the orbital motion of the Earth is admitted, the Earth becomes one of many planets. Thus the others may be inhabited. Kepler is one of the many authors he cites in this connection. "For the Planets, he yields them to be inhabited, he doubts of the stars: and so doth Tycho in his Astronomical Epistles, out of a consideration of their vastity and greatness, break out into some such like speeches, that he will never believe those great and huge bodies were made to no other use than . . . to illuminate the earth. But who shall dwell in these vast bodies, Earths, Worlds, if they be inhabited? rational creatures? as Kepler demands, or have they souls to be saved? or do they inhabit a better part of the World than we do? Are we or they Lords of the World?"

The last few lines may seem familiar; H. G. Wells adopted them as the motto for his famous novel, *The War of the Worlds*.

3. Hans Egede and early astronomy in Greenland

ONE OF THE VERY FIRST astronomical observatories in North America was, surprisingly, at Godthaab, in Greenland. There during the 1780's a Danish astronomer named Linge systematically observed eclipses of Jupiter's satellites, with the aid of a primitive nonachromatic refractor eight feet long. This short-lived establishment at Godthaab was one of a series of observing stations maintained by the Danish government in its overseas possessions; there were others in Iceland and at Tranquebar, in India.

Godthaab had been founded in 1728 by Hans Egede, whose name has been attached to a lunar crater near Aristoteles. He was born in Norway in 1686 when it was part of Denmark. Educated for the church, he was given a pension by King Frederick IV and was sent to Greenland as a missionary to the Eskimos. Landing there with his wife and four children on July 3, 1721, he built a house of stone and earth on Haabets Oe (Hope Island), but later Egede came to Godthaab. Here he was an active missionary until he finally returned to Copenhagen in 1736.

The *Description of Greenland* that Egede published in 1757, a year before his death, contains a long and interesting account of the astronomical knowledge and myths of the Greenland Eskimos. It has special value in dating back to a time before contact with Europeans had made any marked changes in Eskimo culture, so we can recapture their original lore.

To the Eskimos, all the heavenly bodies were originally persons or animals who became transferred to the sky. Anningait, the Moon, was a man who chased his sister Malina, the Sun, into the sky. Now and then the Moon returns to earth in human form to seek amorous adventure. This is why, said Egede, young Eskimo girls are afraid to stare at the Moon. During solar eclipses no man goes outdoors, and during lunar eclipses women stay at home.

As might be expected for an Arctic people who have no dark

nights in summer, the Eskimos' constellation lore was limited to the stars of other seasons. To them Taurus was Kellukturset, a pack of hounds surrounding a bear; this was the star group by which they told the hours of the night. The star Aldebaran was called Nennerroak. Another winter asterism was Siektut, the three belt stars of Orion. According to legend these were three Greenlanders who were lost at sea while seal fishing.

Egede tells us that Nelleraglek was the name for Canis Major, but this attribution may be a slip, for few of the bright stars of this constellation rise in Godthaab's latitude of 64° north. Perhaps the star Sirius was meant, since it can attain an altitude of nine degrees above the southern horizon there. To the Eskimos, Gemini and Auriga formed a single vast constellation, Killaub Kuttuk, the breastbone of heaven. Among the dwellers around Godthaab, Ursa Major was Tugto, the reindeer. Farther north, however, along the shores of Disko Bay, it was called Asselluit, after a wooden implement to which a seal hunter ties the end of his harpoon line.

Egede tells us about the timekeeping of the Eskimos: "They have no calendar or almanacks, nor do they compute time by weeks or years, but only by months; beginning their computation from the Sun's first rising above their horizon in winter; from whence they tell the month, to know exactly the season, in which every sort of fishes, sea animals, or birds seek the land; according to which they order their business."

To this should be added a second account, written nearly two centuries later by the Danish explorer Knud Rasmussen, who in 1903–1904 visited the nomadic Eskimos living in the extreme northwest of Greenland around latitude 77° north. A young man named Maisanguaq told him: "When the constellation of the Great Bear is seen at dawn men are filled with great delight; for then it will not be long till the light comes again. And when at last the Sun comes men call out: 'Joy! joy! the Great Warmer has come; soon we shall be able to seek the sunny side!' And then comes the time when people build sheltering walls of snow and gather round a man's meat at great banquets."

4. The diary of an early American amateur astronomer

When the name of Judge Samüel Sewall (1652–1730) of Boston is remembered, it is more likely to be in connection with witchcraft than with astronomy. He was one of the seven justices named in 1692 by Governor William Phipps of the Massachusetts Bay Colony to try the accused Salem witches. That epidemic of terror led to 19 executions by hanging and one by pressing to death (under heavy stones, for refusal to plead). To his lasting credit, Judge Sewall later acknowledged the wrongfulness of his sentences by doing public penance in 1697 in Boston's Old South Church.

At the close of the 17th century, Sewall was one of the leading citizens of Boston, then a town of perhaps 5,000 souls. One of seven members of the Harvard graduating class of 1671, he had been trained for the ministry. But he soon turned to trade. Sewall married the daughter of the wealthy merchant John Hull, mintmaster of the colony, and (as legend has it) received as a dowry his bride's weight in newly coined silver pine-tree shillings. Later he became the owner of extensive lands, a member of the governor's council, and an overseer of Harvard College, as well as serving 36 years as a magistrate.

The interest that astronomy held for Sewall is made clear from the private diary that he kept for more than half a century. This diary is important to historians for its vivid and detailed descriptions of persons, events, and customs. To general readers it is delightful reading for its intimate and somewhat naive revelation of the personality of a distinguished man; it is also often marked with unconscious humor.

Although by Sewall's time much of Europe had adopted the Gregorian calendar, New England still used the Julian one. In his diary, the Old Style (O. S.) dates up to February 28, 1700, are 10 days behind the Gregorian dates, and thereafter 11 days. The flavor of the diary can be sampled in the following passage from the year 1702,

Among the earliest settlers in New England who can be called amateur astronomers was Samuel Sewall, who was born in England in 1652 and emigrated to America at the age of nine. He became one of the leading citizens of the Massachusetts Bay Colony and served as Chief Justice of its Superior Court. This 1730 portrait by Smibert belongs to the Massachusetts Historical Society.

describing a comet: "Thorsday, Febr. 19 [O. S.]. Mr. I. Mather preached from Rev. 22.16 – bright and morning Star. Mention'd Sign in the Heaven, and in the Evening following I saw a large Cometical Blaze, something fine and dim, pointing from the Westward, a little below Orion.

"Febr. 21 [O. S.]. Capt. Timo. Clark tells me that a Line drawn to the Comet strikes just upon Mexico, spake of a Revolution there, how great a Thing it would be. Said one Whitehead told him of the magnificence of the City, that there were in it 1500 Coaches drawn with Mules. This Blaze had much put me in mind of Mexico; because we must look toward Mexico to view it. Capt. Clark drew a Line on his Globe. Our Thoughts being thus confer'd, and found to jump, makes it to me remarkable. I have long pray'd for Mexico, and of late in those Words, that God would open the Mexican Fountain."

Sewall's "Blaze" was Comet 1702a, a bright object observed in many parts of the world but so roughly that no dependable independent orbit has ever been calculated. A. G. Pingré's *Cométographie* of 1783 (see Chapter 82) gives several reports of it from the logs of Dutch ships, the earliest being on the evening of February 20th (Gregorian) at the Cape of Good Hope. As seen from the Gulf of Bengal on the evening of the 28th, the comet had a tail 43° long. This comet was also viewed about that date from Japan and Louisiana. In recent times it has been suggested by H. Kreutz and by B. G. Marsden that Comet 1702a may be a member of the famous sun-grazing group of comets that includes 1843 I and 1882 II.

In Judge Sewall's diary are brief notes on at least nine lunar and five solar eclipses. Thus: "Friday, Novr. 19, 1686 [O. S.]. . . . so we together viewed the Eclips. As to the time and digits the Cambridge Almanack rightest; had he not unhappily said 'twould not be visible. Clouds hindered between whiles that could not so well see how much the Moon eclipsed, but when near half darkened, and when emerging, had a good view."

This was a partial eclipse of the Moon that lasted 2½ hours according to modern calculations. Mid-eclipse was at 6 p.m. Boston time, when 51 percent of the Moon's diameter was covered.

"Decr. 6, 1713 [O. S.]. Sun is Eclipsed just about the beginning of the Fore noon Exercise; when well enter'd many Guns are Fired by Capt. Brown Going down to Nantasket. Mr. Holyoke observes the Eclipse in the Town House Turret. Very clear day. I saw it plain as I came home at Noon."

By my calculation, based upon Oppolzer's *Canon,* this eclipse was partial at Boston, with a maximum of 66 percent of the Sun's diameter covered, at 11:15 a.m. local time.

In late 1687 Sewall sailed to England on the ship *America,* a two-month voyage, to settle some land claims. He took advantage of this opportunity to visit the Royal Observatory at Greenwich on April 29, 1688 [O. S.]: "Went through the Park to Mr. John Flamsted's, who shewed us his Instruments for Observation, and Observed before us, and let us look and view the Stars through his Glasses."

Two months later, Sewall and his American companions visited the University of Cambridge. One attraction there appears to have been an unusual sundial consisting of trimmed ornamental shrubbery: "Mr. Harwood and I step'd out and saw Queen's Colledge, which is a very good one, in the Garden a Dial on the Ground,

Hours cut in Box." This unusual timekeeper must have been difficult to keep in accurate adjustment from year to year!

Besides astronomical phenomena, there are frequent notes in Sewall's diary of such things as earthquakes, atmospheric halo complexes, and storms. But the context shows that his lively curiosity was only in small part scientific, for he carefully recorded these events as omens indicating God's will. The diary entries tell us less about astronomical happenings than how an intelligent layman reacted to them three centuries ago.

5. America's last King and his observatory

During America's bicentennial celebration King George III got a bad press, for he made a convenient antihero to contrast with the American patriot leaders. On the other hand, his considerable personal merits usually go unmentioned. If the battle smoke of the Revolutionary War were to clear, contemporary Americans might be able to see King George at leisure from state affairs and enjoying his hobbies – farming at Windsor Castle, cutting ivory ornaments on his lathe, botanizing, and using a telescope.

For among George III's personal merits was an enthusiasm for astronomy. He was an avid amateur who welcomed the company of astronomers and had his own observatory at Richmond, near the western edge of London. Usually called Kew Observatory, it was located in the Old Deer Park that adjoins the present-day Kew Botanical Gardens.

Here is a brief description by a German visitor in 1816: "The private observatory of the king at Richmond is particularly distinguished by the beauty of its building, equipment, and location. It was erected at the order of the present king, who himself observed the transit of Venus in 1769. The instruments here are excellent and many: a fine transit instrument, a zenith sector, mural quadrant, several good telescopes including a 10-foot Herschel reflector, a large equatorial under a movable roof, two orreries, and a large collection of physical apparatus. Dr. Demainbray directed the construction of the observatory . . ."

This Stephen C. T. Demainbray, who was born in London in 1710, had as a young man acquired some reputation as a physicist. In 1754 he was selected to tutor the 16-year-old Prince of Wales and his younger brother, the Duke of York, in mathematics, philosophy, and natural history. Two years later, when the prince came of age and his other teachers were dismissed, Demainbray was retained to instruct him in the sciences. Such tutoring continued until George III ascended the throne in 1760. There seems to have been a long-continued friendship between the two, and it is very likely that the king's taste for astronomy was aroused by Demainbray.

It was George's interest in the coming transit of Venus that caused him in 1768 to erect the observatory at Richmond and place it under the care of Demainbray, who received the title of King's Astronomer. The passage of Venus across the Sun's disk on June 3, 1769, was observed by George III with a 6-inch reflecting telescope of two feet focal length. A distinguished group was present with him during the great event. It included Queen Charlotte, Princes Ernest and George of Mecklenburg-Strelitz, Demainbray, and the clockmakers Benjamin and Justin Vulliamy.

The royal interest in Kew Observatory during the following years is described by its first director's great-grandson: "His Majesty frequently attended at the observatory and procured the best clocks and watches that could be made and placed them . . . under the Doctor's care, so that by daily observations of the Sun when passing the meridian, the time was regulated to a second; and for many years the accurate time was taken from the King's Observatory at Kew for the regulation of the clocks in both Houses of Parliament, at the Horse Guards, St. James', and elsewhere [in London]."

The King's Observatory at Kew had only two directors during its existence of 72 years. When Demainbray died in 1782, Sir Joseph Banks (president of the Royal Society) sought to have William Herschel appointed as successor. But George III had already promised the post to his old tutor's son, Rev. Stephen G. F. T. Demainbray, who held it for the next 58 years. His main contribution seems to have been to continue the time service for London until this was taken over by the Royal Greenwich Observatory.

George III died in 1820 after decades of growing blindness and mental illness. George IV and William IV both maintained an interest in Kew Observatory because of their early associations with it, but Queen Victoria, who came to the throne in 1837, did not. Reverend Demainbray was pensioned off, and the astronomical instruments were presented to Armagh Observatory in Ireland, where they remain to this day.

Even more than for founding Kew Observatory, King George III should be remembered as a generous patron to aspiring astronomers. One recipient of his favor was James Ferguson (1710–1776), born a poor farm boy in Scotland, who taught himself to read and who became famous as an astronomical lecturer and a builder of clocks and orreries. He was the author of *Astronomy Explained on Sir Isaac Newton's Principles,* 1756, which went through many editions and served as the textbook from which William Herschel learned the elements of the science.

Beginning in 1762, the king gave a pension of £50 a year to Ferguson, which enabled him to devote himself entirely to astronomy instead of working as an artist. A biographer of the Scottish astronomer tells us: "Ferguson this year, 1768, began, by command of the king, to visit his Majesty at Kew and St. James. Indeed, from this time, till his death, Ferguson had the honour of frequent invitations to converse with his Majesty on Philosophical and Mechanical topics, and on the turning of wood, ivory, etc."

The best-known astronomical protégé of George III is of course William Herschel, who rose from obscurity with his discovery of Uranus on March 13, 1781. It took several months before the new body was proven to be a new principal planet beyond the orbit of Saturn, and in November Herschel received the Copley medal of the Royal Society. Herschel's first audience with the king was in May, 1782. He was asked to bring his telescope to the Queen's Lodge at Windsor Castle, and there on July 2nd the astronomer showed Jupiter and Saturn to the royal family and some scientific personages. "It was a very fine evening," wrote Herschel the next day to his sister Caroline. "My Instrument gave a general satisfaction; the King has very good eyes and enjoys Observations with the Telescopes exceedingly."

It was a successful interview! That month the king granted Herschel a pension of £200 for life, enabling him to give up music and

Through his writings, lectures, and models, James Ferguson spread popular interest in astronomy in 18th-century Britain. He received friendship and financial help from George III.

make astronomy his career. That same month, in a letter to Sir Joseph Banks, Herschel formally proposed *Georgium Sidus* (the Georgian star) as the name for the new planet, in gratitude to his patron. Soon afterward, the king ordered four or five (the sources disagree) 10-foot reflectors from Herschel, presumably at the standard price of 600 guineas each. Nor did royal generosity end there. Herschel began construction of his 40-foot reflector of 48 inches aperture in late 1785. The king had given him £2,000 for the enterprise, later raising this amount to £4,000, in addition to an annual grant of £200 for maintenance.

First views through the giant telescope on February 19, 1787, showed that the mirror was not good, and a second was finished on August 27, 1789. The astronomer wrote: "August the 28th, 1789, having brought the telescope to the parallel of *Saturn,* I discovered a

Sir William Herschel (1739–1822) at age 46. Originally a professional musician, he became famous as an astronomer and won the patronage of King George III by discovering the planet Uranus in March, 1781.

sixth satellite of that planet, and also saw the spots upon *Saturn* better than I had ever seen them before, so that I may date the finishing of the forty-foot telescope from that time."

The clumsy altazimuth mounting of the 40-foot reflector made it difficult to handle, and Herschel preferred to observe with smaller instruments except when the extra light grasp was essential. The last object that the aging astronomer viewed through it was also Saturn, in August, 1815.

That telescope no longer exists, having been taken down in 1839.

Today, the most direct signs of King George III's interest in astronomy are his instruments, originally at Kew, that are on display at Armagh Planetarium. Among them is the 6-inch reflector with which the king saw the 1769 transit of Venus. At one time Patrick Moore, a well-known British popularizer of astronomy, looked at the Moon and Jupiter with this instrument and found the images surprisingly good.

6. The crucial years of Wilhelm Struve

In the year 1895, the elder Otto Strüve had just retired from the directorship of Pulkovo Observatory to spend his remaining years in Germany. As the oldest survivor of the 18 children of Wilhelm Struve (1793–1864), he wanted to preserve for later Struves the personality of his famous father and therefore wrote a little book of family history.

This 79-page volume adds many authentic details to the widely known success story of how Wilhelm as a young student came to Russia in 1808 and soon assumed the directorship of Dorpat Observatory, eventually becoming one of the great astronomers of all time. His later achievements include the establishment of Pulkovo Observatory and fundamental contributions in studies of double stars, stellar parallax, galactic structure, astrometry, and geodesy. But let us turn from the world-famous scientist to the unknown student and see how he succeeded. From his son's account, we can trace how exceptional talents and great opportunities were brought together.

Wilhelm Struve was a north German, born at Altona near Hamburg. His father Jakob Struve had some reputation as a mathematician and was director of the school his son attended. The boy distinguished himself among his fellows both as a student and athlete.

This was during the time of Napoleon, whose armies had overrun Germany and plunged it into political and economic upset. There was also a special hazard then for strong and healthy youths. One day in 1808, as the 15-year-old Struve was walking along a street in the Hamburg suburb of St. Pauli, he fell into the hands of a French press gang seeking army recruits. The kidnapped youngster was locked up in the second story of a house, but he jumped from a window and fled to Altona. There he was temporarily safe, for Altona belonged to Denmark, which as a neutral state admitted no foreign soldiers.

Wilhelm Struve, director of Dorpat and later of Pulkovo Observatory.

But young Wilhelm was in danger of being shot as a deserter if he fell again into French hands. Carrying a Danish passport and a letter of recommendation, he left Altona on July 6, 1808, and after a 35-day journey over poor roads of war-ravaged Prussia and the Baltic provinces arrived at Dorpat in Russia.

This is the city renamed Tartu after Estonia became independent in 1918. Back in 1808, though under the Russian flag, it was in many respects a German community, and the University of Dorpat, reestablished in 1802 by Czar Alexander I, was a German university. There Karl Struve (Wilhelm's older brother) taught philology, and an old family friend, A. C. Gaspari, was professor of history and geography. The day after arriving in Dorpat, Wilhelm Struve enrolled as a student.

The account just given of the flight to Dorpat is what Wilhelm Struve told his son. It contradicts a modern, embellished version that Wilhelm, on escaping from his kidnappers in Hamburg, swam out to a ship in the harbor that merely chanced to be bound for Russia. Instead, it is clear that Dorpat had been deliberately chosen as a haven. In fact, two years earlier father Jakob had written Karl to ask about the feasibility of Wilhelm studying there.

Within a few months of his arrival in Dorpat, Wilhelm was supporting himself by giving private lessons. During his first semester

he attended lectures on philosophy and philology. Apparently the quality of the university teaching was indifferent, and Wilhelm recoiled from both it and the boisterous student life, devoting his time to self-education.

Now came a two-year interlude that did much to shape Struve's future. Without severing his university connection, he accepted the post of household tutor with a noble family named von Berg. The greater part of each year was spent on their country estate of Sagnitz, 70 versts from Dorpat, for the family lived in town only during the coldest months. It was not easy for an ambitious young man to fill this menial post and not clash with the class attitudes of his aristocratic employers, but Struve had the necessary tact and awareness. Despite the 34 hours of lessons he gave each week and his private studies, he had energy and time to take part in all the sports and entertainments of the family. With his social talents and exceptional physical grace, he was a sought-after guest at the dances, hunts, and other activities of the country gentry. This social education gave him a polish that was of utmost value in later years, for it was Struve's personal relations with the Czar and members of the imperial court that made Pulkovo Observatory possible.

Meanwhile, Wilhelm's tastes had turned strongly to mathematics and astronomy. In 1811 he was back at Dorpat, reading all the astronomy texts he could find and learning practical astronomy. At that time the Dorpat professor of astronomy, Johann Huth, was a chronically sick man who could not fulfill his duties either as a teacher or as observatory director. Thus, as far as observing techniques went, Wilhelm Struve was entirely self-taught.

Soon after the 1802 reestablishment of the university, an observatory had been built, and several instruments were secured: a large transit instrument by Dollond, a Baumann repeating circle for measuring star altitudes, clocks, and an 8-foot Herschel reflector. But when Wilhelm Struve first came to the observatory, the larger instruments were still in their packing cases. F. Paucker (later a good astronomer at Mitau) was nominally in charge of the observatory, but he did no work there. He was on friendly terms with Struve, however, and gave him the run of the establishment. Struve unpacked and erected the instruments with his own hands. In particular, the big granite cube that was to support the transit instrument required much time and labor.

Dorpat Observatory was built in 1809–1810 and originally was topped by a dome, which was replaced in 1822 by the cylindrical structure seen here. This photograph, depicting the observatory from the northwest, was taken around the turn of this century; note the open slit in the roof for observing meridian transits. Reproduced from *Publikationen Kaiserlichen Universitäts-Sternwarte, 24,* Part 1, 1914.

During the summer of 1812 Struve undertook a small geodetic survey in the Estonian countryside, using a 10-inch Troughton sextant. This was the summer when Napoleon invaded Russia. During his survey, Struve was surprised by a Russian cavalry patrol, and because of his instrument he was arrested as a spy. Despite his protest, he was escorted under guard 150 versts to Pernau to face a military court. There he managed to clear up the misunderstanding, having lost only a week's time.

The observatory soon became a place of great activity. In 1813 Struve received his doctorate in astronomy for a thesis on the latitude and longitude of the observatory, and that December, at the age of 20, he was appointed assistant professor. At this time he began to observe visual double stars, measuring the relative positions of their components with the transit instrument and with a refractor of 5-foot focus.

Once Struve's position at Dorpat was secure, he returned to Altona to visit his parents and became engaged to marry. On this journey, Struve toured the major German observatories and made the friendship of such leading astronomers as Heinrich Olbers, J. H. Schröter, Carl Friedrich Gauss, and Friedrich Wilhelm Bessel. Now recognition grew steadily: Struve took over the observatory directorship in 1817 and next year became full professor. Also in 1817, the publication of the businesslike first volume of Dorpat *Astronomical Observations* made it clear that a major new astronomical center had come into being.

The more Wilhelm Struve worked with small telescopes, the more he wanted a very powerful instrument for his double star work. In 1820 he visited Fraunhofer in Munich to find how large a refractor he could make. Thanks to cordial support by the university council, Struve was able to place an order for the largest and most modern refracting telescope in the world with the best optical manufacturer of the time. Its aperture was nine Paris inches (9.6 U.S. inches), and it had a sturdy equatorial mounting and a good driving clock. In 1824 the "Giant Refractor" was completed and shipped from Munich. To supervise its transport over the rough roads of the Baltic provinces, Struve sent one of his students to the Prussian border to serve as an escort. This man was Baron Wilhelm Wrangell, later a Russian admiral.

On November 10, 1824, a procession of horse-drawn carts arrived in Dorpat with 22 crates of telescope parts. Everything had been so carefully packed that there was no damage, but by some oversight no instructions for assembly were included. Nevertheless, the next day Struve started to erect the instrument, guided by a drawing he had received earlier. On the 15th he had finished the job.

The Dorpat 9.6-inch telescope was set up temporarily in the west hall of the observatory, where a high window to the south gave a view of the sky to 45 degrees above the horizon. (Not until the next summer was it permanently erected in its own dome.) The accompanying photograph of the instrument shows it much as the delighted Struve saw it during the first days of use. Imagine if you will the wooden pier and the fir-wood tube covered with mahogany veneer that gleamed like polished copper, the shining silver of the

The Dorpat telescope today. When it was installed in 1824, this 9.6-inch refractor was the largest in the world. Its f/18 Fraunhofer objective was of excellent optical quality. With superior mechanical design, reliable driving clock, and convenient micrometer, this was a much more efficient research instrument than any earlier telescope. With it Wilhelm Struve discovered nearly 3,000 double stars, measured accurate dimensions of planets and Jupiter's satellites, and made important physical observations of comets. Courtesy Owen Gingerich.

finely divided declination and hour circles, and the boxes of oculars and micrometers.

"On the 16th, during a clear morning hour, I had a first look through the telescope at the moon and some double stars," Struve reported. "This magnificent work of art was doubly astonishing, both for the excellence of its design and construction, and for its great optical power and the quality of its images." Wilhelm Struve's first use of the Dorpat refractor marked a major milestone in astronomical history. The great, clumsy, altazimuth reflectors of the Herschels' became obsolete on that date, for Struve had begun to measure double stars with the first modern telescope.

7. John Pond: sixth Astronomer Royal

DURING ITS FIRST 300 YEARS, the Royal Observatory at Greenwich had only 13 directors. This sequence of long-lived Astronomers Royal included some of the most famous and honored of British scientists: John Flamsteed, Edmond Halley, and George B. Airy. On the other hand, John Pond, who was director at Greenwich from 1811 to 1835, is seldom remembered, except perhaps as the only Astronomer Royal who was forced to resign. This oblivion is undeserved, for he did some excellent astronomical work despite peculiar difficulties.

Pond was born in 1767 in London, son of a well-to-do businessman. His early interest in practical astronomy was fired by his tutor William Wales (1734–1798), who had been sent by the Royal Society to Hudson Bay to observe the 1769 transit of Venus, and who had sailed as an astronomer on Captain Cook's voyages of 1772–1775 and 1777–1779. Perhaps most of Pond's formal training in astronomy came from his tutor. Although he entered Cambridge University in 1783, astronomical studies were at a very low ebb there, and he withdrew early because of broken health.

After extensive travels, in 1798 he set up a private observatory at

John Pond (1767–1836), director of the Royal Greenwich Observatory from 1811 to 1835. The National Maritime Museum, London.

Westbury in Wiltshire. There he installed an altazimuth instrument with a finely graduated vertical circle 30 inches in diameter for measuring the altitudes of stars. It was the work of the famous instrument maker Edward Troughton (1753–1835) of London, who became Pond's lifelong friend.

The Westbury amateur used this excellent instrument in 1800–1801 to determine the declinations of the same fundamental stars regularly observed at Greenwich. Pond's declinations differed systematically from the Greenwich ones, which were still being measured with an old-fashioned Bird quadrant installed half a century before for James Bradley. In a memoir published in 1806, Pond proved conclusively that the Greenwich quadrant had become warped since Bradley's time and was giving erroneous results. This deformation was verified by Troughton's examination of the Greenwich instrument.

These events gave Pond a scientific reputation, leading to his election to the Royal Society in 1807. They also induced Nevil Maskelyne, the Astronomer Royal, to replace the worn-out quadrant with a new mural circle ordered from Troughton. Pond assisted actively in the design of this instrument, which was delivered to the Royal Observatory in 1812. But Maskelyne died before it arrived, and Pond was appointed his successor in February, 1811.

During the next few years the new Astronomer Royal greatly expanded the staff, the equipment, and the quantity and quality of the observing at Greenwich. There was only one assistant when

The Troughton mural circle that Pond used for observing star positions resembled this one installed at Paris Observatory in 1819. The Greenwich instrument had a 4-inch telescope of 74 inches focal length, limited to the meridian by a horizontal axle supported in bearings by a massive masonry wall. The divided circle, six feet in diameter, turned with the telescope, and its position was read by six micrometer microscopes fixed to the wall face. Although an improvement on the earlier mural quadrant, the mural circle was defective in principle because of its asymmetric support; it was quickly outmoded by the meridian circle, whose axle rests on two piers on either side of the telescope. From J. A. Repsold's *Zur Geschichte der Astronomischen Messwerkzeuge*, 1908.

Pond came, six when he left. The mural circle went into steady use, and in 1816 a 5-inch transit instrument was installed for determining right ascensions.

Pond showed great ingenuity in devising observing methods that would eliminate instrumental errors. In 1824 a second mural circle, made by Thomas Jones and originally intended for the new observatory at the Cape of Good Hope, was erected alongside the Troughton circle. Each was provided with a movable basin of mercury, so that by pointing the telescope below the horizon the reflected image of a star could be observed. Pond used the two instruments for simultaneous observations of the same stars to determine declinations, thereby obviating any need for plumb lines or spirit levels to evaluate corrections to the measured quantities. In one half of his plan, program stars were measured by reflection with one mural circle and by direct vision with the other. In the other half of the plan, simultaneous direct observations were made by the two instruments. Pond could then make a simple algebraic combination of these different sorts of measurements to obtain absolute declinations of the program stars, freed from the index errors. This process was described by Pond to the Royal Astronomical Society on May 12, 1826.

The procedures followed by Pond were perhaps the best in use at any observatory, until in the 1820's positional astronomy was revolutionized by F. W. Bessel, Wilhelm Struve, and Airy. In fact, Pond's star declination observations were so good that the variation of latitude is clearly shown by them, as S. C. Chandler proved in the 1890's. According to Chandler, their freedom from systematic errors was rarely matched in later work.

During Pond's lifetime many astronomers tried unsuccessfully to measure stellar parallaxes. (These slight annual shifts of star positions result from the Earth's motion around the Sun and eventually permitted star distances to be determined.) Among those making the attempt was John Brinkley, a former Greenwich assistant who in 1792 became director at Dublin. There, working with an unwieldy vertical circle eight feet in diameter, he believed he had detected parallaxes of two to three seconds of arc for Altair, Deneb, and some other stars. Pond could not verify such displacements with the Greenwich mural circle observations.

To settle the matter, Pond set up two fixed telescopes 10 feet long and rigidly attached to massive piers. His idea was to measure by transits with one telescope the difference in right ascension between Altair and a star in Pegasus that had the same declination; similarly, the other telescope was used to compare Deneb with Beta Aurigae. The method was sufficiently accurate to show that no parallaxes of the size announced by Brinkley were present. A low-pitched controversy between the two astronomers continued for years, but Brinkley did not see that Pond was right.

However, the scientific work of Pond was achieved only in the face of growing difficulties that finally overwhelmed him. He had become Astronomer Royal on his reputation as an able amateur, but his skill and enthusiasm as an observer could not help in administrative problems. On taking office, he had inherited from Maskelyne responsibility for preparing and publishing the *Nautical Almanac.* He did not share his predecessor's interest in this indispensable aid to navigators, and as a result its usefulness declined quickly. There was much criticism, and in 1818 responsibility for this publication was transferred to Thomas Young. Nevertheless, not until 1831 was this annual ephemeris finally put into satisfactory form.

Further troubles for Pond came in 1821, when the Royal Observatory was assigned the onerous task of rating and testing all marine chronometers offered for government purchase. This chore diverted much of the observatory's limited manpower from astronomical work. In fact, when Airy succeeded Pond as Astronomer Royal in 1835, one of his first acts was to reduce this nuisance to a manageable scale.

By 1825 Pond was in deepening trouble. The observatory's board of visitors received a complaint from S. Lee, assistant secretary of the Royal Society, charging inaccuracies in the published volume of Greenwich observations for 1821 and blaming the Astronomer Royal for failing to do anything about the comet of that year. A committee of the board of visitors finally exonerated Pond of negligence, but it criticized him because "very few observations were made after 10 or 11 at night." Lee returned to the attack, but in such offensive terms that the board rebuffed him. Articles sharply criticizing Pond's work were published in the *Philosophical Magazine,* but these were rebutted firmly by Bessel in the *Astronomische Nachrichten.*

From this time on, Pond's relations with his board of visitors worsened steadily.

Underlying these difficulties was Pond's personality. To judge from the evidence, he was a mild-mannered man who disliked contention, but he was also so awkward and inarticulate that he could seldom please or influence others. In particular, his failure as the superintendent of his growing staff proved disastrous.

In a letter of 1826 Pond wrote: "You know the new methods of observing with the two mural circles . . . evidence of the vast labour which has been bestowed will be found in the journals of 4 quarters of Greenwich Observations . . . containing nearly 10,000 observations. But to carry on such investigations I want indefatigable hard working and above all obedient drudges (for so I must call them, tho' they are drudges of a superior order) – men who will be content to pass half their day in using their hands and eyes in the mechanical act of observing and the remainder of it in the dull process of calculation."

This hardly suggests a congenial atmosphere for master or drudge. Pond's first assistant was Thomas Taylor, who had been appointed during Maskelyne's regime, in 1805 or 1807. He was the principal in the scandal of Groombridge's star catalogue, by preparing an edition so full of errors that Airy in 1832 prevented its publication. Another assistant, William Richardson, and Taylor's son Henry were involved in this business, which Airy declared "a disgraceful affair and an attempt to defraud the Government."

Pond's inability to control his staff was compounded by his deteriorating health, which resulted in frequent absences from the observatory to recuperate at Hastings. The breakdown in the observatory's work forced Pond's resignation in 1835. He died on September 7, 1836, and was buried in the same tomb as Edmond Halley in the nearby churchyard of Lee.

Pond's successor (see Chapter 10), Airy, made it a condition for accepting the post that Thomas Taylor be removed. In applying for a pension for Pond's widow, Airy generously described his predecessor's claims to the gratitude of astronomers: "For certainty and accuracy, Astronomy is quite a different thing from what it was, and this is mainly due to Mr. Pond."

8. Some British amateurs of the Golden Age

TO THIS DAY there are important astronomical consequences of a small dinner party held on January 12, 1820, at Freemason's Tavern in London. The host was Rev. William Pearson, a schoolmaster and able practical astronomer, who had long wanted to found a society to promote his favorite science. At this dinner, Pearson and a group of like-minded friends did just that, arranging to hold their first regular meeting that March 10th. Their group grew and flourished, and in 1831 it became the Royal Astronomical Society. Today the RAS is the leading organization of professional astronomers in Great Britain, and its gold medal is one of the world's most honored scientific distinctions.

But back in 1820, and for some years to follow, the Royal Astronomical Society was composed mainly of amateurs. Pearson himself, although the author of an excellent two-volume treatise on astronomical instruments and their use, was first a teacher and later a country clergyman. Another founding member, Sir James South, was a surgeon who measured double stars. The first president of the society was the octogenarian Sir William Herschel, who with his son John were amateurs of the most distinguished kind. The society's secretary was Francis Baily, after whom are named the Baily's beads seen at total solar eclipses. He was a prosperous stockbroker who retired at an early age to compile star catalogues. His famous measurement of the density of the Earth was conducted with apparatus in his own house in Tavistock Place, close to where the University of London stands today.

The first half of the 19th century in Great Britain was clearly a golden age for amateur astronomers. The science was far less specialized than today, and there was little difference between professional and amateur although this difference was growing. It was possible for a well-to-do and knowledgeable individual to compete on fairly even terms with regular observatories, both in observing equipment and in range of work.

Consider, for example, William Henry Smyth, who upon retiring from the Royal Navy as a captain set up a private observatory at Hartwell House in Bedford. Here in 1830 he installed an equatorially mounted 5.9-inch refractor with an excellent objective by Charles Tully. It was with this instrument that he made the many double star measures published in his well-known *Cycle of Celestial Objects* (see Chapter 12). At the time he began, the 9.6-inch Fraunhofer refractor at Dorpat Observatory in Russia was the largest in the world and was being used by Wilhelm Struve for his double star research. Smyth could enjoy the heady excitement of being a pioneer for some years, until the growing numbers of larger telescopes and better observers began to diminish the significance of his contribution.

One reason for this predominance of private astronomers in Britain about 1830 was that there were very few opportunities for professional employment in astronomy. The Royal Greenwich Observatory was still a small affair staffed mainly by industrious drudges. In the universities the science was at a low ebb, until it was brought to life at Cambridge by George Biddle Airy in 1828. The government sponsored astronomy indirectly through geodetic work, such as the great survey of India. This, too, was a time when the Royal Navy was active in mapping coastlines in distant parts of the world, which required skilled celestial navigators. Many naval officers sought to improve their chances for promotion by inventing new navigational methods and by joining the Royal Astronomical Society.

But there was another kind of employment opportunity for some young astronomers, as workers at a large private observatory. At Birr Castle in Ireland, the third Earl of Rosse had erected a 36-inch reflector by 1839 and a 72-inch in 1850, and a number of men who later gained scientific distinction served as his assistants. Lord Rosse not only designed and built his own telescopes but was an active observer of nebulae with them (see Chapter 76). The Birr Castle assistants were more fortunate in their patron than some others were elsewhere. Sometimes a wealthy man might have a well-equipped observatory built but would lack the time or training to use it himself. So he would hire an astronomer, much as he might hire a gardener or a librarian, in hopes of some discovery that would lend prestige to himself.

One such establishment was George Bishop's observatory in Regent's Park, London, where a 7-inch refractor was put in 1836. Bishop, who had made a fortune in wine, was highly successful in finding a series of very capable aides. With his telescope, the eagle-eyed W. R. Dawes measured double stars (see Chapter 69), J. R. Hind discovered minor planets, and Norman Pogson observed variable stars. Bishop's relationship with his assistants was scathingly described by Dawes in a letter to his friend George Knott: "Mr. B. never did and never *could* observe at all, not even a transit; but after I left his observatory *he put his own name to all my observations!!*"

In this golden age of British amateurs, Lord Wrottesley takes a special place. John Wrottesley, as he was called until he succeeded to his father's title, was born in 1798. Not long after his graduation from Oxford, where he distinguished himself in mathematics, he became a founding member of the Royal Astronomical Society at the age of 22. He was a London lawyer when, nine years later, he began the construction of the first of his observatories.

This was located in the suburb of Blackheath, just south of Greenwich, in the same town where Stephen Groombridge's observatory was active from 1806 to 1827 (see Chapter 68). Wrottesley installed a transit instrument of 3¾-inch aperture and 62 inches focal length and a sidereal clock. He also secured an able and industrious assistant, John Hartnup, who later was director of Liverpool Observatory from 1843 to 1885.

In businesslike fashion, Wrottesley decided upon a single, carefully planned program for his establishment. This was the formation of a catalogue of the right ascensions of 1,318 selected stars from observations with the transit instrument. These stars, all of 6th and 7th magnitude, were ones that had previously been observed by James Bradley at Greenwich or by Giuseppi Piazzi at Palermo.

This program was carried out in full. After tests were made on the instrument, a total of 12,007 transit observations were made between May, 1831, and July, 1835. The bulk of the work was done not by Wrottesley but by Hartnup, who during his patron's frequent absences pushed ahead zealously with the observing and computing.

For this catalogue the Royal Astronomical Society awarded its gold medal to Wrottesley in 1839, and two years later he became its president. So carefully was the work done that it retained some scientific value even into the 20th century, being one of the source catalogues for the Boss *General Catalogue of 33,342 Stars* (1937).

When his father died on March 16, 1841, Wrottesley inherited his title and the family estate in Staffordshire. As Lord Wrottesley, he took up astronomical work again with promptness. He described it rather grandly: "In the commencement of the year 1842, I resolved on erecting an observatory near to my residence in Staffordshire; and on the 29th March in that year the first stone of the building was laid by my youngest son in an elevated position, about 700 yards to the north by west of the mansion. On the 11th May following, the transit piers were fixed, and on the 8th of October in the same year, the 5-foot transit, with which I observed my Blackheath Catalogue of 1318 stars, was finally placed upon them, and ready for work in its new locality."

This second Wrottesley observatory was a stone building with a dome 17 feet in diameter at its east end, housing an equatorial refractor. In addition to the transit room was an 18-by-12-foot computing room from which steps descended to a basement containing a small kitchen and cramped living quarters for the two assistants. The geographical position of the observatory was determined as latitude 52° 37′ 23″ north, longitude 2° 13′ 24″ west. But the location was not a very favorable one, as Lord Wrottesley admitted, for it was within a few miles of extensive coal mines, and the view to the south was partly obstructed by a grove of trees.

The 7¾-inch refractor was intended for double star work, and several lists of measures were published. But an unsatisfactory driving clock must have made it an unhandy instrument for Frederic Morton, the assistant assigned to it. More important was the transit instrument with which Lord Wrottesley desired to repeat his accomplishment of producing a high-quality star catalogue. The result was the publication in 1854 of the right ascensions of 1,009 stars. All of the observations and reductions were made by the first assistant, Richard Philpott.

It therefore seems a trifle ungracious for Philpott's noble employer to have written in his preface to the work: "I have thus brought to a conclusion my . . . last star catalogue, for which I need hardly make an apology for calling that *mine,* which has been produced by a large outlay of my own capital, both material and intellectual." Neverthe-

The overgrown ruins of Lord Wrottesley's second observatory, founded in 1843 and active for about a decade, were photographed in the late 1970's by Malcolm Astley. In 1869 the city of Wolverhampton enacted an ordinance to protect the observatory by forbidding the issue of smoke from furnaces or steam engines within a three-mile radius of it. This act was repealed only around the time this photograph was taken.

less, he goes on to say he does not underrate the value of the services of the able assistant who did all the work!

Lord Wrottesley died on October 27, 1867, full of honors. As a senior statesman of science, he had been president of the Royal Society in succession to Lord Rosse as well as president of the British Association for the Advancement of Science. But the golden age of British amateurs was already waning. One small indication is that his 1854 catalogue didn't share with his earlier one the distinction of having been used in the preparation of Boss' *General Catalogue*. The reason for this was the increased number of regular observatories that were determining star positions with improved instruments and methods. The rapid widening of the gap between amateur and professional had begun.

9. John Herschel's expedition to South Africa

EVEN TODAY the southern half of the sky has been less thoroughly explored than the northern. Despite major new astronomical facilities in Australia and Chile, the most powerful instruments for optical surveys are still concentrated in the other hemisphere. These include the three largest reflectors, two largest Schmidt cameras, and nine largest refractors.

The north-south imbalance was far greater in 1822, the year of Sir William Herschel's death. He had systematically searched the northern heavens with telescopes up to 48 inches in aperture, discovering double stars, clusters, and nebulae by the thousands. In contrast, the far-southern sky was unexplored. The first permanent observatories below the terrestrial equator were just being founded, at Paramatta in Australia and at the Cape of Good Hope. Apart from a few chance finds, the known southern star clusters and nebulae were limited to Nicolas Louis de Lacaille's discoveries with a refractor of only ½-inch aperture.

A wonderful opportunity awaited the first well-equipped explorer of this rich wilderness. Appropriately, it was William's only son, Sir John Herschel, who extended his father's work to the south celestial pole during a memorable expedition to South Africa in 1834 to 1837. At a temporary observatory he erected the 18¼-inch reflector he had previously used in England, and catalogued over 1,700 clusters and nebulae and 2,100 double stars. With remarkable energy he also made numerous photometric measurements of stars, mapped the Magellanic Clouds and the Milky Way, observed double stars micrometrically, and made extensive star counts (see Chapter 75). Halley's comet, the satellites of Saturn, and solar activity

Sir John Herschel photographed in April, 1867, by Julia Margaret Cameron, one of the most famous Victorian photographers. Herschel sat for this picture 30 years after his return from South Africa and four years before his death. Collection of Owen Gingerich.

were also studied. These rich scientific results fill a large quarto volume.

It was a remarkable accomplishment, made possible by John Herschel's special advantages. At a time when there was no academic training for working astronomers in Great Britain, he had been personally instructed by his father in the art of observing. He was a wealthy man, the sole heir of William who had made £16,000 (in purchasing power the equivalent of over $1 million today) from the sale of telescopes. Thus Sir John's four years in South Africa were spent in a comfortable country house with Lady Herschel, their children, and servants. His telescopes were only a few steps from his door. Seldom has a great scientist been able to follow his calling so free from worldly cares.

The intimate history of those days became known only after the auction of Herschel family effects in 1958. A considerable quantity of Sir John's personal papers came into the possession of the University of Texas. There, astronomer David S. Evans, who is also an historian of science, compiled an edition of Herschel's African diaries. *Herschel at the Cape* was a major addition to the limited biographical material about him.

The diaries are the concise and even sometimes cryptic record of a very active social and intellectual life. The man's breadth of interests extended far beyond his astronomical observing. As Evans says: "There are, too, accounts of experiments in many other fields: botany, zoology, ornithology, and chemistry. In meteorology there is every indication that Sir John knew a front when he saw one; a piece of paper asymmetrically stained with graduated pigment which is in the Texas collection, hints that he may, with his strong interest in flower colors, have been on the verge of discovering chromatography. He had his quirks as well, one of which was a half-formed idea that the phases of the Moon were correlated with the weather." African natural history fascinated the visitor from England, to judge from the diary references. It seems he always took his walks gun in hand. His penchant for shooting birds is explained by the need to collect study specimens due to the difficulty in observing birds close up before binoculars were available.

Little more than a month after Herschel landed at Cape Town, his 20-foot reflector was ready for use. The first objects he saw with it, on February 22nd, were the multiple star Alpha Crucis and the Eta Carinae nebula, "a most wonderful object." Watching the full Moon rising, he noted, "The *European* face is quite lost by the reversal of its position."

The 20-foot was mounted as an altazimuth and ordinarily was used near the meridian. Herschel had brought three 18¼-inch mirrors with him, because speculum metal tarnished so quickly that frequent repolishing was needed. In searching for new objects, the telescope was kept stationary while the stars drifted through the 15-minute-of-arc field of a 180x eyepiece. A system of pulleys and a graduated dial allowed the altitude of the telescope to be altered through a range of three degrees during such a sweep. If something interesting were seen, the telescope could be shifted manually to keep the object within sight for a few minutes. When a star was suspected of being a close double, further tests were made with a triangular aperture and with powers of 240 and 320.

Perhaps the typical seeing (steadiness of images) at Feldhausen was not very good, because very few double stars closer than two sec-

John Herschel's 18¼-inch reflector peers skyward from his Feldhausen estate in South Africa, in this drawing he made in September, 1834. In the shelter at right was the 5-inch refractor, formerly the property of Sir James South in London, that Herschel used for measurements of southern double stars. Unfortunately, the micrometer was defective, so that only the determinations of position angles were reliable, not the separations. From Herschel's *Results of Astronomical Observations*, 1847.

onds of arc were logged by Herschel. He suspected with some reason that air currents inside the tube caused this turbulence, and he suggested replacing the tube by an open framework. The optical quality of the 20-foot was not the bottleneck, for the diary entry of September 5, 1834, reads: "Swept till 4 A.M. – The most wonderful astronomical night I ever knew – used 1200 as a good working power – saw well with 2000!"

Nearly everyone in South Africa at that time who had anything to do with science seems to have been acquainted with Herschel. The diary entry for August 21, 1836, recounts a conversation with the explorer Sir Andrew Smith. "Dr. Smith states that he had discovered a mass of Meteoric Iron in Namaqua land beyond the Orange River. It is more than a ton in weight. It was seen to fall and was hot and made a great smoke as described by the Natives. They cut it with chisels & use the Iron. Dr. Smith cut off a piece, a specimen of which he promised to give me." This account of a big meteorite in Southwest Africa suggests the Hoba West iron found near Grootfontein in 1920, except, as Evans points out, Hoba West is extraordinarily tough and hard to cut. Also, it weighs about 60 tons, instead of Smith's "more than a ton." Quite possibly Herschel's diary has recorded a lost meteorite still awaiting rediscovery.

10. The Airy regime at Greenwich

PICTURE A HIGH-CEILINGED OCTAGONAL ROOM with old-fashioned furnishings. On its walls hang some dusty portraits and antique scientific instruments. A great north window has been opened to the bright August sunshine, affording a hilltop view of the trees of Greenwich Park, the winding Thames below, and the Essex countryside in the distance. Standing silently at this tall window is a thin, slightly stooping man in his thirties, whose strong features and piercing eyes indicate a powerful personality. He enjoys a moment of satisfied ambition and then goes downstairs to begin his task of reforming Greenwich Observatory. Thus may we imagine George Biddell Airy on that day in 1836 when he first came to the observatory as its seventh director and Astronomer Royal.

During his 45 years of tenure Airy won laurels that remain fresh to this day. Within a few years he had reorganized, restaffed, and reequipped the observatory. Under his direction, hundreds of thousands of precise positions of stars and solar system objects were observed, reduced, and published on a systematic basis. Also skillful as an optical designer and engineer, he drew up the detailed specifications of effective new observatory instruments; the Airy transit circle, for example, remained in use from 1851 to 1951. Versatile, tireless, and influential, it is no wonder that he was the dominating figure in British astronomy for most of his long life.

The Royal Observatory had been founded to improve the art of navigation by providing better knowledge of the positions and mo-

Sir George Biddell Airy (1801–1892) as pictured in *Men of Mark*, Vol. 2, approx. 1870. Courtesy Owen Gingerich.

tions of heavenly bodies. All of Airy's astronomical work at Greenwich served this utilitarian purpose in one way or another. Because of their importance to navigation, tides, waves, terrestrial magnetism, and deviation of ships' compasses were the subjects of much of his scientific labor. On top of all this, the British government continually used Airy as a scientific consultant in all sorts of engineering matters such as railroad gauges, London sewers, Big Ben, and lighthouse design. It is an interesting question how, even with his enormous appetite for work, he managed to accomplish all this.

"The ruling feature of his character was undoubtedly Order," noted his son Wilfrid Airy. "His accounts were perfectly kept by double entry throughout his life. . . . He seems not to have destroyed a document of any kind whatever: counterfoils of old cheque-books, notes for tradesmen, circulars, bills, and correspondence of all sorts were carefully preserved in the most complete order from the time he went to Cambridge. . . . In everything he was methodical and orderly, and he had the greatest dread of disorder creeping into the routine work of the Observatory, even in the smallest matters.

"As an example, he spent a whole afternoon in writing the word 'Empty' on large cards, to be nailed upon a great number of empty packing boxes, because he noticed a little confusion arising from their getting mixed with other boxes containing different articles; and an assistant could not be spared for this work without drawing him from his appointed duties. His arrangement of the Observatory correspondence was excellent and elaborate: probably no papers are more easy of reference than those arranged on his system. His strict habits of order made him insist very much upon detail in his business with others, and the rigid discipline arising out of his system of order made his rule irksome to such of his subordinates as did not conform readily to it: but the efficiency of the Observatory unquestionably depended mainly upon it."

A contemporary who understood Airy particularly well was the Canadian-born American astronomer Simon Newcomb. As head of the Nautical Almanac Office in Washington and responsible for producing new planetary tables, he was one of the principal users of Greenwich data, and he took an informed interest in Airy's working methods. The manner in which the practical-minded and orderly Astronomer Royal reorganized Greenwich Observatory is told by Newcomb:

"Airy's abilities as a planner and administrator of work were of the highest order. His system was based on the idea that one directing head could work out all the formulae and prepare all the instructions required to keep a large body of observers and computers employed in making and reducing astronomical observations. A few able lieutenants, who would see that all the details were properly carried out, were an adjunct of his system. Acting on these ideas, he reduced the work of the observatory to a system more comprehensive in its details than anyone had ever before attempted in the conduct of astronomical operations."

In short, Airy's reorganization of the work of Greenwich Observatory was closely analogous to the change that had been going on in the British industry – the introduction of the factory system. Before Airy, the staff of an observatory generally consisted of individual observers, each a skilled astronomer working in his own way. His innovation was to put all routine work into the hands of closely supervised subordinates from whom only plodding perseverance was required, not scientific training.

Eight years after Airy's death, a Greenwich assistant gave a vivid picture of the system. "His regulation of his subordinates," wrote E. W. Maunder, "was, especially in his earlier days, despotic in the extreme. . . . For thirty-five years of his administration the salaries of his assistants remained discreditably low, and his treatment of the supernumerary members of his staff would now probably be characterized as 'remorseless sweating.' The unfortunate boys who carried out the computations of the great lunar reductions were kept at their desks from eight in the morning till eight at night, without the slightest intermission, except an hour at midday."

These computers were usually boys of about 18 with a good head for figures, who were set to work 11 hours a day reducing routine observations on printed skeleton forms devised by Airy. The most dreaded of these forms was one for reducing positions of the Moon from measurements with the altazimuth instrument; it required seven-place logarithms and was particularly long. No surer way of stifling an interest in science could have been devised. Usually these youngsters suffered a breakdown within a year or two.

The scheduling of night work deserves some description. For many years two instruments were in regular nightly use, the Airy transit circle and the altazimuth. The four assistants assigned to them had the work divided among them as follows, in a tiring cycle. On the first day, a man had a 21-hour tour of duty with the transit circle, from 6 a.m. to the following 3 a.m. The second day's duty consisted only of two or three hours of computing. Then came a full day's work on the calculations, followed by a night's duty with the altazimuth. This last watch was easy if the Moon was young and the night clear, for an hour could suffice for the required readings of the Moon's altitude and azimuth. But on a mostly cloudy full-Moon night, the hapless observer might have to keep a vigilant lookout from dusk till dawn in order to observe the Moon through any breaks in the clouds. And at 6 a.m. the next 21-hour stint at the transit circle would begin for him!

As Maunder notes, "Such a routine carried on with iron inflexibility was exceedingly trying, as it was absolutely impossible for an observer to keep any regularity in hours of rest or times for meals. . . . It was impossible for a man to be at his best for long under the old *régime,* and from forty-six to forty-seven has been an ordinary age for an assistant to break down under the strain."

Reducing observations made with Airy's altitude-azimuth instrument was especially feared by the drudges who worked for him. The National Maritime Museum, London.

Airy was firm in his conviction that the primary work of the Royal Observatory was determining the right ascensions and declinations of Sun, Moon, planets, and stars, and that no other observations should be allowed to interfere. Studies of asteroids or comets with the equatorial refractors, or timing eclipses of Jupiter's satellites, had distinctly lower priority. How Airy reacted to unauthorized initiative is recalled in Maunder's story of an episode that occurred about 1850, when asteroid hunting was first becoming popular. "Airy found an assistant, since famous, working with a telescope on his 'off-duty' night. That stern disciplinarian asked what business the assistant had to be there on his free night, and on being told he was 'searching for new planets,' he was severely reprimanded and ordered to discontinue at once."

William Ellis was a staff member at Greenwich who, tougher than most of his fellows, outlived Airy. He was on duty at the altazimuth on a night in 1861 when one of the finest of the century's comets unexpectedly appeared above the southern horizon. Torn between his obligation to observe the Moon and his desire to measure the brilliant new comet, he finally chose the latter. When he narrated the event 40 years later, his fear lest he be interrupted by Airy in this unlawful activity was still evident.

Airy's system, wrote Maunder, "militated . . . against the growth of real zeal and intelligence in the staff, and necessarily occasioned labour and discomfort out of proportion to the results obtained. Fortunately, in Airy's later years, the extension of the work of the Observatory, a slight failing in his own powers, and the efforts he was devoting to the working out of the lunar theory, compelled him to relax something of that microscopic imperiousness which had been the chief characteristic of his rule for so long."

This change must have begun some time around 1870. One indication of it is that the chief assistant, E. J. Stone, was allowed to use the 12¾-inch refractor in an attempt to measure with a thermomultiplier the radiant heat from Arcturus. A few years later, in 1873, daily photography of sunspots was begun, and soon afterward spectroscopic observations of bright stars were commenced, in an extensive visual program of radial velocity measurements. A long-overdue salary increase was achieved by the assistants in 1871, with Airy's backing.

The end of the old regime came with his retirement in August, 1881, at the age of 80. The new director, W. H. M. Christie, had been chief assistant and was thoroughly familiar with the positional programs. But he was also an astrophysicist who brought in such refreshingly new lines of work as celestial photography and double star measurements. The old rigid system was over.

One illuminating commentary on Airy's methods appeared later. In 1891 the exhausting 21-hour tour of duty at the transit circle was shortened, and thereafter the observer's responsibility for routine observations was reduced to 19 hours. This improvement in working conditions raised morale and production. In 1870–1880 the average annual number of transit circle observations was 3,930; in 1880–1890 it was 5,100; but in 1894–1904 it jumped to 11,580. Part of the increase must be attributed to a very enthusiastic observer, W. W. Bryant, who joined the staff in 1892. Yet the beneficial effects of the Christie reforms deserve a good deal of the credit.

11. About an astronomer-explorer

AUGUST SONNTAG made only a brief appearance on the American astronomical scene long ago, but his few traces reveal an unusual career with a tragic ending. I have never seen a book or article about him, yet his story can be outlined from scattered bits of information.

Just when Sonntag was born is not known. Our first glimpse of him is in 1848 when he was a youthful aide at the observatory of Altona. This city, of which he was a native, is now a suburb of Hamburg in West Germany, but at that time was in the Duchy of Holstein and under the rule of the King of Denmark.

The small but well-equipped observatory at Altona was directed by a leading Danish astronomer, H. C. Schumacher, who was editor of the *Astronomische Nachrichten,* then the world's leading astronomical journal. Schumacher, from his own pocket, used to hire penniless but talented youths to help around the observatory, and one of these was Sonntag.

Beginning with the July 4, 1848, issue of the *Astronomische Nachrichten,* a progression of notes and papers signed "August Sonntag" shows his growing skills. The first was a series of ring-micrometer positions of the newly discovered asteroid 9 Metis, followed by a calculation of its orbit. During the next four years, sometimes in

August Sonntag made a geomagnetic measurement at Rensselaer Harbor, northwest Greenland. Based on a sketch by Elisha Kent Kane, this picture originally appeared in his *Arctic Explorations* (1856).

collaboration with a fellow assistant, he computed the orbits of seven comets. For comet 1850 I he used nearly 500 measured positions. Such work, in addition to many miscellaneous observations, must have made his name widely known among European astronomers. Significantly, in 1852 he began to send contributions across the Atlantic for publication in the newly founded *Astronomical Journal,* edited by Benjamin Apthorp Gould at Cambridge, Massachusetts. Perhaps this means that Sonntag was already planning to emigrate to America.

Conditions at Altona Observatory had begun to deteriorate in 1848, when the German-speaking inhabitants of Holstein revolted. After the Danish government resumed an uneasy control of the duchy, it reduced its financial support of the observatory and even sought to remove the astronomical instruments for use by the Danish navy. Closing of the observatory was averted by an appeal from scientists in England, France, Belgium, Italy, Germany, Russia, and the United States. Soon after this crisis, Sonntag's patron Schumacher died on December 28, 1850, at the age of 70. The new director, A. C. Petersen, continued to give some financial help to Sonntag. But prospects were so bleak for the assistant that some time late in 1852 or early 1853 he sailed for the New World.

Soon after arriving in America, he volunteered for the post of astronomer on the Arctic expedition being organized by Dr. Elisha Kent Kane (1820–1857) of Philadelphia. This was one of the many missions mounted to search for Sir John Franklin's lost expedition in HMS *Erebus* and HMS *Terror,* last sighted in 1845 in Baffin Bay in an attempt to sail the Northwest Passage.

Dr. Kane was a well-to-do naval surgeon and world traveler who had already sailed to the Arctic as a doctor with Lt. J. De Haven, USN, in 1850–1851 in a vain hunt for Franklin. Now, Kane had obtained the support of Secretary of the Navy J. P. Kennedy and other influential backers to mount his own relief expedition. To secure Sonntag's services, Kane managed to have him put on the payroll of the U. S. Naval Observatory, which together with the Smithsonian Institution and the U. S. Coast Survey provided the scientific instruments he was to use.

Kane's expedition sailed from New York on May 31, 1853, in the 144-ton brig *Advance* with 18 officers and men. Its course northward along the coast of Greenland was dictated by a fantastic theory, then widely accepted, of an open polar sea. There was even hope that on the shores of such a sea some survivors of Franklin's party might still be alive.

The *Advance* left Upernavik, Greenland, on July 23rd and, after encountering much ice, was finally frozen in for the winter on September 10th, in Rensselaer Harbor (latitude 78° 37′ north, longitude about 71° 00′ west). This remote and previously unvisited part of the Greenland west coast is some 150 miles north of the present Thule air base.

Within days, a tiny observatory was set up on an islet 100 yards from the imprisoned ship. Here Sonntag had four chronometers, a theodolite, a transit instrument, and equipment for measuring the Earth's magnetic field. This was to be Sonntag's headquarters for nearly two years. Here, among other observations, he determined his longitude by timing culminations of the Moon (when it was highest above the horizon). Several occultations of Saturn and Mars by the Moon were timed in 1853–1854, as well as a partial eclipse of the Sun.

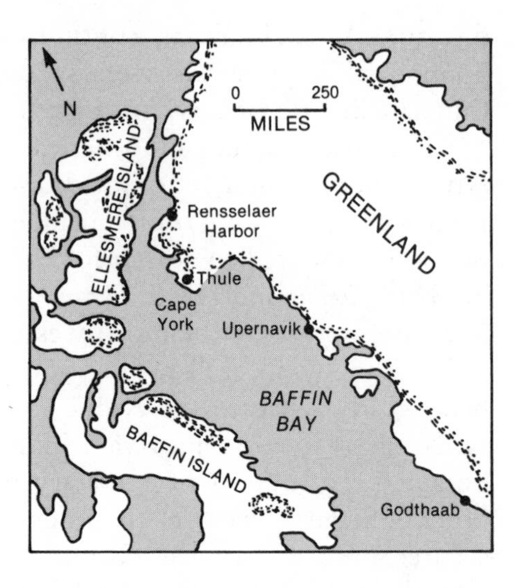

Sites mentioned in the text are identified on this map.

A crisis occurred in August, 1854. The *Advance* was still trapped in the ice, and the half-starved men, already affected by scurvy and fearful of a second Arctic winter, begged Kane to abandon ship. He refused, but on August 28th he allowed Sonntag and 10 others to start southward by sledge. Sonntag demanded and received his pay because, he said, he might return to Europe.

The attempted journey to safety was a fiasco. On December 7, 1854, after three months of extreme hardship, six survivors, including Sonntag, got back to the *Advance* at Rensselaer Harbor and were allowed on board by Kane. That evening the indignant commander wrote in his diary: "These wretched men have reaped the bitter fruit of want of faith. . . . Sonntag, a child and abstractionist, never dreamed that meridian transits and lunars would not teach him to steer a bird line in a raging pack."

Somehow, the remnants of the expedition endured their second winter, and on May 20, 1855, all set out in a desperate escape journey by sledges and open boats to Upernavik, nearly 600 miles to the south. Sonntag was the navigator. Kane's party finally reached this outlying settlement on August 6th, and five weeks later they were picked up by two U. S. Navy relief ships.

In this way Sonntag returned to New York City on October 11, 1855. Kane was hailed as a national hero, and presumably his astronomer too became briefly famous. Early the next year, Sonntag and

Charles A. Schott of the U. S. Coast Survey were working, at Kane's expense, in reducing and tabulating the astronomical, magnetic, and meteorological observations of the expedition.

Later in 1856, Sonntag went to Mexico with the scientific expedition of a Baron Müller and climbed 17,887-foot Mount Popocatepetl. He also made magnetic observations that were published by the Smithsonian Institution.

Finally, in 1859 Sonntag obtained an assistantship at the Dudley Observatory in Albany, New York. A spectacular controversy between its trustees and its director, Benjamin Apthorp Gould, had ended the year before with Gould's removal (see Chapter 13). The acting director and only other astronomer at the observatory was Franz Brünnow. For some reason Dudley Observatory did not turn out to be the haven that Sonntag needed. In 1860 both men left, Brünnow to become professor of astronomy at the University of Michigan, Sonntag to go on another Arctic expedition, with Isaac Hayes in the schooner *United States*.

Hayes' aim was to retrace Kane's track along the west coast of Greenland in search of the mythical open polar sea. In the course of this effort, in January, 1861, Sonntag and an Eskimo companion were traveling over the ice near Cape York in latitude 78½° when the astronomer fell in the water and died soon after of cold and exposure. This restless wanderer whose astronomical career ended so prematurely remains a tantalizing figure, dimly glimpsed.

12. The Sadler-Smyth scandal

A REMARKABLE IMPRESSION has been left on amateur astronomy throughout the English-speaking world by Admiral William Henry Smyth (1788–1865). It stems from his *Cycle of Celestial Objects,* a book heavily drawn upon by later amateur observing handbooks such as Thomas William Webb's *Celestial Objects for Common Telescopes* and William Tyler Olcott's *Field Book of the Skies.* Over 130 years after publication, Robert Burnham, Jr., quoted it extensively in his modern *Burnham's Celestial Handbook.* Nevertheless, the unusual history of Smyth and his book is insufficiently known.

Though Smyth was born in England, his ancestry was American, for his father was a loyalist emigré from New Jersey after our Revolution, and he himself was a direct descendant of Captain John Smith, of Virginia fame. Entering the Royal Navy as a boy, he managed by self-education and strength of character to rise from the ranks, an unusual accomplishment at that time. During the Napoleonic wars he served in the Mediterranean, commanding a brig, making hydrographic surveys, and carrying out diplomatic missions. These duties brought Smyth to Palermo in 1813, where he visited the observatory and met Giuseppe Piazzi, the discoverer of the first known asteroid, Ceres (see Chapter 59). The Sicilian astronomer deeply influenced Smyth, who now decided on an astronomical career; he even named his son Charles Piazzi Smyth.

Retiring from active service as a captain in 1825, Smyth settled at Bedford, England, where he erected an observatory, in which a fine 5.9-inch refractor by Tully was mounted in 1830. During the next nine years he measured several hundred double stars and examined many of the brighter nebulae and clusters. Afterward the instrument was sold to his friend John Lee, who built an observatory for it at Hartwell House.

Smyth published his observations in 1844 as the *Celestial Cycle.* Its first volume is general in character; the double star, cluster, and nebula data are collected in the second volume, which is often known by its subtitle: the *Bedford Catalogue.* This work at once became popular, both as the first manual for the amateur observer and for its delightful style and wealth of curious historical lore. In 1845, the Royal Astronomical Society awarded Admiral Smyth its gold medal for the *Bedford Catalogue,* and in the same year elected him president for a two-year term. His personal influence was great, for his jovial character won him a very wide circle of friends.

Fourteen years after Smyth died in 1865, the *Bedford Catalogue* became the center of bitter controversy. A number of cases had been reported of discrepancies between Smyth's descriptions of double stars and later observations. This led a hitherto little-known young English amateur, Herbert Sadler, to examine the catalogue in detail. He submitted his findings for publication in the *Monthly Notices* of the Royal Astronomical Society. Through accident the paper was

422 THE BEDFORD CATALOGUE.

= − 284″·68; while Mr. Baily assigns the value of its proper motions only, on the following scale:

Æ + 0″·47 Dec. + 0″·02

Flamsteed, in whose day it followed ν Aquilæ to the meridian, seems to have observed it only once, viz. on June 3, 1691; "and then," says Mr. Baily, "at the special request of Sir Christopher Wren." Such a slight has, however, been more than made up of late by the attentions of Messrs. Knorre, Encke, and Stratford, who have given its mean apparent place for every day in the year, in their respective ephemerides. This, however, could only be an accommodation to fixed observatories and grand instruments.

In sharing the somewhat unmerited honour of daily duty with Polaris, in the computations of the *Nautical Almanac*, it will be recollected that this is not the only encroachment which δ has made upon α. This star appears on the Catalogues as Yildun; but the epithet happens to be miscopied from Hyde's notes on Ulugh Beigh for *Yilduz*, the "star," in Turkish, and signifying *Yilduz Shemáli*, the North Star, *i. e.* α Ursæ Minoris.

DCLIV. 22 M. SAGITTARII.

Æ 18h 26m 25s PREC. + 3s·66
DEC. S 24° 01′·4 —— N 2″·31

MEAN EPOCH OF THE OBSERVATION . . . . 1835·57

A fine globular cluster, outlying that astral stream, the Via Lactea, in the space between the Archer's head and bow, not far from the point of the winter solstice, and midway between μ and σ Sagittarii. It consists of very minute and thickly condensed particles of light, with a group of small stars preceding by 3m, somewhat in a crucial form. Halley ascribes the discovery of this in 1665, to Abraham Ihle, the German; but it has been thought this name should have been Abraham Hill, who was one of the first council of the Royal Society, and was wont to dabble with astronomy. Hevelius, however, appears to have noticed it previous to 1665, so that neither Ihle nor Hill can be supported.

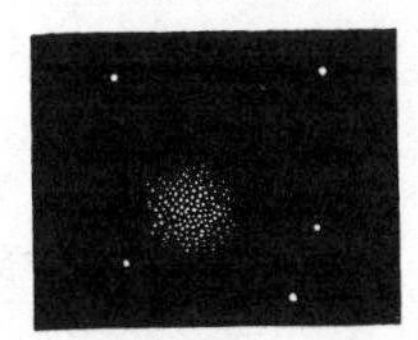

In August, 1747, it was carefully drawn by Le Gentil, as seen with an 18-foot telescope, which drawing appears in the *Mémoires de l'Académie* for 1759. In this figure three stars accompany the cluster, and he remarks that two years afterwards he did not see the preceding and central one: I, however, saw it very plainly in 1835. In the description he says, "Elle m'a toujours parue tres-irrégulière dans sa figure, chevelue, et repandant des espèces de rayons de lumière tout

A sample page from Smyth's *A Cycle of Celestial Objects* indicates something of its antiquarian flavor. At the top of the page are the concluding sentences of the account of Delta Ursae Minoris. The account of Messier 22 that follows includes a drawing he made while observing with the 5.9-inch refractor.

Admiral W. H. Smyth was a prominent figure in English astronomy during the early Victorian era, not only for his observations and books but also for his convivial and warm personality. Assisted by his bowl of punch, he was a favorite guest at scientific dinners. "Smyth was a genial companion and a quaint pleasant writer," according to his friend Augustus de Morgan. Here he is depicted with his 5.9-inch refractor. From *A Cycle of Celestial Objects*.

not adequately examined by referees, so its content was unknown until it appeared in print in the January, 1879, number.

The startled readers of this issue found an elaborate attack on Smyth's honesty, listing many cases where Sadler suggested that Smyth had copied double star measures by previous observers, instead of actually determining the position angles and distances himself. A single example will suffice. The star Gamma Persei has a distant companion first measured by Sir John Herschel, who gave its position angle (number of degrees counted eastward from north in the sky) as 224°.9, which later measurements showed to be a misprint for 324°.9. Nevertheless, the position angle was cited by Smyth as 226°.0 for the epoch 1837.6.

Sadler's enthusiastic attack gave deep offense to many. The surviving friends of the admiral bitterly and loudly resented the imputation of dishonesty, and the members of the Royal Astronomical Society were antagonized by the tactless criticism of its award of the gold medal to Smyth. At a tumultuous meeting of the society, censure of the council was urged for having allowed the offending paper to be printed, and Sadler's resignation was demanded.

It was not until a year later, in 1880, that the true nature of Smyth's double star observations was finally clarified and the dispute settled. S. W. Burnham, already the leading American expert on double stars, in 1879–1880 re-examined the stars of the *Bedford Catalogue* with the 18½-inch refractor at Chicago (now at Dearborn Observatory). This revealed that Smyth's observations fell sharply into two classes: well-known, moderately close pairs for which Smyth's measurements were uniformly of fair quality; and wide pairs not measured by anyone before Smyth, for which his data were often grossly in error. A satisfactory explanation of this was promptly given by E. B. Knobel's thorough study of Smyth's original records. First, for wider pairs Smyth often had merely estimated position angles and distances roughly, but unfortunately printed them in such a way that they could be mistaken for measurements. Next, many pairs had been measured with a double-image micrometer of unusual construction that did not fix the position angle uniquely, and Smyth was led to choose one of several possible values by comparison with data by earlier observers. In addition, Knobel found many arithmetical errors in Smyth's reductions.

The result was that Smyth's honesty was entirely vindicated, but his double star work lost its reputation for reliability. It is now also known that the unusual colors that Smyth ascribed to faint stars are in general fallacious – a point worth mention as they are sometimes quoted even today.

Despite its defects – many of which were removed in the 1881 second edition – Smyth's *Celestial Cycle* remains a classic that is still delightful reading for its quaint style and out-of-the-way information. It will long be a cloudy-night solace to the amateur observer – if he knows where to get a copy!

13. The adventures of C. H. F. Peters

In any gathering of leading American astronomers around 1870, amid such worthies as Simon Newcomb, S. P. Langley, and Joseph Winlock you might also see a slender man of about 60. His light hair and accent would suggest a north German or Danish origin. On seeking an introduction, you would meet Christian Heinrich Friedrich Peters, the celebrated asteroid hunter of Hamilton College.

On further acquaintance, you would realize that this soft-spoken man was an extremely versatile and capable astronomer. And if you won his confidence, he could tell fascinating stories of his earlier life in many countries as a scientist, soldier, political refugee, and astronomical controversialist. Peters was also a gifted linguist who made several trips to European libraries to study Greek and Arabic manuscripts of Ptolemy's star catalogue. "Of his personality it may be said that it was extremely agreeable so long as no important differences arose," wrote Simon Newcomb. "Those traits of character which in men like him may be smoothed down to a greater or less extent by marital discipline were, in the absence of any such agency, maintained in all their strength to his latest years." These traits of Peters' led to several dramatic episodes.

C. H. F. Peters served as principal astronomer during the active era of Litchfield Observatory, Hamilton College. Courtesy Peter J. Millet, Hamilton College.

Peters was born in Schleswig, then Danish but now German, on September 19, 1813, the son of a clergyman in the little village of Coldenbüttel. He attended the University of Berlin, studying mathematics and astronomy under the famous J. F. Encke, and received his doctorate in 1836. But the 23-year-old astronomer found jobs scarce, and after failing to get a position at Copenhagen Observatory he went to Göttingen University for further studies.

There Peters made friends with the geologist Sartorius von Walterhausen and accompanied him to Sicily, where until 1843 they engaged in a detailed scientific examination of Mount Etna. The young Dane made so good a reputation that he was offered the directorship of Catania Observatory, but he accepted instead an appointment as head of the geodetic survey of Sicily. Some of his time was spent in Naples, where in 1845–1846 he observed sunspots systematically with a 3½-inch refractor and discovered a comet on June

26, 1846. This object (1846 VI) had an estimated period of 13 years, but it has never been detected on any later return.

In 1848, when Sicily revolted against the tyrannical misrule of King Ferrante II of Naples, Peters sided against the government and was deprived of his post. He escaped on an English ship to Malta but quickly returned to Sicily and joined the patriot army. As a major of engineers under Gen. Ladislaw Mieroslawski (a soldier of fortune who led other insurrections in Poland and Germany), Peters fortified Catania and Messina. But the Neapolitan army quickly overran the island, and in mid-1849 he again had to flee, first to France and then to Turkey.

At Constantinople Peters learned Turkish and Arabic and made many friends, including the American ambassador, George P. Marsh. The Sultan Abdul-Mejid II proposed sending the astronomer on a scientific expedition to Syria and Palestine, but the plan was abandoned with the outbreak of the Crimean War in 1854. At Marsh's suggestion, Peters decided to seek his fortune in the United States.

Arriving in this country later that year with letters of recommendation from European astronomers, he first visited Harvard Observatory. He met other astronomers in 1855 at the Providence, Rhode Island, meeting of the American Association for the Advancement of Science. There he read a paper on the solar rotation period as derived from his sunspot observations at Naples.

Peters then obtained a position with the U. S. Coast Survey at Washington, D. C., as a protégé of Benjamin Apthorp Gould. He quickly showed his ability as a practical astronomer. A year or two later, when Gould became director of the newly founded Dudley Observatory in Albany, New York, he brought Peters as his assistant.

In this way Peters soon became embroiled in the notorious dispute between Gould and the trustees of Dudley Observatory. This group of Albany businessmen had financed the observatory and wanted to retain control of it. Gould, on the other hand, regarded it as a research institution to be run by a scientist. The misunderstanding deepened into a bitter and complex dispute, aggravated by the fact that Gould was dividing his time between the Coast Survey and the observatory. Irked by Gould's slow progress, the trustees proposed that Peters take over the observatory's operation. The European

Benjamin Apthorp Gould (1824–1896) was founder and first director of Cordoba Observatory. Photograph from the National Observatory of Argentina.

astronomer compounded the offense to Gould by discovering a comet at Dudley Observatory on July 25, 1857, and naming it Olcott's Comet after one of the most prominent trustees. A complete rupture ensued between Gould and Peters, and the latter resigned from his Coast Survey post. This job had paid him only $540 a year, hardly enough to live on even then.

For a few months the trustees allowed Peters to live in an apartment at the observatory, until the indignant Gould forced him out. But meanwhile Olcott and his friends had made arrangements by which Peters finally found a safe haven only a short distance away.

In 1858 Peters accepted a call to become professor of astronomy at Hamilton College in Clinton, New York, a few miles southwest of Utica. The year before the first steps had been taken to build an observatory. Enough gifts and subscriptions were collected to erect a two-story building 27 feet square with a cylindrical dome 20 feet in diameter. Inside was a 13½-inch refractor (some references say a 13-inch), one of the largest in the United States at the time. (This telescope was the work of Charles A. Spencer (1813–1881) of Canastota, New York, a small town east of Syracuse. Although famous

The 13½-inch refractor at Litchfield Observatory. Courtesy Peter J. Millet, Hamilton College.

for his excellent microscope objectives, Spencer is almost forgotten as a maker of astronomical optics; he is not even mentioned in H. C. King's standard *History of the Telescope.)*

It was an insecure life at Hamilton College Observatory for the first few years, as may be seen from the correspondence between Peters and his close friend George P. Bond, second director of Harvard Observatory. Bond wrote to Peters on January 7, 1863: "What you say of the financial prospects with which you begin the new year, nearly completes the list of twenty-five observatories *started* (not *founded*) within the past twenty years in the United States and left to die of want." In his reply of February 1st, Peters told Bond: "Lately for a day I was in Albany to speak with a lawyer about the payment of my last year's salary. The trustees here, too, will find that there are 'fighting' astronomers."

An improvement in Peters' fortunes came in 1867 when a railroad magnate, a Mr. Litchfield from nearby Delphi Falls, gave the college funds to pay its astronomer a modest but regular salary. The director of the renamed Litchfield Observatory also became Litchfield professor of astronomy. Peters used the 13½-inch by day to measure the positions of sunspots and by night to hunt for new asteroids. The sunspot work was continued assiduously for a decade, but the results remained unpublished until 1907, long after Peters' death.

His first discovery of a minor planet was 72 Feronia, which he at first supposed was 66 Maia. Peters became the most successful asteroid hunter of his generation, eventually achieving 48 discoveries. All these finds were made in the old-fashioned visual way, the observer laboriously comparing a star chart (usually plotted by himself) with the heavens to pick out interlopers.

Some of the names Peters selected for his new minor planets throw interesting sidelights on his tastes and activities. For example, 102 Miriam was discovered August 22, 1868. E. S. Holden explains: "The name of his asteroid Miriam (sister of Moses) was chosen in defiance of rule, and of malice aforethought; so he could tell a theological professor, whom he thought to be too pious, that Miriam also was a 'mythological personage.' "

Peters' 21st and 22nd minor planets were discovered on the same night, June 3, 1875. Their names Vibilia and Adeona are those of minor Roman goddesses of journeyings and homecomings, respectively. They were chosen to commemorate Peters' return from New

Zealand, where he had led the United States expedition to observe the transit of Venus on December 8, 1874. His station was in the South Island at Queenstown, where he arrived on October 16th and stayed for three months; although the weather was bad on the day of the transit, he managed to make a good timing of internal tangency as Venus entered upon the solar disk. It is characteristic of Peters' energy and stamina that he found his last new asteroid, 287 Nephthys, in 1889 when he was almost 76 years old.

His planet hunting was done in combination with the preparation of a new series of star charts with a limiting magnitude of 11. He made over 100,000 approximate determinations of star positions during this work. On a scale of about two inches to a degree, each chart covers 20 minutes in right ascension by five degrees in declination. Peters' plan called for 182 charts to cover the ecliptic zone of the sky, but he published only 20 of them. They appeared in 1882, just when astronomers in several countries were beginning to realize the vast potential of stellar photography for mapping the sky. It is no coincidence that large visual star-charting programs at Paris and Marseilles were abandoned at about the same time as was Litchfield Observatory's.

Peters had now achieved a prominent and highly respected niche among American astronomers. Although about 70 he was still hale and active, in an era when there was no compulsory retirement age for professors. He could reasonably anticipate a serene old age, divided between astronomical studies and tending his famous roses, which grew all over the walls of the observatory.

Actually, his misadventures were not over. Yet to come was his 1879 controversy with James C. Watson (1838–1880) and Lewis Swift (1820–1913). Watson was then the brilliant director of the University of Michigan's observatory at Ann Arbor, discoverer of 23 minor planets, and author of a famous textbook on orbit theory. Swift was the celebrated comet hunter whose story appears in Chapter 15. At the time in question he was still a self-trained amateur living in Rochester, New York.

Ever since 1859—when U. J. J. Le Verrier had predicted the existence of a planet circling the Sun inside Mercury's orbit, and Dr. Lescarbault thought he had observed it moving across the Sun – the elusive planet Vulcan was eagerly sought by both professional and amateur astronomers. Here is Agnes Clerke's summary:

"The next announcement of the discovery of 'Vulcan' was on the occasion of the total solar eclipse of July 29, 1878. This time it was stated to have been seen at some distance south-west of the obscured sun, as a ruddy star with a minute planetary disc; and its simultaneous detection by two observers – the late Professor James C. Watson, stationed at Rawlins (Wyoming Territory), and Professor Lewis Swift at Denver (Colorado) – was at first readily admitted. But their separate observations could, on a closer examination, by no possibility be brought into harmony, and, if valid, certainly referred to two distinct objects, if not to four; each astronomer eventually claiming a pair of planets. Nor could any one of the four be identified with Lescarbault's and Leverrier's Vulcan, which, if a substantial body revolving round the sun, must then (as Oppolzer showed) have been found on the *east* side of that luminary. The most feasible explanation of the puzzle seems to be that Watson and Swift merely saw each the same two stars in Cancer: haste and excitement doing the rest."

The most drastic demolition of the Watson-Swift claims was in a long article by Peters, "Some Critical Remarks on so-called Intra-mercurial Planet Observations," which appeared in *Astronomische Nachrichten* in 1879. This paper is a strange blend of sharp insight and utter tactlessness. "It is therefore quite apparent to every unbiased mind," wrote Peters, "that Watson observed θ and ζ Cancri, nothing else. His well known ability in the method of searching corroborates in the same time what others did find, viz. that there was no object of any extraordinary kind in the surroundings of the Sun." Peters' main argument was that Watson had grossly overestimated the accuracy of the setting circles with which his portable equatorial refractor was furnished. These were used to determine the positions of any objects swept up during the 2.8 minutes of totality.

Peters wasted little time on Swift: "Where we have to judge from internal evidence, the first statements must have the greatest weight, since additions given later under the title of explanations are too often influenced by afterthoughts. In Mr. Swift's successive publications is perceivable so singular a gradation in the statements, that Prof. Watson's communications alone can be the subject of a scientific discussion." Both Watson and Swift published indignant rejoinders in *Astronomische Nachrichten* and reiterated their claims. But Peters was right; no Vulcan had been observed.

It is not very surprising that Peters' personality sometimes brought him into court as a litigant. Earlier I described how during his early days at Hamilton College he went to law to recover arrears in salary. The last of his misadventures was a lawsuit that achieved national notoriety.

At Litchfield Observatory, when Peters was observing comets and asteroids with the 13½-inch refractor, he was continually obliged to search through the pages of astronomical literature to find precise positions for the comparison stars he used. Clearly, it would be worthwhile to hunt through all the astronomical journals and miscellaneous collections of observations to compile a catalogue of star positions and to publish them in a single volume for convenient reference. The task involved only routine search and calculation, but it required effort on a large scale. Some progress was made on this project by Peters and his assistant Jermain G. Porter. (The latter received his master's degree from Hamilton in 1876 but left two years later and eventually became director of Cincinnati Observatory.) Then the right man came along in the form of Charles A. Borst, a graduate of the class of '81 and a favorite pupil of Peters. Simon Newcomb described Borst: "He was a man of extraordinary energy and working capacity, ready to take hold in a business-like way of any problem presented to him, but not an adept at making problems for himself." Peters hired Borst as an assistant at $500 a year (later raised to $600), first using him to do miscellaneous reductions.

In May, 1884, Borst began on the star catalogue with great vigor, taking the work home with him for his sisters Emma and Lucy to help in the calculations. By their joint efforts, mostly at home, the catalogue was completed in manuscript early in 1888. Borst reported the end of the job to his chief and submitted a proposed title page, which said that the work was performed by Charles A. Borst under the direction of Christian H. F. Peters. According to the young assistant's account, Peters tore up the paper, threw the fragments into the stove, and ordered, "Bring me the catalogue!"

This Borst refused to do, and the angry director instituted a suit *in replevin* to obtain possession of the manuscript. *Peters* v. *Borst* was held before the Supreme Court of New York, Oneida County, in 1889. The famous Elihu Root, who later was secretary of war under

Presidents McKinley and Theodore Roosevelt, was present as counsel for the plaintiff.

Both Judge Williams and the lawyers were badly confused by the astronomical technicalities in the case. In addition, the plaintiff and the defendant had totally divergent recollections of the terms under which Borst had begun work on the catalogue. Peters appears to have regarded Borst as his servant, looking upon his assistant's work as done for himself personally. Borst, on the other hand, said that Peters had advised him to undertake some special project to gain a reputation, suggesting the catalogue in particular. The confusion never was cleared up. Unfavorably impressed by the youthful Borst's sweeping criticisms of Peters' professional stature, Judge Williams decided for the plaintiff and awarded him the manuscript catalogue as well as damages.

A number of astronomers attacked the verdict as unfair to Borst in denying him any right to the work he had done. Particularly critical was Newcomb, who expressed his dissent in newspaper and magazine articles both in the United States and Europe. Newcomb was probably correct in maintaining that the dispute should never have gone to court but should have been submitted to a board of astronomers for arbitration.

Peters never did get the catalogue. On the morning of July 19, 1890, he was found dead on the doorstep of the college building where he lodged. His observing cap was on his head and a half-burned cigar in his fingers. Presumably a heart attack had struck while he was going to the observatory to commence his nightly work.

This was not quite the end of the legal tangle. Edwin R. Root, the administrator of Peters' estate, had to face an appeal by Borst. *Root v. Borst* was heard by three judges of the New York Supreme Court in September, 1892, who upheld by two to one the earlier decision in favor of Peters. Finally, in April 1894, the Court of Appeals of New York reversed the judgment, on grounds that improper evidence had been admitted, and granted a new trial. It was never held. Quite possibly neither Borst nor Peters' modest estate could afford to continue.

After Peters' death, no one was appointed to succeed him as director of Litchfield Observatory, which was allowed to fall into decay.

According to the Utica *Observer* of May 5, 1917, the building was demolished. All the instruments of value were removed and placed in storage, but the refractor's granite pier was left and still stands in front of the Sigma Phi fraternity house.

G. Harvey Cameron, who became head of the physics department at Hamilton College in 1932, later recalled that the 13½-inch Spencer objective at that time laid wrapped in purple velvet in a large iron safe in his office. Some years later, the American Optical Co. (successors to Spencer Lens Co.) asked for and obtained the lens for their museum as an example of Spencer's early work.

14. John Tebbutt, his observatory, and a probable nova

NOT FAR FROM WINDSOR, New South Wales, Australia, stands the remains of an observatory erected by the remarkable amateur astronomer John Tebbutt (1834–1916). It was sited at his lifelong home, on a low hilltop just east of town in the middle of a 250-acre tract of rich farmland, which was known as the Peninsula Estate. The name derived from the fact that the parcel was nearly surrounded by the Hawkesbury River and South Creek. Tebbutt's observatory was unusual. Its hard-working proprietor eschewed such ordinary amateur pleasures as looking at the Moon and planets or spectacular star clusters, but devoted himself to measuring comet and asteroid positions, running a time service, keeping systematic meteorological records, and from time to time making observations to determine his latitude and longitude with high accuracy. So much work of professional quality was done by Tebbutt single-handedly that it filled some 300 papers in scientific journals. His Windsor Observatory even issued a businesslike, printed, 24-page annual report. One gets the impression that this zealous amateur was trying with some success to run a one-man Greenwich Observatory in the Southern Hemisphere.

Tebbutt was also eager to spread public interest in astronomy. He

Although John Tebbutt was an amateur, his contributions to astronomy in Australia were so great that in 1984 that country honored him by placing his portrait on the back of its $100 bill.

was a prolific writer on celestial topics for Australian newspapers and magazines, and when in 1895 the New South Wales Branch of the British Astronomical Association was organized, Tebbutt became its president.

The humble beginnings of his career go far back. As a teenager, his mechanical bent led him to study steam engines, then clocks, and finally the heavens. "It dawned on me that the universe was really mechanism of the highest order," he wrote in his autobiography. So in 1853, at the age of 19, he purchased a sextant and a copy of J. W. Norie's *Epitome of Navigation*. His other equipment consisted of a 1⅝-inch telescope and a star atlas.

Comets attracted Tebbutt from the start. He observed the one of 1853 by alignments with stars, obtaining positions from which he calculated its orbit. He also determined the orbits of Comet 1858 VI (Donati) and 1860 III from his sextant observations. His enthusiasm was much enhanced by his discovery on May 13, 1861, of the great comet of that year, when it was still a faint and nearly stationary object in Horologium. So isolated was Australia in those days that Tebbutt's find, 1861 II, went unobserved in other countries until it

John Tebbutt's observatory in 1880, seen from the northwest. At left is the transit house he built in 1863, topped by the cupola of the 3¼-inch refractor. The squat pyramid next to it is the original hut (1874) for the 4½-inch telescope, which in 1879 was remounted over the office building at far right. In 1886 the 4½-inch was dismounted to make room for an 8-inch refractor. Behind the gate is the fireproof library building. From *History and Description of Mr. Tebbutt's Observatory*, by himself, Sydney, 1887.

was rediscovered by C. W. Moesta in Chile on June 10th and by E. Liais in Brazil the following night. But Tebbutt continued his observations and calculations, and he announced in the Sydney *Morning Herald* on June 13th that the Earth would pass near the end of the comet's tail on the 29th. This prediction was strikingly verified by the observations of J. F. J. Schmidt at Athens on the 30th.

Despite Tebbutt's zeal, his equipment remained modest for many years, perhaps an indication that he did not become financially independent until middle age. In November, 1861, he acquired a 3¼-inch refractor with which he made ring-micrometer observations of comet positions. In 1864 a 2-inch transit was obtained; it was used both in the meridian for time determinations and in the prime vertical to find the geographical latitude. In 1872 he bought a 4½-inch refractor for £100, which he mounted equatorially and outfitted with crossbar and filar micrometers. But it was not until 1886 that he acquired his main instrument, an 8-inch Grubb refractor that cost him £400.

These instruments were in steady use, timing phenomena of Ju-

The head of Donati's comet nearly covered the star Arcturus on October 5, 1858, when George P. Bond made this drawing. The Big Dipper is at right. From *Astronomical Engravings . . . of Harvard College Observatory*, 1876.

piter's satellites and occultations of stars by the Moon, measuring double stars, and tracking comets and minor planets. For example, Tebbutt measured positions of Comet 1898 VII (Coddington-Pauly) on 103 nights! One suspects that on a clear night any casual visitor to the observatory at Windsor was politely but firmly turned away by its proprietor, with the explanation that program work had to be done.

In fact, cometary astronomy in Australia was virtually Tebbutt's private preserve for some years. During the two decades 1880–1899, the government observatories at Sydney, Melbourne, and Adelaide produced 95, 174, and 20 comet positions, respectively, whereas this amateur's total was 700.

"In the year 1882," writes Tebbutt in the third person, "a request was made by the Scientific Society of Boston that he should use his influence for the formation of an Australian Corps of Comet Seekers, similar to that in the United States. A club of this kind has existed for some years past in that country, and nearly all the honours in recent comet discovery have been carried off by it. . . . With kind words of encouragement from friends in England he proceeded to the task, but it was unfortunately one which ended in disappointment. Much time was spent in correspondence and in the preparation of regulations for the work, but although many gentlemen were written to, only two promised to join. . . . Although there are several gentlemen in the Colonies possessed of telescopes adapted for ordinary amateur work, the idea of systematic observation, which alone can be of any benefit to science, does not present itself to their minds." This episode sheds as much light on Tebbutt's personal standards as it does on the state of amateur astronomy in Australia nearly a century ago.

What has been told about Tebbutt will help the reader form his own opinion of the problematical celestial object Tebbutt discovered in 1862, which modern variable star catalogues list as the nova V728 Scorpii. He did not suspect that he had seen anything out of the ordinary until 15 years later, when he reexamined his observations of Comet 1862 III. On October 4, 1862, he had used his tripod-mounted 3¼-inch refractor with a ring micrometer to determine the comet's place relative to two faint comparison stars. Because his charts did not show these, he resorted to the following means of identifying them, which seems to have been his standard practice. Leaving the telescope stationary, he waited until a naked-eye star – in this case estimated by him as 5th magnitude – entered the field. By timing the elapsed interval, he could tell the approximate difference in right ascension between the 5th-magnitude star and each of the comparison objects. Tebbutt repeated this procedure on the nights of the 5th, 6th, and 9th. As his 5th-magnitude star was not in any of the few star catalogues at hand, he determined its position on the last night by sextant distances from Altair, Antares, Theta Scorpii, and Epsilon Sagittarii. Tebbutt obtained coordinates that, precessed to 1950, are $17^h\ 35^m\ 31^s$, $-45°\ 27'.2$.

Reexamining these records in November, 1877, Tebbutt was surprised that none of the considerably larger number of catalogues now in his library contained this star. His curiosity aroused, he turned his 4½-inch refractor to the spot on November 13th and found that the 5th-magnitude object was missing. The only stars seen in the field were of magnitudes 10 and 11.

Did the Australian amateur make a mistake of some kind, or was he the sole observer, on four nights, of a nova in 1862? From Tebbutt's own businesslike report of the affair, published in the *Monthly Notices* of the Royal Astronomical Society for March, 1878, it appears that he reexamined his sextant observations in 1877 and was convinced that the approximate position deduced was correct. Furthermore, his journal of 1862 stated that the problematical 5th-magnitude star was visible with the unaided eye about one degree northeast of Sigma Arae and was brighter than the latter. The journal also noted that there was a "small round nebula" to the north of the star and in the same field of view. All this checks out, the "nebula" being the globular cluster NGC 6388 at $17^h\ 32^m.6$, $-44°\ 43'$ (1950). The only change needed in Tebbutt's description is to call his star 4th magnitude instead of 5th, as the visual magnitude of Sigma Arae, which it outshone, is 4.6.

15. Lewis Swift and the lives of a 16-inch refractor

PERHAPS LEWIS SWIFT would never have become one of the best-known American astronomers of his time if he had not fractured his left hip when he was 13. This mishap in 1833 permanently lamed him, and gave him the opportunity to gain an education instead of laboring from dawn to dark on the family farm in Monroe County, western New York state. For three years he attended the little academy in Clarkson, every school day trudging two miles on crutches.

During this formative time he gained a taste for science, and he saw the epochal Leonid meteor shower of 1833 and later the great comet of 1843. But it was not until about 1855 that Swift became an avid amateur astronomer. Inspired by some astronomical books of

Lewis Swift with his 4½-inch refractor on the roof of Duffy's cider mill. This picture, originally published in the Rochester *Union and Advertiser* for August 13, 1892, was based on a photograph taken in 1880. At right is the market basket in which Swift carried the optical parts.

Thomas Dick that he had purchased from a peddler, he acquired a damaged 3-inch objective for five dollars. He made a brass mounting and an eyepiece for this telescope and began a survey of the sky. About 1860 he bought a 4½-inch refractor by Henry Fitz. This he used on a platform attached to his barn in Marathon, New York, where he had moved.

Swift discovered his first comet, 1862 III, with the 4½-inch on July 15, 1862. He did not at first recognize it as a new object, thinking he had seen 1862 II. It was independently found three days later by Horace P. Tuttle at Harvard Observatory in Cambridge, Massachusetts. Comet Swift-Tuttle became a fine, bright object that had a tail 25 degrees long. With a period of 120 years, it travels in the same orbit as the Perseid meteors.

Swift was now well launched on his life's work of comet hunting.

But to support his family, he moved to Rochester in 1872 and opened a hardware store. He soon attracted attention among his fellow townsmen for his unusual avocation and for his public lectures on astronomy.

To have an unobstructed view of the sky when searching for comets, he observed from the flat roof of Duffy's cider mill on White Street, half a mile from his home. "One cannot discover comets lying in bed," Swift used to say. Roused by his alarm clock, he would walk to the mill, carrying the telescope lenses in a market basket. To get to his observing place, where the telescope tube and mounting were kept, he had to climb three ladders and then walk some 100 feet across a sloping roof. Especially on an icy night, it was quite a trip for a middle-aged man with a lame leg.

Despite these difficulties, Swift was able for a while to average a new comet each year by discovering 1877 III, 1878 I, 1879 II, 1880 IV (an unexpected return of Tempel's comet 1869 III), 1881 II, and 1881 VIII.

Rochesterians were now aware that the enthusiastic amateur astronomer and popular lecturer was an internationally known figure. Civic pride swelled. A wealthy patent medicine manufacturer, H. H. Warner, offered to build a substantial observatory for Swift if local citizens would raise the money for a 16-inch refractor. Warner was a vigorous promoter, and soon the telescope was ordered from Alvan Clark and Sons of Cambridge, Massachusetts.

Thus, at a cost of $100,000 the Warner Observatory was built on a lot at the corner of East Avenue and Arnold Park, not far from the mansion of the patent-medicine king. Swift was made director in 1882, enjoying the use of the fourth largest telescope in the United States.

Warner in 1881 announced that he would award a gold medal and $200 in cash for the first American discoverer of each new comet. Furthermore, if less than five "Warner Safe Remedy Prizes" were awarded in any one year, their value would be increased to total $1,000. Swift was to judge the awards. Soon after, the appearance of a naked-eye comet in August, 1881, brought him a flood of 3,000 letters claiming the prize. Warner solved the problem by offering a substitute prize for the best essay on comets.

The next few years were happy ones for Swift, who for the first time in his life was financially secure. He found comets 1883 I, 1889

Swift and the 16-inch Clark refractor at Warner Observatory, photographed about 1890. Near the base of the observing chair is an elaborate spectroscope; on the shelf at the near side of the pier are a filar micrometer and comet eyepieces. At lower left is what may be Swift's 4½-inch. Courtesy Blake McKelvey.

VI, and 1892 I and also discovered about 900 new nebulae. The latter achievement was of considerable astronomical importance in those days when celestial photography was still primitive.

Probably Swift also enjoyed his role as a showman. "Every Tuesday and Friday evening, from eight until ten, the observatory was open to the public," we read in an article by two Rochester historians, Ralph Bates and Blake McKelvey. "Visitors could gain admit-tance on other days (except Wednesday and Sunday, reserved for research) by applying first for a twenty-five-cent ticket at Warner's patent medicine establishment on St. Paul Street. It was the first observatory of any size in the world to which the public was welcome. . . . Visitors to the city were invariably driven out to see it. . . . In recognition of the public service of the Warner Observatory, the state legislature passed a bill exempting it from taxation."

This paradise began to fray. By the late 1880's Warner had lost much of his interest in the observatory. He continued to pay Swift's bills but stopped the comet awards. There had been much new construction in downtown Rochester, and city lights and smoke started to interfere seriously with the work of the 16-inch. Also, Swift had little contact with the local scientific community. He never joined the Rochester Academy of Science or had much to do with the University of Rochester. The largely self-taught old man probably felt uncomfortable among the professionals.

Warner's business empire collapsed during the financial panic of 1893, and no one else offered to support the observatory. Still eager to continue his astronomical work, Swift prepared to go to California at the invitation of one of 19th-century America's more colorful and versatile characters.

The many careers of Thaddeus Lowe, a New Hampshire man born in 1832, included meteorology, aviation, chemical engineering, and real-estate development. He became wealthy from his important inventions in the manufacture of artificial ice, illuminating gas, and coke, and he eventually moved to California. About 1889 he formed a plan to build a cog railway up Mount Wilson and erect a hotel near the present site of Mount Wilson Observatory, but he was thwarted by the interests then controlling the mountain. Lowe floated a company to build an electric railway from Altadena into the San Gabriel Mountains west of Mount Wilson. By 1893 the line extended from downtown Altadena up Rubio Canyon to Echo Mountain, elevation about 3,500 feet. Lowe built two resort hotels on Echo Mountain and added a zoo, a museum, bridle paths, and an observatory to attract tourists. It all sounds rather like a 19th-century Disneyland.

This was the observatory to which Lowe invited Swift, who came to California in 1893, bringing with him his 16-inch Clark refractor. It was reerected on the slope above the Echo Mountain House. Here the veteran astronomer divided the nights between public demon-

The elegance of Thaddeus Lowe's Echo Mountain resort is suggested by this photograph, showing the observatory, Echo Mountain House, and a chalet; the San Gabriel Valley lies in the distance. From *Publications* of the Astronomical Society of the Pacific, 7, 1895.

strations for visitors and searching for new comets and nebulae in the marvelously clear skies.

To his list of cometary discoveries were added 1895 II, 1896 III, and 1899 I. The last of them was found when Swift was 79 years old. About 300 more nebulae were added to his lists. Assisting the old man was his teenage son Edward D. Swift, who himself discovered Comet 1894 IV.

Meanwhile Lowe extended his railway four miles farther into the San Gabriel Range to Mount Lowe Springs. There he built another even more luxurious hotel, the Alpine Tavern, for $450,000 in 1895. But at this very time a major economic depression set in, and Lowe encountered serious financial difficulties. The railway went into receivership, and he had to mortgage his home in Pasadena.

Lewis Swift stuck it out at Echo Mountain Observatory until 1900, after his eyesight had started to fail. He went back to his old home in Marathon, where he died on January 5, 1913, at the age of 92. In his later years he was practically deaf and almost blind.

The great astronomer E. E. Barnard, who knew and admired Swift, described his disposition as genial and happy, and praised him for his generous aid and advice to struggling amateurs. Swift's valuable contributions to astronomy, despite his adversities, won him many honors from scientific organizations at home and abroad.

On August 11, 1900, a new director (and sole staff member) took charge of Echo Mountain Observatory. This was Edgar Lucien Larkin, who had been born in a log cabin in La Salle County, Illinois, on April 5, 1847. Like Swift, Larkin was a self-made and largely self-educated man. In 1879 he built a private observatory at New Windsor, Illinois, equipped with a 6-inch Clark refractor. This telescope in 1888 was transferred to Knox College at Galesburg, Illinois, where Larkin was in charge of the observatory until 1895.

At Echo Mountain, he seems to have devoted his time chiefly to public nights for the thousands of visitors who came on the scenic railway. After Lowe's failure, the line and hotel were bought in 1902 by the Pacific Electric Railway Co., which continued to operate them.

One well-known astronomer, William H. Pickering, spent several months at Echo Mountain Observatory in 1904, using the 16-inch to continue his studies of the Moon and Jupiter's satellites. But such episodes seem to have been exceptional.

About that time, Larkin acquired a volunteer assistant in Charles S. Lawrence, a photographer for the railway who sold pictures to visitors. Lawrence took an active part in the public demonstrations, and after Larkin's death in 1924 he became the last director of Echo Mountain Observatory.

Fire is a major hazard in the arid mountains that fringe greater Los Angeles, and the observatory had narrow escapes from destruction in February, 1900, and December, 1905. On the latter occasion all the buildings on Echo Mountain burned down except the observatory. The flames came so close that the 16-inch objective was removed from the telescope and lowered into a water tank for safety.

However, the end of the observatory came not from fire but during a windstorm in 1928 that blew the building away. Lawrence was inside the observatory at the time. He escaped without a scratch although timbers fell around him and the lecture room sailed away from him several hundred feet through the air.

The observatory was never rebuilt, and hardly any trace of Lowe's development remains. By 1936 the last of the hotels had burned down. Sightseers no longer were taking "The Greatest Mountain

The remains of Echo Mountain Observatory as they appeared in 1971. This view, toward the northwest, shows the dome support wall and pier for the 16-inch refractor. Courtesy David Kaplan.

Trolley Ride in the World," which ceased to run in November, 1939.

Phoenixlike, the 16-inch refractor has survived. In 1941, the University of Santa Clara bought it from the Pacific Electric Railway Co., and this telescope was reerected in its third observatory.

16. Garrett P. Serviss and some Brooklyn amateurs

IF YOU WERE an American amateur astronomer in the mid-1800's, perhaps your interest in the sky was first aroused by a view of the great comet of 1861, or the nova of 1866 in the Northern Crown, or the total solar eclipse of 1869. You would have acquired a few books like Thomas Dick's *The Practical Astronomer* or Elijah Burritt's *Geography of the Heavens,* and, if your means permitted, you might have bought a small Clark or Fitz refractor.

But in many ways your hobby would have been very different from today. Telescope making was a difficult and mysterious art restricted to a relatively few gifted practitioners, while celestial photography was virtually nonexistent in those days of slow and cumbersome wet plates. Since there were few or no regional societies, you would very probably be following your hobby in isolation, with the newspapers as your main source of facts about current astronomical events and discoveries. There were no popular magazines in America specifically devoted to astronomy (one founded in Cincinnati by O. M. Mitchel in 1846 lasted barely two years, and there may have been a few others still more obscure and ephemeral). In fact, amateurs were starved for information until the *Sidereal Messenger* began publication in March, 1882, edited by Prof. W. W. Payne of Carleton College in Minnesota. The appearance of this monthly, which survived as *Popular Astronomy* until 1951, quickly upgraded amateur astronomy and spread it enormously.

In this context we can see the special importance of a long series of well-written, informative articles on astronomy and other sciences that were published on the editorial page of the New York *Sun* beginning in the late 1870's. These excellent unsigned articles continued for over a decade. They attracted much attention nationwide because of their juxtaposition to the *Sun's* politically influential editorials.

Several years later it became known that the anonymous "*Sun's* astronomer" was Garrett P. Serviss, a copy editor on that newspaper who had been keenly interested in the heavens since boyhood. Born in 1851 at Sharon Springs, New York (some 35 miles west of Schenectady), he graduated from Cornell in 1872 and studied law at Columbia, being admitted to the New York State bar in 1874. Immediately afterward he turned to journalism as a career, first with the New York *Tribune* and later with the *Sun,* at which he became night editor in 1882.

Serviss lived in Brooklyn. He had a small observatory containing a 3⅜-inch refractor by John Byrne at his home at 8 Middagh Street not far from the eastern approach to the Brooklyn Bridge, which

Garrett P. Serviss (1851–1929) was an immensely popular astronomical author, editor, and lecturer. Sketch by Steven Simpson.

opened in 1883. At that time night skies must have been quite dark there, for electric lights were few; the first in Brooklyn was installed in a Fulton Street store in 1878. Looking southwestward across the Upper Bay toward New Jersey, Serviss could watch from 1884 to 1886 the advancing construction of the Statue of Liberty on Bedloe's Island (now Liberty Island).

Not far from Serviss another well-known amateur, Stephen Van Culen White, lived in the Columbia Heights section. This prosperous broker owned a 12-inch Fitz refractor with an objective that had been refigured by Alvan Clark and Sons in 1867. The fine telescope had previously belonged to Jacob Campbell, a New York City banker, whose telescope, observatory, and house were bought by White.

One user of White's 12-inch later became a prominent professional astronomer and educator. Sarah Frances Whiting used to bring her pupils at Brooklyn Heights Seminary to look through this telescope during the late 1870's. Later, when she was professor of astronomy at Wellesley College in Massachusetts, she persuaded Mrs. John

By profession a shorthand expert, Henry Parkhurst used his spare time to invent astronomical gadgets, observe variable stars, edit a magazine, print books, and construct an artificial language. From *Popular Astronomy,* April, 1908.

Whitin, a trustee of the college, to buy the telescope in 1898. It is still in use today at Whitin Observatory on the Wellesley campus.

Probably the most active observer in Brooklyn was Henry M. Parkhurst (1825–1908). This talented if slightly eccentric court stenographer was a prolific inventor of astronomical equipment, especially visual photometers, with which he made many observations of variable stars. During the 1880's Parkhurst worked mainly with a 9-inch Fitz refractor at his home.

This geographical concentration of persons sharing an interest in astronomy led in 1883 to the formation of a society of which Serviss was the secretary. It must have been one of the first groups of its kind in the country and "was originally designed to be a National Society. . . ." An editorial note in the *Sidereal Messenger* for April, 1884, tells us: "In Brooklyn, N. Y., and vicinity, there is an amateur astronomical society. Mr. S. V. White, the broker, of 210 Columbia Heights, Brooklyn, is president, and William T. Gregg, of the same place, is treasurer. One Washington astronomer seriously doubts whether an amateur society can possibly carry so large a name as the 'American Astronomical Society,' and says it has no right to if it can."

This Washington astronomer was Simon Newcomb, director of the Nautical Almanac office. He was a very able mathematician and administrator who had risen at an early age to the top of his profession without ever having been an amateur himself. He evidently wanted the name reserved for a national professional society, such as he was to help found in 1898.

By 1885 the American Astronomical Society in Brooklyn (as it actually called itself) had 79 members, but only a handful were very active, since published references repeatedly give the same few names but no others. A view of the society in action can be gleaned from the account that Serviss wrote for the *Sidereal Messenger* of the meeting of October 5, 1885. This was little more than a month after the news of the discovery of a bright new star, now known as the supernova S Andromedae, close to the nucleus of the great Andromeda nebula (see Chapter 77). The meeting was largely devoted to this startling event.

Parkhurst was there and described how the new star had faded in a seemingly irregular manner from magnitude 8½ in early September to about 12. Similar observations were reported by John H. Eadie of Bayonne, New Jersey, a well-known amateur specializing in variables and meteors. Both men alluded to a curious behavior of the new star: occasional large oscillations in brightness that happened almost instantaneously and were not mimicked by other stars in the field. This same flickering was also seen by the famous E. E. Barnard with a 6-inch refractor at Nashville, Tennessee. In an excellent description published in the same issue of the *Sidereal Messenger,* he pointed out that this phenomenon was most marked after the new star had faded greatly. Barnard convincingly explained it as a familiar effect of poor seeing upon a faint star lying in bright nebulosity.

The little magazine that the Brooklyn society published is quite a bibliographical rarity among astronomical Americana. Issued in August, 1885, the first number of *Papers Read Before the American Astronomical Society* contained 32 printed pages of short popular articles on why the Moon has no air or water, variable stars, the structure of the universe, the Sun's temperature, sunspots and the weather, asteroids, a comet, and telescopic ghost images. The second installment, dated March, 1887, was largely occupied by a mathematical study of the Sun's motion relative to the solar system's center of gravity. The article was penned by George W. Coakley, professor of mathematics and astronomy at the University of the City of New York. In the Harvard Observatory library's Xerox copy of the *Papers,* there is only one later number, dated January, 1888, and entirely filled with a technical account by Coakley of P.-S. Laplace's nebular hypothesis. Frankly, it is dull stuff.

Only in 1978 was the later history of the society revealed, by Trudy Bell in the *Griffith Observer.* A crisis developed in 1888 when the Brooklyn Institute, in the process of reorganizing, suggested that the American Astronomical Society affiliate with it. The membership had to decide whether to continue attempting to unite the nation's astronomers or become merely a local group. On May 17th the society chose the latter course, voting to become the Astronomical Department of the Brooklyn Institute. Under this new banner the society flourished, and by 1910 it had 283 members. This reconstituted organization continued at least into the 1950's.

The one amateur in the old Brooklyn circle who made a lasting name for himself was Serviss. For over half a century he lectured and wrote superbly lucid books and articles that perhaps reached a wider reading public than any astronomer's who lived before him. Two of the best and most famous of his books are *Astronomy with an Opera Glass* (1888) and *Pleasures of the Telescope* (1901). The latter is a charming account of the deep-sky objects visible in a small telescope, complete with old-fashioned constellation charts. Although full of information, it breathes the relaxed air of a comfortable age when astronomy was much simpler than today.

If you have access to a large public library, these and any other Serviss books that you may encounter are well worth looking into. In addition to astronomical works, he also wrote several volumes of science fiction, including *The Conquest of Mars,* 1898; *The Moon Metal,* 1900; and *A Columbus of Space,* 1911.

Serviss died at Englewood, New Jersey, on May 25, 1929. His close friend, Clyde Fisher, of the American Museum of Natural History, wrote in *Popular Astronomy* for August-September 1929: "We believe that Garrett P. Serviss has done more to popularize astronomy than any one in America, and perhaps in the entire world."

17. The legacy of S. W. Burnham

TOGETHER WITH Edward E. Barnard, Maria Mitchell, and George E. Hale, the name of Sherburne Wesley Burnham (1838–1921) is prominent among the folk heroes of American astronomy. His fame is secure for his discovery of over 1,300 visual double stars, which include some of the most interesting and most difficult binaries; for his great *General Catalogue of Double Stars;* and for his many thousands of accurate micrometer measures at Dearborn, Washburn, Lick, and Yerkes observatories. One well-qualified authority, Robert G. Aitken, dates the beginning of the modern period of double star astronomy from Burnham's entry into the field. Nevertheless, these were the accomplishments of a self-trained amateur who did not become interested in astronomy until he was grown up, and who remained a nonprofessional nearly all his life.

Shorthand writing was the other thread in his life. He taught it to himself before leaving his native Thetford, Vermont, in 1858 for New York City, where he earned his living by it. During the Civil War, after New Orleans surrendered to Admiral Farragut in April 1862, Burnham served there as a shorthand reporter at General Butler's Union army headquarters.

The story is often repeated how in New Orleans Burnham chanced upon a book auction and out of curiosity bought a copy of Burritt's *Geography of the Heavens.* Its star charts are said to have turned his attention to astronomy. However, some prior interest is suggested by Burnham's own statement that "when in London, about 1861" he purchased a cheap 3-inch refractor on a table tripod, which was good for landscape use but disappointing astronomically.

He settled permanently in Chicago in 1866, about the time he obtained a 3¾-inch portable refractor, "just good enough to be of some use, and poor enough, so far as its optical power was concerned, to make something better more desirable than ever." For some reason, the young amateur concentrated his attention almost exclusively on double stars, using as his reference a copy of the first edition of T. W. Webb's *Celestial Objects for Common Telescopes.*

Sherburne Wesley Burnham was 19th-century America's outstanding authority on visual double stars. He remains famous for his discovery of 1,340 new pairs, for his many thousands of accurate measurements, and for his monumental *General Catalogue of Double Stars*. During most of the 40 years in which he used large telescopes at several American observatories, he was an amateur astronomer who earned his living as clerk of the U. S. District Court in Chicago.

His astronomical appetite was whetted by the fact that his house was only two blocks from the stone tower of the old Dearborn Observatory, with its 18½-inch Clark refractor erected in 1866. The director, T. H. Safford, gave Burnham free access to the observatory library, limited though it was.

Alvan G. Clark, of the Massachusetts firm of telescope makers, went to Iowa to observe the total eclipse of August 7, 1869, and stopped in Chicago on his way home. Burnham happened to meet him, and as a result ordered a 6-inch refractor from Alvan Clark and

Sons for $800, stipulating that "its definition should be as perfect as they could make it, and that it should do on double stars all that it was possible for any instrument of that aperture to do."

Early in 1870 the 31-year-old amateur mounted his new 6-inch refractor in a small observatory he had prepared in the backyard of his house at 36th Street and Vincennes Avenue. That telescope's optical excellence matched the keenness of its owner's eyesight, and his career of discovery was fairly launched.

Working eight hours a day as a shorthand writer and court reporter, Burnham spent his clear evenings hunting for doubles that were not listed in Webb or the Dearborn Observatory's books. The first installment of the results appeared in the March, 1873, *Monthly Notices* of the Royal Astronomical Society, as "Catalogue of Eighty-one Double Stars, Discovered with a Six-inch Alvan Clark Refractor."

Because the 81 pairs in this list are in order of right ascension, the first of them to have been discovered, on April 27, 1870, is numbered β40. The discoverer described it as 52 minutes of arc north of 12 Scorpii and a little west, with estimated position angle and separation of 350° and 4″, respectively, magnitudes 8 and 10½.

As received by Burnham, the 6-inch had no driving clock, but he improvised a satisfactory one. Having no micrometer, he had to estimate position angles and separations instead of measuring them. Measurements for many of the early Burnham pairs were made by Baron Ercole Dembowski, a veteran double star observer of the first rank who worked in Italy.

Within a year of his first catalogue, Burnham published a second and a third in the *Monthly Notices,* together with a paper on William Herschel's double stars. In addition, he began active correspondence with Webb and other astronomers interested in doubles. In 1874 he became a fellow of the Royal Astronomical Society, on Webb's nomination. That same year Burnham spent a summer vacation in New Hampshire, and on 10 nights he had the use of the 9.4-inch refractor at Dartmouth College Observatory for double star discovery. The most notable part of this trip was his brief stop at Washington, D. C., where he was allowed to use the U. S. Naval Observatory's new 26-inch refractor for one night. Of this visit we are told by Capt. W. de Abney, in his address to the Royal Astronomical Society when Burnham was awarded its gold medal in 1894:

Edward S. Holden (1846–1914) was the first, though controversial, director of Lick Observatory. In 1889 he organized the Astronomical Society of the Pacific, which continues today as an almost unique body supporting interests of both amateur and professional astronomers. Courtesy Mary Lea Shane Archives of Lick Observatory.

"The fifth catalogue has 71 new pairs, and brings out a peculiar characteristic of your medallist. If a star disc deviated an almost infinitesimal quantity from the circular, his eye detected it at once. In 1874, in Washington, on the night of August 11, he scanned some of his old discoveries, with the result that he [added] 14 new pairs to his list. I give one instance. No. 291 in the catalogue had on some occasion offended his critical eye when he looked at it through the 6-inch, so he turned the 26-inch on it and found that it consisted of two 8½-magnitude stars separated by a distance of only 0″.2."

On that visit to the Naval Observatory, Burnham gained a highly influential friend in Edward S. Holden (1846–1914). It was the support of Holden and also of G. W. Hough (1836–1909), who succeeded T. H. Safford as director of Dearborn, that supplied the final ingredients for Burnham's career.

In 1881 Holden left the Naval Observatory to become director of

The Lick 12-inch refractor as it appeared in 1886, two years before Burnham began using it for his double star work. This etching is from *Publications* of Lick Observatory, *1*, 1887.

Washburn Observatory at Madison, Wisconsin. Burnham was invited to join the Washburn staff, and he worked there for five months with the 15.6-inch refractor before going back to Chicago. Holden later became president of the University of California and first director of Lick Observatory. As before, Holden asked the double star expert to join him there.

Because living conditions on top of Mount Hamilton were not suitable for a family, Burnham settled his wife, six children, and brindle dog Hector in San Jose, near the western foot of the mountain. Burnham was working at the observatory by August 5, 1888, when he took measurements of the close pairs Beta Delphini and Zeta Sagittarii with the 12-inch refractor. E. E. Barnard, who was one of the younger members of the original Lick staff, told many years later of his first impressions of Burnham: a thin, rather small man with light blue eyes and a cordial greeting, whose gentle manners and keen wit won friends quickly.

The exceptionally favorable observing conditions on Mount Hamilton were already known to Burnham, for in 1879 he had visited it with his 6-inch refractor to make seeing tests for the Lick

The great Lick 36-inch refractor, also depicted about the time Burnham worked at the observatory. The telescope's objective was made by A. Clark & Sons, the mounting by Warner & Swasey. From *A Text-Book of General Astronomy* by Charles A. Young, 1900.

trustees. He had made a second visit in 1881 with Holden to observe that year's transit of Mercury with the 12-inch telescope, already erected in a temporary shelter.

The four years Burnham spent at Lick were the most fruitful of his astronomical life. With the 36-inch and 12-inch refractors he made enough double star measurements to fill a sizable volume of the Lick *Publications*. The number of pairs discovered by him increased to 1,274. Some of these were of types hitherto virtually unknown: double stars with separations under 0.2 second of arc and close doubles with components of very unequal brightness.

Barnard gives this vignette of Burnham as an observer: "In the early days at Mount Hamilton, when conditions permitted, he worked through the entire night. Near midnight, when one's vitality always gets low, we would meet in his office or in mine and have a little lunch consisting of coffee and crackers and cheese, the coffee being made by Mr. Burnham over a lamp chimney by a little device that would hold a small pot. With this stimulant the rest of the night would go by in work on the measurement of double stars. . . . His rapidity in measuring was remarkable. This was partly due to the fact that he always made three settings of the wires before recording the readings, keeping them in his memory in the interval."

Meanwhile, Burnham pursued a favorite activity in the fine library of Lick Observatory. This was the compilation of a manuscript catalogue of the observations of all double stars north of declination $-31°$, which he had begun two decades before when his 6-inch showed him many more doubles than he could conveniently identify. Thereafter, he spent much effort in keeping this catalogue up to date and in trying to get it published. Holden had long encouraged this project, which appealed to his bibliographical bent.

But there were also many relaxations. Wiry of build and physically tireless, Burnham loved to go for a day's tramp with Barnard in the rough country around Mount Hamilton. He was an enthusiastic photographer of landscapes and animals, but curiously he avoided celestial photography, except for sharing in a lunar program of Holden's. Barnard tells of one occasion when Burnham was taking a picture of the 36-inch refractor, using a very small aperture and an exposure of an hour or longer to get maximum sharpness. "Having arranged everything for the photograph with the camera on the balcony, he started the exposure and went down to his office until it was finished. In the meantime some one, showing a visitor round, came into the large dome. . . . He moved the telescope about and raised and lowered the floor and turned the dome, the usual program for the edification of visitors. When he got through he left everything in a different position from that in which he had found it. Finally Mr. Burnham came back onto the balcony, removed his plate-holder and went into his dark room. . . . Here he had rigged up an oscillator for very long development. At that time he believed in developing very slowly – an hour or more – with a very weak developer. The result of the above exposure was that when the finished negative was examined there was one pier but on it were several 36-inch telescopes and the floor and the inside of the dome were in an awful mix-up."

All this came to an end when Burnham resigned in June, 1892. His departure attracted much attention in the San Francisco newspapers, which blamed Holden, who had chronically bad relations with the press. However, Holden has been absolved of blame in F. J. Neubauer's well-informed history of the observatory: "The fact was that Burnham could not see any sense to his looking after the two establishments he had to maintain, one on the mountain, the other for his family in San Jose where his children attended school. In short, it was a money question. He could always go back to work as clerk of the U. S. District and Circuit Court in Chicago at a much higher salary than he received at the Lick Observatory as an astronomer."

That was just what Burnham did. But in Chicago he still maintained his interest in astronomy. In 1894, when Yerkes Observatory was under construction, he accepted a staff appointment. After the great 40-inch Clark refractor was finally ready for use in 1897, he was assigned two nights a week with it. His custom was to take the Saturday afternoon train from Chicago to Williams Bay, observe Saturday and Sunday nights, and then on Monday to go back by the early morning train to Chicago and his courtroom duties. Burnham continued to commute after 1902, when he resigned the clerkship to devote full time to astronomy.

During this period, he was deeply engaged in updating the manuscript catalogue of double stars. For this work, he used the 40-inch to obtain thousands of measures of neglected pairs. At last, with the support of George Ellery Hale, he was able in 1906 to have the Carnegie Institution of Washington publish his *General Catalogue of Double Stars Within 121° of the North Pole,* two volumes. This monumental work contains data on 13,665 double and multiple systems, with a meticulous observational history for each. It was a great stimulus to double star astronomy, for now observers could draw up working lists of pairs that were most in need of measurement, to avoid wasting telescope time on unnecessary measures. Orbit calculators too found it a boon.

The analysis and interpretation of this treasury of observations were left by Burnham to other astronomers. Singlemindedly he confined himself to a narrow specialty in which his keen eye and taste for orderly record-keeping brought golden rewards. He had no taste for astrophysics or mathematical complexities. A deep mistrust and even a hostility to new astronomical ideas and techniques can be seen in some of Burnham's scientific writings.

Burnham continued to work at Yerkes Observatory until 1913, but his interest shifted to faint optical companions of naked-eye stars. Maintaining that such companions were background objects without appreciable proper motions of their own, he used them as reference points from which the proper motions of the bright stars could be determined. But the great labor that the old man put into this program was too late, because even at that time photographic methods were supplanting the filar micrometer for differential proper-motion work.

Burnham died after a long illness in Chicago early on the morning of March 11, 1921, and all that day the flags at Yerkes Observatory hung at half mast.

18. Harvester of the skies

IMAGINE AN EARLY EVENING STROLL about a century ago, along a dirt road winding through the rich farmlands and orchards of Ontario County in western New York. As you approach the village of Phelps there is a red cottage set near the road. Glancing into its garden, you see a tall, bearded man about 40 years old. He is looking at the western sky through what is unquestionably a telescope of fair size. You walk over to him and make the acquaintance of the village photographer, William R. Brooks. The telescope, he explains, is a 9¼-inch reflector on an altazimuth stand. Yes, he made the mirror himself, on a home-built grinding and polishing machine. On this clear evening he is busily searching for comets. Although Mr. Brooks is obviously glad to meet someone else interested in astronomy, he seems eager to resume his observing, and so you bid him good evening.

This preoccupied man in the garden, you later learn, happens to be known around the world as a discoverer of comets. By the time of his death in 1921, he will have joined Charles Messier, Jean-Louis Pons, and Edward Emerson Barnard as one of the most successful comet hunters of all time.

Today, though the name Brooks is attached to about two dozen objects moving around the Sun, it is hard to find traces of him as a man. Working in comparative isolation, he made few personal contacts with other scientists. Red House where he worked in Phelps has long been torn down, and perhaps the most tangible remaining symbol of his astronomical career is the collection of his record books preserved at Hobart College. The fullest source of information about Brooks is a series of four newspaper articles by E. Thayles Emmons in the Geneva (New York) *Times* for March 14–17, 1960. Mr. Emmons, who knew the comet hunter personally, kindly sent me clippings of these articles, from which the following account has been borrowed extensively.

Brooks was born at Maidstone, England, on June 11, 1844, and seems to have first become interested in astronomy as a small boy during a voyage to Australia. The family moved to New York State when he was 13, and the next year he saw Donati's comet through a homemade spyglass. The young man worked as a draftsman before settling in Phelps in 1870. There he constructed a 2-inch f/18 achromatic refractor and a 5-inch f/10 silver-on-glass reflector. With the latter telescope, Brooks discovered his first comet in October, 1881, sharing the catch with W. F. Denning in England. This was Comet 1881 V, a particularly interesting object of short period (8.69 years). Success whetted the ardor of the New York amateur, and he built his 9¼-inch telescope for more effective searching. On

William R. Brooks (1844–1921) discovered comets at the rate of nearly one a year for three decades. Beginning as a self-taught amateur, this skillful observer eventually became professor of astronomy at Hobart College. Postcard courtesy Paul Luther from a private collector.

October 21, 1882, he chanced upon a companion to the great September comet of that year, some eight degrees east of the main object. Next, about the time of full-Moon rising on February 23, 1883, Brooks picked up 1883 I near Beta Pegasi. It was independently discovered by Lewis Swift at Rochester, New York. Particularly important was an extremely faint nebulosity in Draco, swept up on September 1st of the same year. It turned out to be the long-awaited Comet Pons of 1812 on its first observed return.

Those were the days when a persevering and fortunate amateur could earn a modest living by finding comets. A wealthy patron of astronomy, H. H. Warner of Rochester, New York, regularly rewarded each new discovery with a $200 prize (see Chapter 15). Perhaps spurred by this source of income, Brooks achieved the notable feat of sweeping up three comets in five weeks, on April 17, 30, and May 22, 1886. The story is often repeated how another impoverished photographer of that same era, E. E. Barnard in Nashville, Tennessee, built for his new bride a "comet house" financed in part by several Warner prizes.

By now the amateur of Phelps had become a famous figure, not only among the world's astronomers but also among his neighbors. In nearby Geneva lived a well-to-do nurseryman and amateur astronomer, William Smith. In April, 1888, Smith brought Brooks to Geneva to take charge of a fine new private observatory equipped with a 10⅛-inch refractor. Here further continual patient searching met with renewed success. Comet 1888 III was discovered in Ursa Major on August 7, 1888. It was notable for a tail pointing toward the Sun, not away as is typical of comets. At that time Smith Observatory, as it was called, was not quite completed. Therefore, Brooks had set up his telescope on a platform in the garden of his house next door.

One example of the uncertainties in keeping straight the statistics of comet discoveries is provided by 1889a. Brooks found this cometlike object in Sagittarius on January 14th, but it was not seen by anyone else despite a special search at Lick Observatory. No such ambiguity is attached to 1889 V, which turned out to be a fine periodic comet that returns about every seven years. When Brooks first saw it, this comet was in Cetus, moving so slowly that it remained nearly six days in the same telescopic field.

In 1900, Brooks became professor of astronomy at Hobart College in Geneva. I am not sure whether his later hunting was done at Smith Observatory or with Hobart instruments. At any rate, his record of finds continued to grow. The last discovery claimed was on October 20, 1912. Brooks may be said to have died in harness, for Emmons relates that the elderly man's labors in photographing

One of Brooks' most spectacular discoveries was Comet 1911 V. Its long, straight gas tail is evident in this photograph taken October 19th of that year at Yerkes Observatory.

Comet Pons-Winnecke contributed to the physical collapse that ended in his death.

William Brooks' unusual career brought him some gratifying rewards during his later years, such as the Lalande medal of the Paris Academy of Sciences. Yet he seems to have outlived a large part of his fame, for his passing on May 3, 1921, received only brief mention in astronomical magazines.

It is not easy to say exactly how many comets William Brooks discovered. Emmons gives the total as 27, but he includes some objects seen earlier by other astronomers. Published lists of comets differ among themselves. For a safe minimum, we may follow Brian G. Marsden's 1979 *Catalogue of Cometary Orbits*. It contains 17 different comets named after Brooks alone, and five others bearing his name linked with some other. Thus the total is at least 22.

19. The many Moons of Dr. Waltemath

AMONG THE WORLD'S popular astronomical magazines is *Die Sterne,* published in East Germany. It contains much news about what European professional and amateur astronomers are doing. Also, while browsing through old issues, you will encounter curiosities such as the note "Discovery of a Second Earth Moon?" in 1926.

The author of this item, Cuno Hoffmeister, mentioned that a German amateur named W. Spill was convinced he had seen an Earth satellite cross the Moon's disk on May 24, 1926. But Hoffmeister went on to recall some much more sweeping claims made a quarter century earlier by "a Dr. W . . . in Hamburg (the name does not deserve being saved from oblivion)."

This worthy, it appears, had "published a series of *Astronomical Reports* with the subtitle *Organ of the Union for Investigation of the Dark Moon of the Earth*. His abusive attacks on leading astronomers such as W. Schur and Wilhelm Förster can hardly be equaled. All the discoveries he announced were his own. He discovered not only a third moon but asserted the existence of a whole system of midget moons, concerning which he made strange speculations. Since his moons were not seen as bright objects, he maintained that they absorbed all incident sunlight. His foolishness at that time found few converts; the time was not yet ripe."

The identity of Dr. W . . . is given in *Popular Astronomy* for 1898, which chides another popular journal for giving nearly a column to Dr. G. Waltemath's supposed second satellite of the Earth. "That which will interest astronomers about this second Moon is the alleged fact that its orbit has been computed. Its distance from Earth is 640,000 miles; its diameter is 435 miles. It is faint generally, and can be seen only with the aid of a large telescope. 'Sometimes,' says Dr. Waltemath, 'it shines at night like a Sun,' and he thinks Lieut. Greely, when in Greenland in 1881, saw this little Moon, and by mistake regarded it as the Sun, though it was ten days after the time the Sun should have been invisible. . . . It is not probable, at all, that *anyone* having *any* knowledge of the elements of astronomy would be misled by such statements."

Behind *Popular Astronomy's* caustic note was a flood of newspaper publicity given at the start of 1898 to Waltemath's prediction that his second moon would pass in front of the Sun on February 2nd, 3rd, or 4th of that year. Surprisingly, the professional journal *Astronomische Nachrichten* published a detailed observational report of this event only a few weeks later: "On a Phenomenon Observed February 4, 1898, in Greifswald," by Martin Brendel.

That article told how a dark body was seen crossing the Sun by 12 persons at the Greifswald post office (Herr Postdirektor Ziegler, members of his family, and postal employees). They observed by naked eye, without protection against glare. It is easy to imagine a preposterous scene: an imposing-looking Prussian civil servant pointing skyward through his office window, reading Waltemath's prediction aloud to a knot of respectful subordinates who solemnly agree with everything he says.

On being interviewed, these witnesses spoke of a dark object having one fifth the Sun's diameter that took from 1:10 p.m. Berlin time to 2:10 to traverse the solar disk. The editor of the *Nachrichten* added a footnote expressing his disbelief of the story. And in fact it was soon proved that there had been a mistake. During the very hour in question, the Sun was being scrutinized by two experienced astronomers, W. Winkler in Jena and Baron Ivo von Benko of the Austrian naval observatory at Pola. Both men reported to the *Astronomische Nachrichten* that only a few ordinary sunspots were on the disk.

The failure of this and later forecasts did not discourage Waltemath, who continued to issue predictions and ask for verifications. H. H. Turner, director of Oxford University Observatory, expressed the understandable impatience of a busy scientist when he wrote in August, 1898:

"Waltemath might be tolerated as comparatively harmless were it not that one has to answer so many times the genial query, 'Oh! by the way, how about these new Moons?' and even to write letters about them. The time may come when people with new theories will have to be made to deposit a substantial sum, to be forfeited in case the bubble bursts; and those who want to have questions answered must pay a consultation fee of two guineas. It would probably make them value the replies a little more; and one could buy some nice piece of apparatus with the money obtained. But I fear this will not be arranged before the next century."

20. A South American tragedy

AS THE 20TH CENTURY OPENED, one of the bright lights among young German astronomers was Friedrich Wilhelm Ristenpart. He was already well known in 1891, when at the age of 23 he obtained his doctor's degree at Strassburg, then the best graduate school of astronomy in Germany. His thesis was a study of fundamental astronomical constants and the Sun's motion through space, as indicated by the proper motions of stars. For the rest of his life, star positions and motions remained Ristenpart's favorite astronomical subject.

This able young man was also an enthusiastic observer. At the observatory in Karlsruhe, where he became W. Valentiner's assistant in 1891, he would often during the same night make observations

F. W. Ristenpart at age 28. He tried to build in South America the world's largest observatory; his failure led to suicide. Courtesy Mary Lea Shane Archives of Lick Observatory.

with the meridian circle, use the zenith telescope for latitude determinations, and finally measure comet positions with the refractor.

In 1898, when he moved to Kiel Observatory, Ristenpart began a huge project: collecting all the hundreds of thousands of star positions from all the precise catalogues of the 18th and 19th centuries, and publishing these positions, reduced to the same epoch, in a single master catalogue. Starting in 1900 the Prussian Academy of Sciences supported the enterprise financially. Today the many volumes of this *Geschichte des Fixsternhimmel* fill a large shelf in observatory libraries, and they have been of very great value to astronomers concerned with stellar positions and motions.

Ristenpart had moved to Berlin in 1900 to supervise the bureau of the Prussian Academy where this work was being carried out. But even this labor did not sate his seemingly boundless capacity for work. In 1904 he also became lecturer in astronomy at Berlin University, making a name as a very effective teacher and public speaker. In addition, he found time for occasional observations of comets with the 12-inch refractor of the Urania Observatory in Berlin.

From this one can see something of the intense energy of the man, and also indications of restlessness and a desire to observe rather than spend all his efforts among the books and papers of his office. Ristenpart was ambitious and wanted a wider sphere of activity. With this in mind, we can picture the mood in which he studied an invitation that reached him early in 1908 from the government of Chile, offering him the opportunity to establish a major astronomical observatory.

In the first years of the 20th century, the Chilean observatory at Santiago had fallen into relative inactivity. Its director then was Jean Albert Obrecht, an Alsatian who had studied astronomy at Paris, but whose astronomical work in South America was minor. Obrecht seems, however, to have been a prominent and well-liked public figure. In 1906, Pedro Montt became president of Chile. Being interested in astronomy, he took steps toward a thorough reorganization of the national observatory. This resulted in the invitation to Ristenpart to become director.

Sometime in July, 1908, the German astronomer decided to accept a five-year contract with the Chilean government. He left Berlin on August 20th and arrived at Santiago during the first days of October. Ristenpart immediately set to work with his characteristic enthusiasm and vigor to raise Santiago to the level of a first-class observatory. A new site was chosen, spacious and farther from the city lights, a greatly enlarged staff was enrolled, the construction of new buildings and the erection of large, modern instruments were pushed. Much money was needed, but it could be obtained through the influence of President Montt.

The reorganized observatory began intense activity. Its staff numbered 40 at times during Ristenpart's administration, and the personnel were organized in three sections. The meridian department secured within a few years almost 20,000 transit-circle observations of stars. The equatorial department, whose chief was the able German astronomer Richard Prager, made many measures of the positions of comets and asteroids on a systematic plan, including a fine series of 139 positions of Halley's comet during its 1910 return. Another German, Walter Zurhellen, directed the astrophotographic department and began work on the sky zone between declinations $-17°$ and $-23°$, which had been allotted by the Paris conference of 1887 as Santiago's share of the ambitious *Carte du Ciel* mapping project (see Chapter 81). Ristenpart undertook many other programs. There were observations to follow the variation in latitude, work on star charts of the southern sky down to magnitude 10, and total solar eclipse expeditions to Argentina in 1908 and Brazil in 1912. The latter expedition failed doubly; it rained during the eclipse, and the instruments were lost in a shipwreck of the returning party!

When Ristenpart left his native land, the Prussian government had given him a two-year leave, so that he could return to a secure position if affairs in Chile did not work out. A few months after his arrival in South America, Ristenpart felt that conditions were so propitious that he decided with characteristic optimism to stay in Chile for the rest of his life.

This optimism doomed him. His loyal friend, President Montt, died suddenly on August 16, 1910. From this time on, the Chilean government lost interest in the new observatory, which began to appear more of a financial burden than a cultural asset. There was growing resistance to Ristenpart's requests for construction funds until, finally, the ambitious plans (which had called for the erection of no fewer than 29 buildings on the 27-acre site) had to be postponed to the indefinite future.

There were other troubles. Ristenpart was a hard man to get along with: forceful and self-assured, he made immediate decisions and held to them vigorously. His extraordinary capacity for work led him to make great demands on his staff – too great, they thought. Many of his people were government appointees who had little or no real interest in astronomy, and they could not understand the director's motivation. Ristenpart, on the other hand, did not understand people.

Thus there arose an ever-deepening rift between the director and his staff, until the differences could no longer be kept inside observatory walls. He was forced to accept the government's plan that he turn over the administration to a Chilean while he retained direction of the scientific work. No solution to these problems came and, finally, Ristenpart's contract expired on February 15, 1913. The failure of his glowing plans depressed him deeply, and the strains of the last few months had undermined his health and destroyed his ability to work. After deciding to return to Germany, he was momentarily buoyed with hopes for the future. But on the early morning of April 9th he took his own life.

The German and Chilean accounts of Ristenpart's final crisis differ. The former suggests intrigues by subordinates, while the latter cites government action against an unbalanced autocrat. It is difficult to know what part politics and antiforeign feeling may have played. But undoubtedly the seed of the disaster was in Ristenpart's own personality.

Upon Ristenpart's death, the leadership of the observatory was temporarily given to the same man who had been appointed to investigate the charges against him. Then, on June 1, 1913, the popular but ineffective Obrecht again assumed the directorship, and the great dream was over.

21. The curious career of Leo Brenner

THE NAME of Leo Brenner is seldom heard today except, perhaps, in connection with the large but inconspicuous lunar crater that his German selenographer friend Philipp Fauth (1867–1941) called after him. This present obscurity clashes sharply with the sudden fame that Brenner gained around 1895 as a gifted amateur observer of the Moon and planets. But this international reputation soon began changing to notoriety, and eventually he made a dramatic end to his astronomical career.

This strange man and his actual accomplishments are hard to disentangle from the legends that he invented. Nearly everything that Brenner said or wrote about himself was intended to deceive. A small example of the resulting confusion may be found in H. P. Wilkins and P. Moore's *The Moon,* 1955. The six-line biographical sketch of Brenner on page 356 contains not a single correct statement except for his birth year (1855), his place of birth (Trieste, then in

Leo Brenner (Spiridion Gopchevic), from a photograph taken in 1890 and reproduced in M. Heim's *Spiridion Gopchevic,* 1966.

Austria), and that he built an observatory at Lussinpiccolo where he and his wife observed the Moon and planets.

"Leo Brenner" was one of the pseudonyms of a remarkable Balkan political adventurer, Spiridion Gopchevic, whose life was carefully researched by the German historian Michael Heim. The biography by Heim, together with a judicious use of Brenner's own extensive astronomical writings, makes it possible to view Brenner's career more clearly, and to glimpse some of the demons that drove him.

At the end of the last century, the Austrian flag flew over the chain of picturesque islands along the eastern coast of the Adriatic Sea. Lussinpiccolo, the main town on Lussin, one of the northern islands of the chain, was at that time a fashionable health resort. It was here that Gopchevic settled in October, 1893, prosperous because he had married a well-to-do wife a few years before and because he was receiving a political subsidy from the Austrian government. It was

Manora Observatory, where Leo Brenner and his wife observed from 1895 to 1909. In his astronomical writings he always referred to her by the pseudonym of Frau Manora. From M. Heim, *Spiridion Gopchevic*.

here that he adopted the name of Leo Brenner and with great enthusiasm took up astronomy as a new career.

He settled in a villa on 20 acres of gardens with a 15-year lease. Upon the roof of this house he erected his Manora Observatory, whose wooden dome contained an excellent 178-mm (7-inch) refracting telescope made by Reinfelder and Hertel. A filar micrometer, star spectroscope, camera, and other accessories were purchased, together with the beginnings of an astronomical library that eventually contained over 4,000 volumes. Regular observing with the 7-inch began on May 9, 1894. From the start, Brenner devoted himself primarily to the study of the Moon, Venus, Mars, and Jupiter. Lussinpiccolo was evidently a very favorable site for astronomical work. Brenner extolled its mild climate, clear, dust-free air, and steady seeing. The last quality he associated with the small day-to-night temperature variation.

Brenner spared no pains to publicize Manora Observatory and its work. For example, prominent amateurs and professional astronomers were often invited to visit. He began to publish many observational articles, at first chiefly on Mars and Jupiter, in such periodicals as *Astronomische Nachrichten, Observatory,* the *Journal* of the British Astronomical Association, and the *English Mechanic and World of Science.*

The good initial impression that Brenner made on his contemporaries is not hard to understand. In 1895 the systematic study of plantetary surfaces was confined to a few workers and was still mainly descriptive, limited to visual inspection and drawing. The shining examples of G. V. Schiaparelli and Percival Lowell offered exciting prospects of discovery by a sharp-eyed observer working at a first-rate site.

But the findings that Brenner so confidently announced became more and more extreme. From his numerous drawings of dusky markings on Venus, he proclaimed in 1895 that the rotation period of that planet was 23 hours 57 minutes 36.2396 seconds. In the following year he revised the seconds to 36.3773. Also in 1896, he triumphantly announced the rotation periods of Mercury and Uranus to be 33¼ hours and 8 hours 17 minutes, respectively. Today we know that Brenner's value for Mercury is clearly wrong. But, tantalizingly, his period for Uranus is almost exactly half of the best, albeit uncertain, value available. When a definite period is established, perhaps the changing features shown in his drawings can be reconciled with it.

Likewise remarkable was Brenner's map of Mars from his observations in 1896–1897; it showed no fewer than 164 canals, most of them new. Fully 18 of them radiated in various directions from one dark spot, Trivium Charontis. Brenner was later to claim that he had detected 34 Martian canals using only a 3-inch refractor.

And there was also the matter of the companion of Sirius, which had passed through the periastron point of its orbit in 1894. Between March, 1891, when S. W. Burnham last measured it with the Lick 36-inch refractor, and August, 1896, when R. G. Aitken again saw

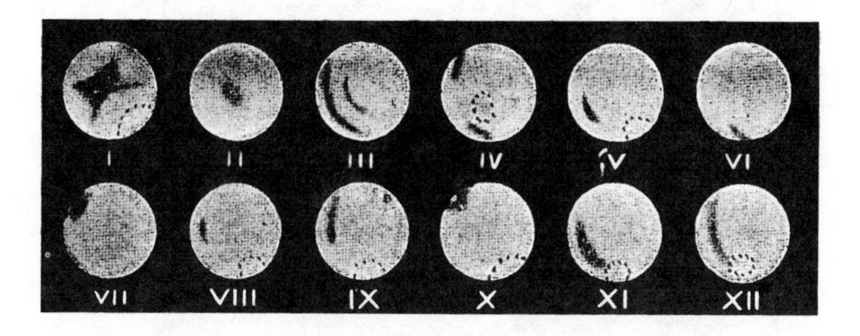

Leo Brenner's observations of Uranus from April 28, 1896, to June 9th. From A. v. Schweiger-Lerchenfeld's *Atlas der Himmelskünde,* 1898.

it with the same telescope, the faint companion was lost in the glare of its brilliant primary. Nevertheless, Brenner published in the *Astronomische Nachrichten* what purported to be micrometer measurements of Sirius B with his 7-inch on two nights in March, 1897.

The kindest interpretation of these varied marvelous observations by Brenner is that he could not distinguish between what he imagined and what he saw. The astronomers whose plaudits Brenner sought turned away from him, and Heinrich Kreutz, editor of the *Astronomische Nachrichten,* finally declined to print any more of his contributions.

Brenner aggravated his worsening relations with the astronomical world by vicious abuse of anyone who disagreed with him. For example, Percival Lowell in early 1896 had visited Lussinpiccolo and carried out Venus observations with him. Lowell, however, did not subscribe to Brenner's 24-hour rotation period and therefore came under heavy attack for both his Venus and Mars work.

Another vendetta was conducted against E. M. Antoniadi and Camille Flammarion at Juvisy Observatory in France, who had the temerity to differ with some of Brenner's 1896–1897 Mars work. Antoniadi was from 1896 to 1917 the director of the British Astronomical Association's Mars section, and the wanton alienation of this influential figure must have damaged Brenner's repute in French and British amateur circles. A third target of Brenner's scurrilous attacks was the Vienna Observatory. How they began I do not know, but for years he libeled its equipment and staff.

Astronomische Rundschau 1899 Tafel I.

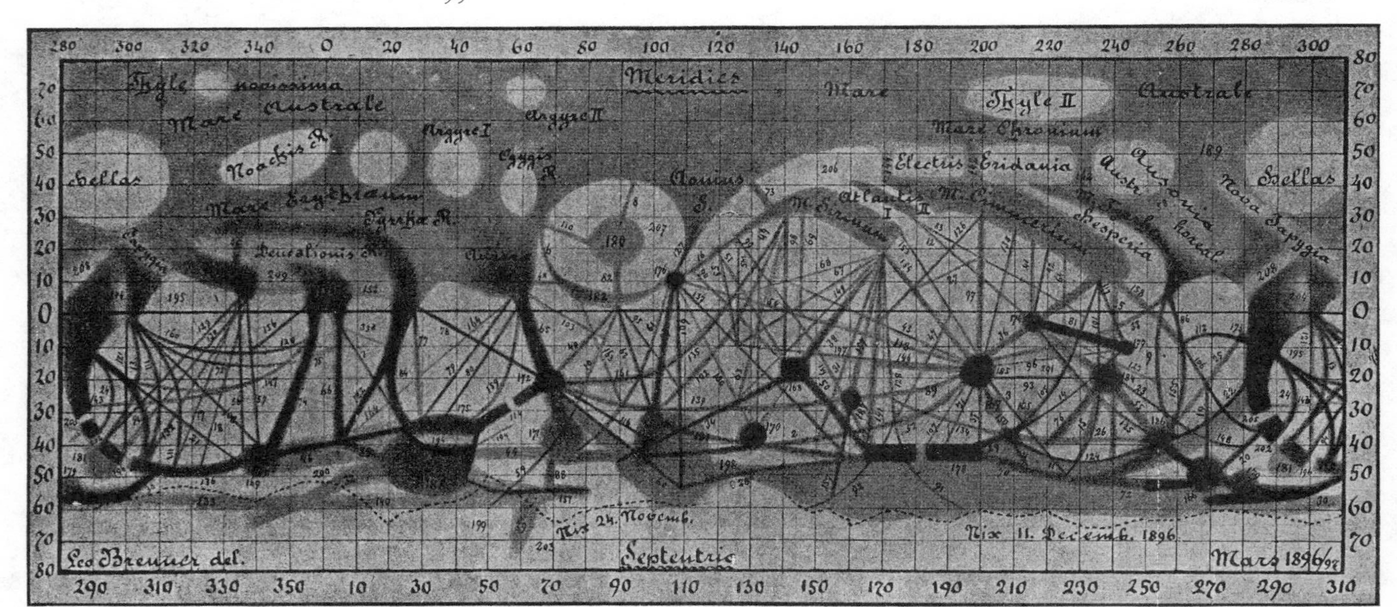

From observations in 1896–1897, Leo Brenner prepared this chart of Mars' surface markings. The illusory pattern of extremely narrow "canals" resembles that depicted by many of his contemporaries (such as Giovanni Schiaparelli and Percival Lowell). From *Astronomische Rundschau*, 1899.

Because he was no longer a welcome contributor to scientific journals, Brenner in 1899 started his own monthly magazine, *Astronomische Rundschau* (Astronomical Review). It was, however, much more than a place to publish his own observations. Each issue contained information about new astronomical developments culled from the professional literature. There were articles by well-known astronomical figures including Fauth (see Chapter 52), T. J. J. See (see Chapter 22), E. E. Barnard, and Simon Newcomb, but I suspect that often these features were copied rather than contributed. One *Rundschau* article of lasting value is that by Fauth (*3*, 172, 1901) on the lunar formation Linné. It demolishes conclusively the old misunderstanding that Linné had changed physically from a crater to a bright spot sometime around the middle of the 19th century (see Chapter 53).

Each issue had an extensive section of printed replies to Brenner's correspondents. Some were routine, as when R. F. in Vienna was told of a book where he could read how to determine the height of a lunar mountain from a measurement of its shadow length. But others are clearly fakes, inserted to further Brenner's personal enmities. For example, the distinguished Hungarian astrophysicist N. von Konkoly-Thege was treated as follows. In Vol. *1*, No. 7, of the *Rundschau,* correspondent H. N. in Vienna was informed: "The judge Konkoly-Thege in Komarom who was suspended for scandalous misuse of office is named Ludwig, and hence is not the same person as the scientist, who is named Nikolaus. However, they are brothers." The next issue added two more insults. A reply to H. N. said: "Dr. N. von Konkoly assures us that the judge in question is not his brother, although a relation." The other reply, addressed to Dr. N. v. K. himself, read: "Until the judicial inquiry is complete and the verdict published, we will not comment on the degree of guilt. The error arose from your own words. We are delighted that the *Astronomische Rundschau* has come to your attention and meets with your approval."

The *Rundschau* continued to appear until March, 1909. But Brenner's interest in astronomy had evidently been waning for a long time as his observing became less and less frequent. There appear to have been money problems, too. He states that financial aid from the Austrian government ceased. The contents of Manora Observatory were sold. In the final issue of the *Rundschau* Brenner dramatically announced that he was really Count Spiridion Gopchevic, with achievements that included command of 30 languages. Because of the shameful way that the government and scientific cliques had treated him, he said, he was going to disappear from the world of astronomy, and he did.

According to Heim's inquiries, Gopchevic was not of noble lineage, despite his claims to be a count and to be a descendant of the early kings of Serbia. His father, also named Spiridion, was a wealthy shipowner and merchant of Trieste. He was bankrupted by the Crimean War and committed suicide in 1861 when his son was six. The widow took young Spiridion and his sister to Vienna for their education but died soon after.

The picture that emerges of Gopchevic as he grew into manhood is a highly intelligent school dropout, embittered by the decline in his family fortunes and ambitious to win distinction. An opportunity came in 1875, when political tensions in the Balkans came to a head and war broke out between Montenegro and Turkey. Gopchevic went to Montenegro to seek command of a military unit on the strength of a fancied relationship with the ruler, Prince Nikola. But the prince paid little attention to the 20-year-old Austrian, who returned disappointed to Vienna where he wrote a book about the war. In this work he gave the impression that he had actually served in the field. The book became a best seller, and Gopchevic was fairly launched as a journalist.

During the next decade he was a voluminous author of polemical articles and books on Balkan politics. This is not the place to detail how he actively championed and abandoned in turn such causes as Serbian nationalism, Albanian independence, a South Slav state, and the integrity of the Austro-Hungarian Empire. These frequent changes of allegiance seem related to his tendency to alienate his backers.

At one time about 1890 he was publishing an antisemitic newspaper in Vienna. He became a military commentator, writing knowingly about strategy, torpedoes, and fortifications. He was also the author of a travel book or two and some now-forgotten novels.

Undoubtedly Gopchevic was a man with great mental curiosity and powers of comprehension, combined with an intense if somewhat unsystematic capacity for work. Yet, as Heim remarks, his was an uncreative and rather shallow mind, and his personality prevented cooperation with others. Eventually he must have realized that authorship did not give the satisfaction he was searching for. Thus he decided to become an astronomer and took the new identity of Leo Brenner.

After Manora Observatory was sold in 1909, Gopchevic went to America and lived obscurely in San Francisco for several years. According to Heim, in 1912 he composed two operas there, *The Paris September Days* and *The Life Saver.*

Returning to Austria, he worked as a propagandist during World War I in a fruitless effort to convince Serbia to change sides and become an ally of Austria against Russia. After the Austrian collapse he eked out a living in Berlin by writing. He was in such poverty that (if tradition be believed) he obtained electricity for his lodgings by bypassing the meter. Gopchevic had dropped so far out of public sight that even the year of his death is not known, various sources citing anything from 1909 to 1936.

The last known publication by this complex and tragic figure was a 1922 article about Atlantis and Lemuria. Heim aptly remarks that the search for sunken lands is symbolic of Gopchevic's restless life.

22. The sage of Mare Island

On July 5, 1962, the New York *Times* carried a long obituary of Thomas Jefferson Jackson See, who had died the day before at the age of 96 in Vallejo, California. The *Times* story identified him as an internationally known astronomer, mathematician, author, and lecturer whose career included a controversy with Einstein. American astronomers, however, had all but forgotten See except as a colorful eccentric. Yet press accounts and oral tradition alike tend to overlook several small but unquestionable claims he has to astronomers' gratitude. The actual See has been lost within the personal legend he himself created.

Let us pick up See's career on Christmas Day, 1892, as the German liner *Saale,* covered with snow and ice, steams into New York harbor after a slow and stormy voyage from Bremen. Among the pas-

T. J. J. See photographed in Berlin at the start of his doomed career. Courtesy Mary Lea Shane Archives of Lick Observatory.

sengers at the rail is a 26-year-old Missourian, returning to the United States with a newly earned doctorate in astronomy from the University of Berlin. In his pocket is a letter from president William R. Harper of the University of Chicago, offering a teaching job.

This was See at the outset of a career filled with high promise. At Berlin he had studied for three years under some of Europe's finest astronomers, mathematicians, and physicists. There he had enthusiastically specialized in double stars, measuring them with the 9-inch refractor, calculating their orbits, and writing a weighty dissertation on the origin of binary stars.

The University of Chicago to which See came had been founded the year before, but its Yerkes Observatory was not to be ready until 1897. From the beginning, the astronomy department was ably staffed by George Ellery Hale in astrophysics and Kurt Laves in celestial mechanics. Joining as an instructor, See published a useful volume of orbit determinations of 40 visual binaries, compiled with the aid of his students. The handsome young astronomer with bushy moustache was ambitious, self-confident, and articulate. He was also outspokenly convinced that he had ability unmatched by his colleagues. When he asked for and failed to receive a promotion from instructor to associate professor – the rank held by Hale – he left Chicago in the spring of 1896.

Percival Lowell invited See to come to Flagstaff, Arizona, to measure double stars with the 24-inch refractor there. Because Martian work had top priority at Lowell Observatory, the night hours when the red planet was well placed were reserved, but the remaining time was used by See and his assistant, W. A. Cogshall. During the winter of 1896–1897 Lowell temporarily relocated his observatory near Mexico City, in hopes of even better seeing than Arizona offered. The low latitude permitted See to search for new doubles as far south as declination $-65°$.

This work had merit, for southern double stars were little known at that time. But the quality of the results was affected by the principal observer's overconfidence, which led him into carelessness. One by-product was many misidentifications among the new pairs, even of naked-eye stars. W. H. van den Bos has further pointed out that all of See's position angles for about a month have the same large systematic error; evidently See did not bother to check the zero point of the micrometer anew every night.

Serious illness of Percival Lowell forced him to abandon astronomical work in 1897, an interruption that lasted two or three years. See was left temporarily in charge at Flagstaff, but his arrogant behavior toward the observatory staff created an explosive situation that ended in his being fired in July, 1898. See thereupon joined the staff of the U.S. Naval Observatory in Washington, D.C., after being appointed a Navy professor of mathematics in February, 1899. The next three years were mainly devoted to extensive micrometer work at the 26-inch Clark refractor, on double stars, faint planetary satellites, the minor planet Eros, and planetary diameters.

That year See's pride was deeply wounded in a controversy with Forest Ray Moulton, a mathematical astronomer of exceptional ability. See had proclaimed that certain apparent irregularities in the orbital motion of the binary star 70 Ophiuchi were explained by the companion having a massive dark satellite moving about it in a 36-year period. Moulton published a paper proving that such a three-body system would be highly unstable and hence that the postulated dark body could not exist. See refused to accept defeat, and wrote an ill-considered letter to the *Astronomical Journal* that led to his lifelong disbarment as a contributor.

The somewhat formal atmosphere of the Naval Observatory and its highly organized program of work could not have been a congenial environment for See. Tension and overwork contributed to a breakdown he suffered in 1902. Six months later he was transferred to the U.S. Naval Academy as an instructor in mathematics, but this arrangement lasted only one semester. A second transfer sent See to the naval shipyard at Mare Island, California, where he remained as officer in charge of the chronometer and time station until his retirement in 1930.

The catastrophic collapse of See's hopes led him to seek fame as a discoverer of the laws of cosmic evolution. Working in isolation from other astronomers, he issued a stream of publications on the internal structure of the Sun, the cause of earthquakes, the origin of the solar system, the Milky Way's size, and more.

The most comprehensive of these writings is his 1910 publication, *Researches on the Evolution of the Stellar Systems, Vol. II, The Capture Theory of Cosmical Evolution.* (Vol. I was his 1896 collection of 40 binary star orbits.) This enormous quarto of almost 750 pages contains unorthodox views on every aspect of astronomy. Its author described his task: "For after long and careful meditation I have concluded that unless some one has the courage to brush aside the erroneous doctrines heretofore current, as one would the accumulated dust and cobwebs of ages, we shall never be able to cut loose from antiquated traditions and make lasting progress in reducing Cosmogony to a scientific basis. . . . The necessity for getting rid of this dull tread-mill of stationary effort, has appeared to justify a stand not one whit less resolute than that which was taken by COPERNICUS when he laid the foundations of the true system of the world."

See's opinion of his success is quoted in the original italics: "*The Capture Theory is so overwhelmingly indicated by the most diverse phenomena of the Starry Heavens, that I cannot doubt that it represents an ultimate truth of the very first order of importance.*" Few astronomers read such books from beginning to end, yet there are some points of interest here and there in the text, which is otherwise generally unrelated to modern astronomy. Worth a look even today are See's speculations on the origin of stars from condensing dust clouds and the formation of comets.

The cool reception of this work doubtless increased See's isolation from the world of astronomers. The separation was deepened by Moulton's strongly-worded article in *Popular Astronomy* for February, 1912, entitled "Capture Theory and Capture Practice." It demonstrated that an important section of See's book had been taken without credit from Moulton's *Introduction to Celestial Mechanics,* becoming garbled in the process. After this stinging attack, See never again wrote for *Popular Astronomy,* though until then he had articles in nearly every issue.

In 1913 appeared the 298-page *Brief Biography and Popular Account of the Unparalleled Discoveries of T. J. J. See.* "Professor See is universally recognized as the most intrepid and indefatigable of the explorers of Nature," states the preface, and this adulation continues throughout the text. The author's name on the title page is W. L. Webb, but the book is clearly autobiography told in the third person. Style and viewpoint are characteristically See's, as can be verified by comparing the text with reprints of his articles included in the same volume. The evidence suggests that either he wrote the book himself or drastically revised a Webb manuscript.

Viewed in long retrospect, the book did See a grave disservice. It distracted attention from his useful activities, such as arranging for the publication in 1912 of Sir William Herschel's collected works in the magnificent London edition of two volumes. We need not regard him as another Hipparchus, but we can sympathize with the unusually painful fate of a man who saw his career collapse and who remained trapped for more than half a century in its ruins.

23. Some invisible astronomers

THE SMOOTH RUNNING of every observatory depends on certain "invisible astronomers" – who are not formally trained astronomers at all, who are never seen at professional meetings, and who never write journal articles. These are the veteran night assistants,

Godfrey Sykes and wife at home. From *A Westerly Trend,* by Godfrey Sykes, 1944.

the telescope maintenance men and instrument makers, and certain secretaries with decades of service.

Godfrey Sykes of Lowell Observatory was an outstanding example. He went to Arizona in the 1890's to help Percival Lowell erect and maintain his telescopes and stayed at the observatory in charge of its shop until about 1950. Born in England, he had been a cowboy in Kansas before meeting Lowell. Sykes was a skilled craftsman from a bygone age, who learned to do with a file precision work for which a modern mechanic would need power tools. The wooden dome of the 24-inch Clark refractor at Lowell Observatory was designed and built by Sykes. Much of the older equipment around the observatory bears signs of his ingenuity and craftsmanship. In addition, he did some work for the Steward Observatory in Tucson, being a close friend of its first director, A. E. Douglass. When I visited Lowell Observatory in 1948 I met Godfrey Sykes and re-

The dome for Lowell Observatory's 24-inch refractor (seen inside) shortly after it was built with the aid of Godfrey Sykes. Courtesy Lowell Observatory; supplied by Arthur A. Hoag.

member him as a small, very alert white-haired man, who had taken up painting as a hobby.

Another invisible astronomer was Frank Bowie at Harvard Observatory. From 1904 to about 1945 he was the night observer in Cambridge, personally exposing and developing a good fraction of the several hundred thousand sky patrol photographs in the Harvard plate stacks. All of the patrol cameras were run unguided, the usual exposure time being one hour. On a clear night Bowie would make his rounds of these instruments from dusk to dawn, changing plate-holders, turning the cameras to new sky areas on a predetermined schedule, and filling in the record books. About midnight he would find time to sit down with coffee and a newspaper in his little office and select his horses and numbers for the next day's bets. He always

Some of the earlier cameras of the Harvard Observatory patrol program. They were mounted on this special platform at the Oak Ridge station in Harvard, Massachusetts.

carried in his pocket a long folding sheet of paper that listed each day's winning number for the last decade.

About 1935 deteriorating sky conditions in Cambridge caused transfer of most of the observatory's instruments to a country site 27 miles away, the Oak Ridge station. Bowie stayed in Cambridge with the remaining instruments, while Henry Sawyer became the night assistant at Oak Ridge. Among the tens of thousands of high-quality plates that Sawyer took in the next 15 years was a series with the 16-inch wide-angle Metcalf refractor. It covered the northern sky to about magnitude 18 and was used by Harlow Shapley for his studies of the distribution of galaxies. Every graduate student and visiting astronomer who worked at Oak Ridge during those years remembers Henry Sawyer as the stogie-smoking gray-haired Yankee in the fur cap. He owned five elderly cars in varying states of

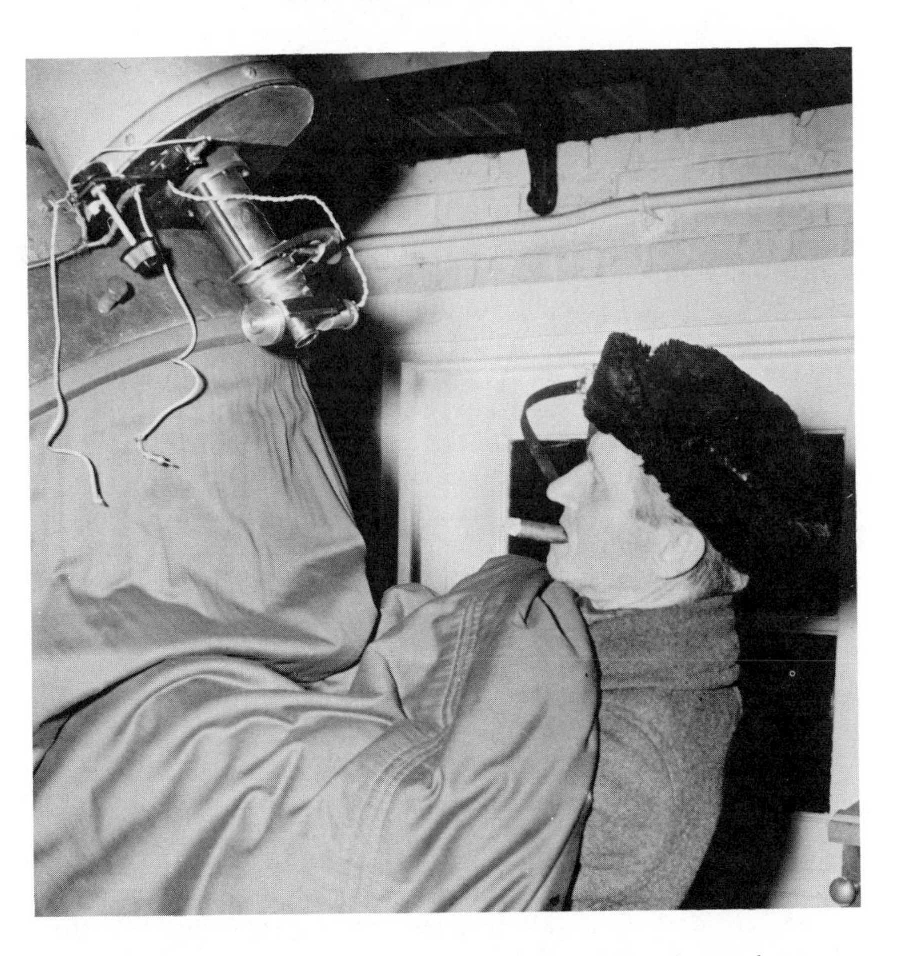

In a picture of a quarter-century ago, Henry A. Sawyer loads a plate into the 16-inch Metcalf refractor at Harvard's Oak Ridge station in Massachusetts. The lightproof cloth shroud was used because the plate was put directly into the camera back, without a holder. A partial vacuum bent the glass against a metal template that matched the curved focal surface. The eyepiece in the guide telescope may well be one of the Coddington oculars Sawyer used to make of lucite. Photograph by P. Southwick.

disrepair, but by transferring parts from one to another he could always manage to drive one of them from his home in West Acton to the station. An unusual feature of this home was a milling machine in the middle of the parlor.

For many years a large part of the astronomical research done at

Harvard Observatory used the photographs Bowie and Sawyer had taken, and it is proper to call them silent partners in the work of many famous astronomers.

Because such people seldom appear in printed records, invisible astronomers are soon forgotten. The *Observatory* for 1896 has preserved an account of one of these men who was associated with the 72-inch reflector built in 1845 by the third Earl of Rosse at Birr Castle, in Ireland (see Chapter 76). It reads: "On March 7th the death occurred in Parsontown of William Coghlan, who had been for 55 years the mechanical assistant attached to Birr Castle Observatory, and had assisted at the casting and polishing of the 6-foot speculum, and erection of the mounting of the telescope. He was the last survivor of the staff of the workshop, and gave up work only a year before his death. He was by trade a locksmith in a neighboring village, and showed much aptitude for mechanical work. After coming into the Birr Castle workshop he learned iron and brass founding and moulding, turning, etc.; many such jobs of necessity having to be done on the spot in the midst of an agricultural country."

Lastly, there is the shadowy figure of John Stone, who for many years was Sir John Herschel's observing assistant and factotum. At Slough, England, Stone helped Herschel during his survey of the northern heavens from 1825 to 1833. This effort resulted in the discovery of 2,300 new double stars, clusters, and nebulae. Later, when Herschel went to South Africa to extend this work over the entire sky, he brought along Stone (see Chapter 9). The latter supervised the erection of the 18¼-inch reflector and the 5-inch refractor at Feldhausen and took a share in the observing. But hardly any more information about this mysterious helper can be found.

TELESCOPES AND TECHNIQUES

24. The era of long telescopes

IN THE YEAR 1659, when the invention of the astronomical telescope was still within living memory, Christiaan Huygens published his book *Systema Saturnium.* It told of the fulfillment of the secret dream of many an amateur astronomer today: To use powerful homemade instruments to settle decisively some famous astronomical problem, and to top it off with the discovery of a noteworthy new object in the sky. Both these feats were reported by the 26-year-old Dutch astronomer in his book.

Saturn had been a perplexing problem to astronomers ever since Galileo first viewed it in his optic tube in 1610. With the primitive telescopes of that time, the planet was variously seen as triple, single, oblong, or embellished with semicircular or elliptical arcs. Huygens began to observe Saturn in 1655 with a nonachromatic refractor 12 feet long made by himself. Later that year he pressed a 23-foot refractor into service, and in 1656 one 123 feet long. His systematic observations enabled him to prove conclusively that the curious phenomena of Saturn were due to "a thin, flat ring, nowhere attached, and inclined to the ecliptic." During this work, moreover, he discovered Saturn's brightest satellite, Titan, on March 25, 1655, with the aid of the 12-foot telescope.

This instrument was one of the very first Huygens made after he took up lens grinding with his brother Constantyn early in 1655. Its objective was a single plano-convex lens, and its ocular, giving a power of 50, was a simple lens of a little less than three inches focal length. Fortunately, the objective of this telescope has been preserved. It was discovered in 1867 by P. Harting among the possessions of the physics department of Utrecht University in Holland. It is 2¼ inches in diameter, and is inscribed with the date February 3, 1655, and a Latin sentence, which is the same anagram Huygens had used to safeguard the priority of his discovery of Titan. (The anagram reads *Admovere ocvlis distantia sidera nostris, vvvvvvvcccrrhnbqx,* which can be rearranged to yield *Saturno lvna sva circunducitur diebus sexdecim horis quatuor:* Saturn has a moon revolving about it in 16 days four hours.) Harting carefully compared the inscription with

Christiaan Huygens (1629–1695) as portrayed in an engraving by Edelinck. From A. v. Schweiger-Lerchenfeld's *Atlas der Himmelskünde,* 1898.

Huygens' manuscripts preserved at Leiden Observatory and found that they were in the same handwriting.

Huygens spoke in his book of a 12-foot telescope, whereas the focal length of this lens is 10.6 feet. The explanation seems to be that the old Dutch astronomer described his instruments by their overall length, including dewcap and eyepiece. There seems to be no doubt that this is the actual objective used for the discovery of Titan.

Here was an interesting opportunity for modern tests of the performance of a famous 17th-century telescope. In 1883, J. A. C. Oudemans fitted this objective with a 3-inch eyepiece from another Huygens instrument, thereby obtaining a telescope practically identical with the original, as far as the optical parts were concerned. With its aid, the rings of Saturn were just recognizable, and the star Mizar was readily shown as double. More extensive tests of the same object glass were made about 15 years later by the Utrecht astronomer, A. A. Nijland. He added a modern ocular, providing a mag-

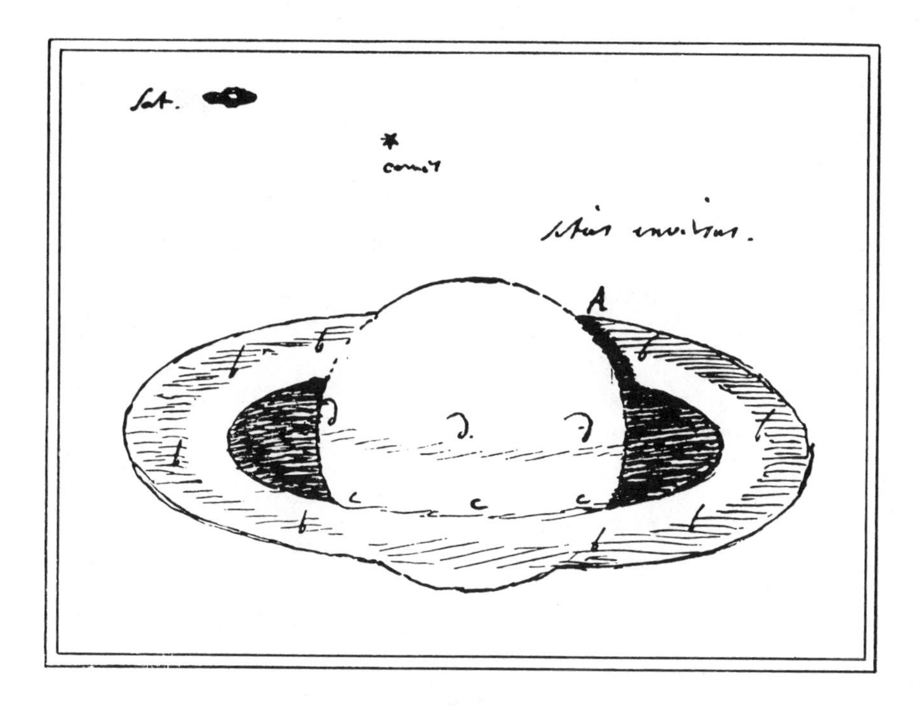

Saturn as observed by Christiaan Huygens on December 8, 1675, with a telescope 36 feet long, at Paris in company with G. D. Cassini. The upper part of the sketch shows the location of Titan with respect to Saturn. In the large Saturn drawing, *A* is the shadow of the planet on the ring, and *ccc* the shadow of the ring on the disk. The outer part of the ring was seen darker than the inner, the boundary being labeled with *b*'s. On the planet, a dark belt is marked *ddd*. This pen-and-ink sketch in Huygens' observing book is reproduced from Volume 15 of his collected works.

nification of 88, and placed the assembled instrument on an equatorial mounting.

The results give us some insight into the difficulties under which the observer of 300 years ago worked. Even with the most careful adjustment, Nijland found that the telescope gave diffuse images. A 1st-magnitude star showed a bright, irregular central patch from which extended colored rays marked by fragments of several diffraction rings. Only double stars with separations over five seconds of arc could be split. Jupiter and Saturn showed some surface features, but the Moon alone, easiest of all telescopic objects, yielded him a pleasing view. The performance, according to Nijland, was not as

Christiaan Huygens' 123-foot aerial telescope had its objective lens mounted on a ball-and-socket joint attached to a slider that could be moved up and down a tall pole. The lens was aimed by pulling on a long wire attached to the eyepiece carrier. Today, it is difficult to imagine how such a cumbersome instrument could be used. From A. v. Schweiger-Lerchenfeld's *Atlas der Himmelskünde,* 1898.

good as that of another 17th-century objective he tested at the same time. This was made by Giuseppe Campani of Rome and had a clear aperture of 1.65 inches and a focal length of 10.4 feet. It revealed the 1st-magnitude star Procyon as a perfectly round yellow disk surrounded by neat diffraction rings.

Later in his career, Huygens made and used much larger and more effective instruments than the 12-foot. With single-lens objectives of small aperture, very great focal lengths minimize spherical aberration and render chromatic aberration less noticeable. This led to the construction of enormously long telescopes, whose wooden tubes, hung by ropes from poles, were very difficult to handle. This problem was eased by the Huygens brothers' invention of so-called aerial telescopes. For these tubeless instruments the objective was placed on a high tower and pointed by means of long cords held by the observer.

Three objectives for such telescopes, ground by Constantyn Huygens in 1686, are in the possession of the Royal Society of London. They have diameters of 7½, 8, and 8½ inches, and focal lengths of 123, 170, and 210 feet, respectively. The first of these is probably the same one mentioned in a revealing note in the *Philosophical Transactions* of the society for 1718: "Much about the same time the excellent M. *Christian Huygens* of Zulichem, made the Society a present of the Glasses of a Telescope of 125 Foot length, with the *Apparatus* for using them without a Tube. . . . But those here that first tried to make use of this Glass, finding for want of Practice, some difficulties in the Management thereof, were the occasion of its being laid aside for some time. Afterwards it was designed for making perpendicular Observations of the fixt Stars passing by our Zenith, to try if the Parallax of the *Earths* annual Orb might not be made sensible in so great a Radius, according to what Dr. *Hook* had long since proposed: but in this we miscarried also, for want of a place of sufficient height and firmness, whereon to fix the Object Glass, so that it lay by neglected for many Years." It is clear from this excerpt why short, convenient reflecting telescopes were so popular after they became generally available in the mid-18th century.

25. Herschel's "large 20-foot" telescope

WILLIAM HERSCHEL's career as a telescope maker extended over 45 years. He was still an obscure musician in the autumn of 1773 when he began to grind and polish metal mirrors for telescopes. The last entries in his "polishing journal" are dated December, 1818, when the great astronomer at the age of 80 was instructing his son John in the art. By Herschel's own statement in 1795 he had then made 430 mirrors, including two that were 48 inches in diameter.

Herschel's achievement can be seen by a comparison with the instruments that served his French contemporary Charles Messier. These are listed in Kenneth Glyn Jones' book, *Messier's Nebulae and Star Clusters* (1969), as little more than a few old nonachromatic refractors, a 3.3-inch Dollond achromatic refractor, and some small reflectors, such as a Gregorian of six inches aperture. All were inefficient by modern standards. It is little wonder that Messier's life list of clusters and nebulae totaled only about 100, while Herschel discovered some 2,500.

The way in which Herschel became an amateur telescope maker at the age of 34 is told in his journal for the year 1773, when he was living at Bath, England:

"I was . . . informed that there lived in Bath a person who amused himself with repolishing and making reflecting mirrors. Having found him out he offered to let me have all his tools and some half-finished mirrors, as he did not intend to do any more work of that kind. The 22nd September when I bought his apparatus, it was agreed that he should also show me the manner in which he had proceeded with grinding and polishing his mirrors, and going to work with these tools I found no difficulty to do in a few days all what he could show me, his knowledge indeed being very confined."

"About the 21st October I had some mirrors cast for a two feet reflector, the mixture of the metal was according to a receipt I had obtained with the tools. It was at the rate of 32 copper, 13 tin and one of Regulus of Antimony, and I found it to make a very good, sound, white metal. In the beginning of November I had other mirrors cast, among them was one intended for a 5½ feet Gregorian reflector, and as soon as they were ground and figured as well as I could do them, I proceeded to the work of polishing. About the middle of December I got so far as to give a tolerable gloss to some metals, and having advanced considerably in this work it became necessary to think of mounting these mirrors."

In January, 1774, Herschel put this 5½-foot mirror into a square wooden tube, but found the Gregorian adjustment so troublesome that he converted the instrument to a Newtonian. That March 1st he began his observing journal with the note that at 40x Saturn's rings looked "like two slender arms."

Although not the largest, the most useful of all William Herschel's telescopes was his "large 20-foot" of 18.7 inches aperture, built in 1783. (It was so-called to distinguish it from his previous "small 20-foot" that had only a 12-inch mirror.) The picture on the next page of the large 20-foot telescope is from an engraving of 1794, which was little known until recently when the original copperplate was discovered in the archives of the Royal Astronomical Society by J. A. Bennett. The engraving is labeled "Published by Dr. W. Herschel 1 Feb. 1794," and therefore may be accepted as a reliable record of the structural details of this famous telescope.

The large 20-foot came into being as a result of Herschel's strong desire to resurvey the heavens with a much larger instrument than the 7-foot Newtonian (aperture 6.2 inches) with which he discovered Uranus on March 13, 1781. His small 20-foot reflector was not suitable for the purpose. This must have been a most troublesome telescope to use, for the upper end of the tube hung by ropes from a mast, and the observer had to perch precariously on a ladder to reach the focus. Then he had to climb down again to record what he had seen. Despite the inadequacy of this mounting, Herschel began to build a 30-foot telescope of the same design; he abandoned it only after repeated failures in casting the mirror.

Evidently Herschel reconsidered the mounting problem thoroughly, for the altazimuth stand adopted for the large 20-foot was a great advance. The picture shows this instrument not in its original form but after a decade of improvements. For example, the telescope was at first used as a Newtonian, but after 1786 as a front-view reflector with the eyepiece fixed in the mouth of the tube. "The machinery of my twenty feet telescope is so complete that I have been able to take up the planet [Saturn] at an early hour in the evening and to continue the observations of its own motion, together with that of its satellites, for seven, eight, or nine hours successively," wrote its maker. When the telescope was operated in this way, the muscles of at least two helpers were needed, to shift the mounting in azimuth, to change the altitude of the tube by turning a crank, and to raise or lower the observer's gallery. Herschel noted that it took about 15 minutes to point the telescope at a new object.

The so-called large 20-foot reflector that Sir William Herschel used from 1783 to 1814 was the world's most effective telescope for deep-sky surveys. Note the circular track on which the entire instrument could be moved in azimuth. Turning the crank at the lower end of the tube raised or lowered the telescope. The observer stood on the gallery and looked through an eyepiece down the tube. This is a reproduction of the 1794 engraving described in the text.

A simpler and more convenient mode was used in discovering nebulae and clusters. The telescope was kept in a fixed azimuth (usually the plane of the meridian) and was moved continually up and down through a two-degree angle by a workman at the crank. Remaining at the eyepiece, Herschel could thus sweep a strip of sky two degrees wide in declination and, due to the diurnal motion of the sky, extending several hours in right ascension. For these sweeps he customarily used a 157-power eyepiece with a field 15 minutes of arc in diameter. As set up at Slough, England, where he had moved in April, 1786, the 20-foot telescope stood in a garden close to the house. To avoid the necessity of taking notes by lamplight, he stationed his sister Caroline in a room with an open window, so that he could call out his observations for her to write down.

Herschel could determine approximate coordinates for new objects discovered during a sweep by treating the 20-foot telescope as a transit instrument. When an object passed a vertical wire through the middle of the field of view, he called to Caroline who would note down the clock time, from which the right ascension could be calculated. To obtain the declination, she could read a dial whose pointer was turned by a cord attached to the tube. Of course both the right ascensions and declinations had to be calibrated by star observations during each sweep. Herschel stated in 1786 that when he observed the same object in different sweeps, the positions seldom differed by as much as one minute of arc in declination or three or four seconds of time in right ascension.

It was during sweeps with the 20-foot that Herschel carried out his counts of stars in different parts of the sky (see Chapter 75). Also, with this instrument he discovered the two outer satellites of Uranus, Titania and Oberon, on January 11, 1787. Saturn's moons Enceladus and Mimas were both first seen with it.

Over the years four mirrors were made for the 20-foot telescope. One of the evils of speculum metal – in addition to its brittleness and relatively low reflectivity – is its liability to tarnish. This degradation made frequent repolishing necessary, so Herschel usually kept spare mirrors on hand. He had two 18¾-inch blanks cast in the autumn of 1783 and had one mirror sufficiently finished to begin observing that October 23rd with the 20-foot.

Many of Herschel's mirrors show a quality of figure and freedom from zonal irregularities that is surprising, considering that none of our modern tests were available to him (see Chapter 29). "His method of trying out his mirrors was to examine the stonework of Windsor Castle with the peripheral and central portions of the mirror alternately exposed. When the foci of both parts agreed, the mirror was judged to be good," according to the British astronomer W. H. Steavenson, who made a careful study of Herschel's surviving equipment in the 1920's.

William Herschel's last sweep with his 20-foot telescope was made

Three deep-sky objects drawn by John Herschel using the "large 20-foot" reflector. At left is the galaxy NGC 4565, at middle the galaxy Messier 64 and at right the diffuse nebula NGC 6589-90. From *Philosophical Transactions* of the Royal Society of London, *123*, Part 2, 1833.

in 1802, and he looked through it for the last time in August, 1814. But soon thereafter his son John, born in 1792, began preparing in earnest to continue his father's work. Two new specula were cast in June, 1817, and one of these was polished by John under his father's direction late the next year. The other mirror was later completed by John himself. Meanwhile, the much-decayed wooden fabric of the 20-foot telescope was restored, and the instrument was again operational by 1820. John's own sweeps began in 1823, the year after his father's death, and by 1835 he had discovered 2,300 nebulae and clusters of which about 1,800 had never been seen by earlier astronomers.

It was this same rebuilt 20-foot reflector that John Herschel took with three mirrors to South Africa for his famous exploration of the southern heavens in 1834-38. This work brought the total of his nebulae and clusters to 4,105 and his double stars to 5,449. But after John's return to England, he never reerected the 20-foot telescope and its career ended. One of the mirrors for the 20-foot telescope ended up at the Royal Observatory at the Cape of Good Hope, and two are in England at the National Maritime Museum, Greenwich.

26. A harem of telescopes

QUITE UNEXPECTEDLY I finally met up with Dr. Kitchiner. It was at the Cambridge, Massachusetts, headquarters of the American Association of Variable Star Observers. My eye was traveling over the bookshelves of the association's library when it chanced on a small dark green volume, gilt-lettered on its spine *Telescopes: Kitchiner: 1825*. The title page of this rare book explains why generations of amateur astronomers have delighted in it: *Of Telescopes; Being The Result of Thirty Years' Experiments with Fifty-one Telescopes, of from One to Nine Inches in Diameter, in the Possession of William Kitchiner, M.D.*

The author of the book lived in England from about 1778 to 1827. He was an amateur astronomer at a time when the avocation was still rare but gaining esteem from the shining example of Sir William Herschel. Son of a prominent London coal merchant who died worth £2,000 a year, Kitchiner was educated at Eton and obtained his medical degree from Glasgow. As the law then stood he could not practice medicine in London, but his father's death in 1794 left him a wealthy man, free to follow a surprising combination of hobbies.

To his contemporaries he was famed as a gastronome and amateur chef, whose select dinners made his house in Fitzroy Square a social center. To share his skill, Kitchiner wrote a best seller, *The Cook's Oracle,* a collection of 600 recipes that ran through at least five editions. The odd variety of the subjects on which he wrote books suggests an eccentric jack-of-all-trades: *Observations on Vocal Music; Peptic Precepts To Prevent and Relieve Indigestion; The Pleasure of Making a Will; Loyal, National, and Sea Songs;* and *A Companion to the Telescope,* to name some of them.

He took up astronomy at about the age of 18, as he tells us: "When I first began to play with Telescopes, in the Year 1796, I purchased a 3½ feet Achromatic with a double Object-glass of 2.7 inches aperture . . ." This was the start of the collection of 51 instruments

One of the many curiosities in William Kitchiner's book is this anthem for amateur telescope makers. He presumably composed it himself, rather than Galileo, Huygens, Ramsden, or Dollond, whom he suggests offhandedly as possible authors. Courtesy American Association of Variable Star Observers.

listed on page 14 of his book. This private hoard of telescopes included achromatic refractors ranging from one of 4.6 inches aperture and a five-foot focal length to two 1.1-inchers a foot long. He had many Gregorian reflectors, from a 9.3-inch 27 inches in length to a 2-inch only four inches long! Curiously, there was only a single Cassegrain in the lot. At least two of his telescopes were made by William Herschel himself: a 7-inch f/12 Newtonian and a 6.3-inch f/13.3 Herschelian or front-view reflector. Altogether, he spent about £2,000 for his telescopes.

The reflectors, of course, all had speculum-metal mirrors, for in 1825 the art of depositing silver on glass still lay in the future. Kitchiner quotes with approval the composition chosen by Rev. John Edwards, who in 71 melts found that "the hardest, whitest, and most reflective" speculum metal was 32 parts copper, 15 tin, and one each of brass, silver, and arsenic.

Telescope making in those days was an empirical art. "*The Practical Optician* can employ his skill in producing suitable Specula for counteracting each other's errors with respect to the united effect of their separate aberrations, better than *the calculating Theorist* can pretend to direct," Kitchiner quotes with manifest approval from Dr. Rees's *New Cyclopaedia*. His views on optical testing have an engaging simplicity: "You cannot judge accurately of the excellence of any Telescope, by observing any object which you are not acquainted with . . . . The Dialplate of a Watch, placed about 100 feet from the Telescope, is *an Excellent Test* of the distinctness and Achromaticalness of a Glass, especially when the Sun shines strongly upon it – so is a Weathercock, or any object in the Day, with a bright light behind it, and the Moon, and best of all the Planet Jupiter, when near to the Meridian." Another test object, recommended to him by a correspondent, is sunlight reflected from "the ball of a Mercurial Thermometer at about one hundred yards distance."

Stars, however, were far too stringent a test of early 19th-century optics. "When a Telescope is pointed at a Star, the least Defect in the figure, or adjustment of the metals in a Reflector, or of the object-glass in an Achromatic, immediately stares in your Eye, – the Star not appearing round, but surrounded by false lights, radiating points, and little flitting luminous accompaniments . . . If I was an Optician, – I think that I would about as willingly *Waltz blindfold and barefoot among 9 Red-hot Ploughshares laid at unequal distances from each other,* as have all my Telescopes tried by that truly troublesome test, a Fixed Star."

Kitchiner is thoroughly insular. He seems not to have heard of the wonderful advances under way among continental telescope makers, and his book contains no mention (as far as I can tell) of the giant 9.6-inch refractor Fraunhofer had completed for Dorpat Observatory in Russia the year before. To him Sir William Herschel's large altazimuth reflectors were the summit of optical excellence. The book is full of excerpts from Herschel's papers.

But the doctor has sound ideas about the superior convenience and manageability of medium-size instruments and moderate powers. On none of his 51 telescopes is a magnification of more than about 400 recommended, save for an occasional experiment. "I do not believe that Art has yet produced a Reflector beyond 7 inches, or an Achromatic beyond 3.8 inches in diameter, of which the Defining power has proved quite perfect . . ."

Evidently Dr. Kitchiner did not think much of the Earl of Stanhope's proposal, sometime before 1820, to construct a monster reflector 384 feet long and of 72 inches aperture. "Immense TELESCOPES, *are only about as useful, – as the* Enormous SPECTACLES *which are suspended over the doors of Opticians.*"

There is something for everybody in this pleasant old book. Modern telescope users can pick up practical hints; many old drawings and descriptions of Saturn await the planetary buff; and the historian of astronomy will find a rich harvest of information about Kitchiner's contemporaries who made, sold, or used optical instruments. And there are occasional little puzzles for 20th-century readers. For example, what were the "Telegraph Telescopes" casually mentioned on page 336? The context suggests small refractors, but the use of the word *telegraph* is odd for 1825, long before the telegraph we know had been invented. Were they perhaps terrestrial telescopes, used in connection with the lines of semaphore signals that connected London with certain naval bases such as Portsmouth?

27. The telescopes of William Lassell

IF YOU TRACE THE DEVELOPMENT of reflecting telescopes from Sir William Herschel (see Chapter 25) to the present, you will find that the direct line of descent passes not through Lord Rosse but through the English telescope maker William Lassell (1799-1880). His versatile and effective reflectors with their equatorial mountings look more modern than the unwieldy 72-inch leviathan of the Irish peer, which was slung by ropes from two great masonry towers (see Chapter 76).

As in Herschel's case, Lassell's astronomical career was facilitated by beer. The former's marriage to the widow of a wealthy brewer gave him financial independence, while Lassell founded his own brewery in 1825, and from it secured the fortune that eventually permitted him to devote his entire time to building and using large telescopes.

His career as an amateur telescope maker started at the age of 21, when he made a Newtonian and a Gregorian, each of 7-inch aperture. With this experience, he erected a private observatory containing a 9-inch Newtonian. Thanks to Lassell's great mechanical ability, both his mountings and his speculum-metal mirrors were of unusual excellence for that time.

William Lassell photographed in 1874. Courtesy Mary Lea Shane Archives of Lick Observatory.

About 1844 he decided to construct a 24-inch reflector. The first step was to visit Birr Castle in Ireland, where he inspected the Rosse workshops and instruments. Work was begun with a straight-stroke grinding machine like Lord Rosse's, but soon Lassell replaced it with a curved-stroke machine of his own invention. Lassell was also fortunate to have the cooperation of James Nasmyth, another amateur astronomer, who was a gifted mechanical engineer with extensive foundry experience (see Chapter 28). Such knowledge was invaluable in casting large metal mirrors.

With this fine instrument, Lassell discovered Triton, the brightest satellite of Neptune, on October 10, 1846, and two years later he found Hyperion, the eighth satellite of Saturn. The latter was simultaneously detected in America by W. C. Bond with the 15-inch refractor of Harvard Observatory. In 1852 the English amateur moved his 24-inch telescope to the island of Malta for the winter. Under clear Mediterranean skies he was able to observe satisfactorily

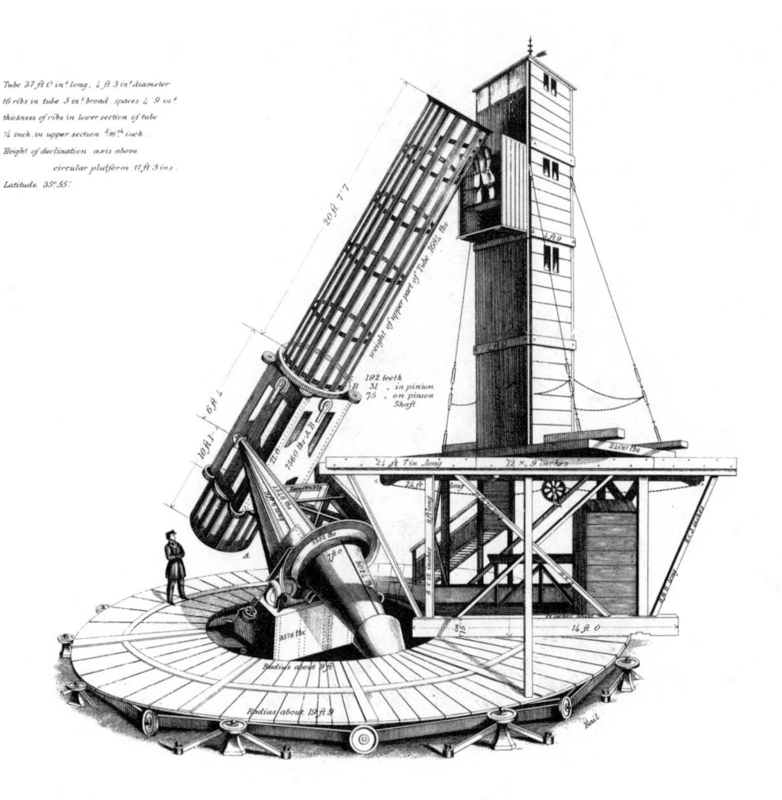

Lassell's 48-inch reflector had no driving clock but was moved by turning the crank seen at the base of the pier near the attendant. Lassell used the instrument mainly to observe faint nebulae and planetary satellites. From *Memoirs* of the Royal Astronomical Society, *36*, 1867.

all four of the satellites of Uranus known before G. P. Kuiper's discovery of a fifth moon, Miranda, in 1949.

Nothing contributes more to the obsolescence of a good amateur-built telescope than its success, which habitually leads to the desire for a still larger and more effective instrument. Lassell therefore undertook the building of the 48-inch reflector pictured here.

Erected in 1859 on the grounds of his villa near Liverpool, this telescope was an f/9.4 Newtonian with a tube 37 feet long made of flat iron bars. "There is no roof or covering over the telescope," wrote its owner, "but the observer or observers are protected by being placed in one or other of the storeys (according to the altitude of the object to be viewed) of a Tower, which affords a means of getting conveniently at the eye-piece, which, when the telescope points to the zenith, is about 39 feet from the ground.

"A staircase within the tower leads to the different storeys, which are about 4 feet 6 inches square, and afford abundant room for papers, micrometers, eye-pieces, lamps, and any other small apparatus required; beside furnishing to the observer a most grateful shelter from the dew, and occasionally from an inclement wind. During observation, however, the size of the storey in use becomes practically much larger, by the opening of the folding-doors and letting down of the platform, as shown in the engraving; the available space being then about 6 feet 9 by 4 feet 6 inches. The tower is carried round on a circular railway, and has besides, a revolution on its axis, and a radial motion to and from the telescope: so that at most altitudes and hour-angles the eye-piece is easily accessible. It has been usual, however, for the most obvious reasons, to observe within three hours of the meridian, east or west."

The telescope did not have a driving clock but was moved through a gear train by an assistant turning a crank once a second, in synchronism with the beats of a clock. With this primitive arrangement, a star could be kept within the field of view for several hours, even though the moving parts of the telescope weighed about eight tons.

Nebulae were favorite targets for the 48-inch reflector, the largest in England. A fine drawing of the Dumbbell nebula (Messier 27) in Vulpecula was published in October, 1860, in the second of a series of circulars. (The first had been devoted to the Ring nebula, Messier 57.) Lassell regarded this picture as a faithful representation of the Dumbbell, except that its brighter parts, which prompted its name, were not sufficiently prominent compared to the fainter extensions. The English amateur correctly commented that there was no indication of resolution of the nebula, and the stars in the field were totally unconnected with it, since "the sky around is quite as full of Stars as the space occupied by the Nebula."

In 1861 Lassell moved this large telescope to Malta, where during the next four years its main scientific work was done. Besides numerous measurements of faint satellites, examination was made of many of the brighter nebulae. The most important result of this

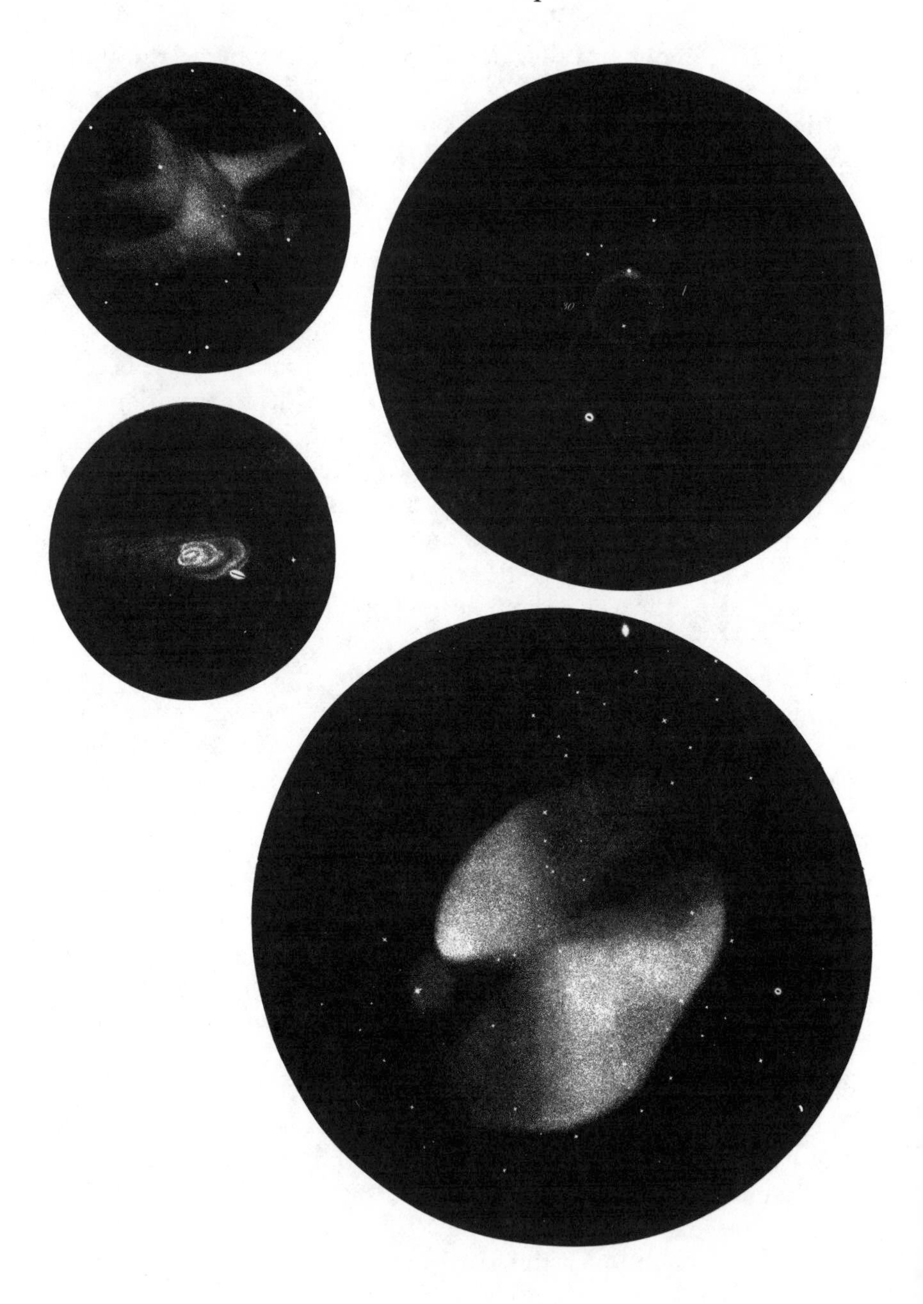

survey was the recognition of spiral form in a number of galaxies. Also, a catalogue of 600 previously unknown nebulae was compiled. Most of the actual observing during this second Malta expedition was not done by Lassell but by his assistant Albert Marth. An able and hardworking German astronomer who had emigrated to England, Marth has never received adequate recognition for his varied contributions to astronomy.

Lassell's return to England in 1865 marked the end of the usefulness of the 48-inch reflector, for he never re-erected it. Shortly before he died in 1880, it was sold for scrap metal. "When witnessing the breaking up of the specula," wrote Lassell, "I was not without a pang or two on hearing the heavy blows of sledge-hammers necessary to overcome the firmness of the alloy."

28. James Nasmyth's telescopes and observations

SOMEONE SHOULD WRITE A BOOK about famous amateur telescope makers. There are some highly interesting personalities among them, from Galileo and William Herschel to Russell W. Porter, Horace Dall, and Anton Kutter. It would be enlightening to read not just what each optician innovated, but what his background and special skills were, and what motivations directed him. After all, a William Lassell or a Bernhard Schmidt telescope is an extension of the craftsman who made it; thus telescope and constructor each tells us something about the other.

Four drawings made by William Lassell with his 48-inch telescope at Malta. At upper left is Messier 1 (the Crab nebula, a supernova remnant); below it is Messier 88 (a galaxy); and at bottom is Messier 27 (the Dumbbell nebula, a planetary). In all these renditions, south is at top, east to the right. From *Memoirs* of the Royal Astronomical Society, *36*, 1867.

James Nasmyth, in a photograph believed to have been taken about 1870. From *Men of Mark,* Vol. 2, approx. 1870, courtesy Owen Gingerich.

In the gallery of distinguished telescope makers, the Scottish engineer James Nasmyth (1808–1890) deserves a prominent place. His family was highly gifted. Son of the painter Alexander Nasmyth, he was younger brother of the landscapist Patrick Milner Nasmyth, and between 1829 and 1866 no fewer than six women artists of this family exhibited their works in London.

As a youngster in Edinburgh, Nasmyth spent many hours in iron foundries and chemical laboratories, and he became known as a maker of beautiful working models of steam engines. In 1827, at the age of 19, he constructed a "road steam-carriage" that was actually a primitive automobile. "Many successful trials were made with it on the Queensferry Road, near Edinburgh," Nasmyth recalled many years later in his autobiography. "The runs were generally of four or five miles, with a load of eight passengers, sitting on benches about three feet from the ground."

After working for three years with the famous engineer Henry Maudslay in London, Nasmyth started his own factory at Patricroft, near Manchester. His firm acquired a high reputation for manufacturing machinery of all kinds, steam engines, and especially improved machine tools. He became a rich man from his many inventions, which included steam hammers and pile drivers, a hydraulic punch, and flexible shafting for driving small drills such as dentists use. Thus it was that Nasmyth could retire from business at the early age of 48 and settle at Penshurst, Kent, to devote himself to astronomy.

He had made his first telescope in 1827, a 6-inch reflector with a mirror of speculum metal. His first experiments in casting speculum-metal disks failed, because the conventional closed sand molds gave an excessively brittle product. He therefore adopted an open iron mold to which a pouring pocket was fitted, so that the melt entered the bottom of the mold. This technique yielded a brilliant alloy and a mirror that was much freer from inhomogeneities and thermal strains.

Maudslay, too, was interested in astronomy, and Nasmyth cast and polished for him a mirror variously described as an 8-inch and a 10-inch. He was planning a 24-inch for the private observatory that Maudslay was about to build at Norwood, when the latter died in 1831.

After moving to Patricroft, Nasmyth resumed his telescope making. "As I had all the means and appliances for casting specula at the factory I soon had the felicity of embodying all my former self-acquired skill in this fine art by producing a very perfect casting of a ten-inch diameter speculum. The alloy consisted of fifteen parts of pure tin and thirty-two parts of pure copper, with one part of arsenic. It was cast with perfect soundness, and was ground and polished by a machine which I contrived for the purpose. The speculum was so brilliant that when my friend William Lassell saw it he said 'it made his mouth water.' "

This episode occurred about 1840, when the meeting between the two amateur telescope makers began a close friendship that lasted 40 years. Nasmyth quickly began casting for a 13-inch speculum. Looking back as an old man on his enthusiasm, he wrote in his autobiography: "I know of no mechanical pursuit in connection with science that offers such an opportunity for practising the technical arts as that of constructing, from first to last, a complete Newtonian or Gregorian reflecting telescope . . . Buy nothing but the raw material, and work your way to the possession of a telescope by means of your own individual labor and skill. If you do your work

The 20-inch reflector and its maker, from *James Nasmyth: An Autobiography*, edited by Samuel Smiles, 1884.

with the care, intelligence, and patience that is necessary, you will find a glorious reward in the enhanced enjoyment of a night with the heavens. . . ."

Nasmyth's ambition expanded, and he was soon at work on a 20-inch reflector. Hitherto his telescopes had been Newtonians, but now he chose a novel modification of the Cassegrain design. Here is his description:

"In order to avoid the personal risk and inconvenience of having to mount to the eye-piece by a ladder I furnished the telescope tube with trunnions, like a cannon, with one of the trunnions hollow, so as to admit of the eye-piece. Opposite to it a plain, diagonal mirror was placed, to transmit the image to the eye. The whole was mounted on a turn-table, having a seat opposite to the eye-piece, as is shown in the engraving [reproduced here].

"The observer, when seated, could direct the telescope to any part of the heavens without moving from his seat. Although this arrangement occasioned some loss of light, that objection was more than compensated for by the great convenience which it afforded for the prosecution of the special class of observations in which I was engaged, namely, that of the sun, moon, and planets." Although this type of mounting was neglected for decades, it became widely adopted for telescope designs in the latter part of the 20th century.

Systematic lunar observations with the 20-inch began in 1842. On every favorable opportunity, large drawings in black and white chalk on gray paper were made of "the most characteristic and instructive features" of the lunar surface. The observations led Nasmyth to the conviction that the Moon's surface features had been produced by volcanic action, no doubt in the remote past, since all agencies of change seemed to have ceased. He argued that the enormous scale of lunar craters, as compared to terrestrial volcanic formations, could be explained by the lesser surface gravity of our satellite.

These ideas were expounded in the book *The Moon Considered as a Planet, a World, and a Satellite,* which was published in 1874 by Nasmyth and James Carpenter (1840–1899), a former assistant at Greenwich Observatory. This work is famous for its photographs of Nasmyth's plaster-of-Paris models of lunar features. These models show the Moon's surface as far more jagged than it really is, because they were based upon drawings rather than actual measurements of shadows. The book was widely read, and was translated into German by Hermann Klein of Cologne. Fittingly, both Nasmyth and Carpenter have had lunar craters named after them, near Phocylides and Anaximander, respectively.

Nasmyth made many observations of the fine structure of sunspots, which directed his attention to the apparent texture of the solar photosphere. His interpretation of what he saw gave rise to the once-celebrated "willow-leaf" controversy. In 1861 Nasmyth announced to the Literary and Philosophical Society of Manchester that the entire luminous surface of the Sun is composed of small

This plaster model of the lunar mountain Pico was photographed in strong sunlight to make an illustration for Nasmyth and Carpenter's *The Moon*. In reality, the mountain is not nearly this steep.

elongated features, which he called willow leaves from their shape. These were described as crossing each other in all directions and constantly moving. Most of the leading solar observers of the time tried to verify his results, and a lively discussion ensued, in which the existence of the willow leaves was discredited. Particularly critical was the very able English amateur W. R. Dawes (1799–1868), who is best remembered for the Dawes limit, which specifies the resolving power of a given size telescope (see Chapter 69). Evidently Nasmyth was observing but misinterpreting the granulation pattern of the Sun. (The term "granulated" was introduced by Dawes in 1864.) There is a large subjective element in most visual studies of granulation, since the true granules are so small as to resist resolution. The coarser pattern ordinarily seen is due to the unequal spac-

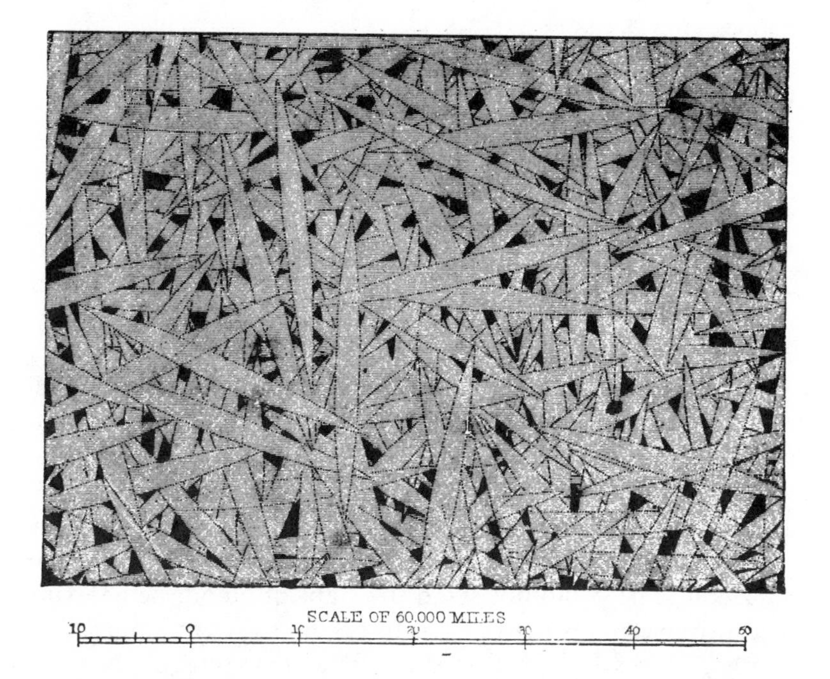

Nasmyth's "willow-leaf" pattern on the Sun. In a bizarre interpretation of these spurious features, John Herschel proposed that they were not composed of gases but of "sheets, flakes, or scales, and having some sort of solidity." He then went on to "regard them as *organisms* of some peculiar and amazing kind . . ."! From *The New Astronomy,* by Samuel Pierpoint Langley, 1888.

ing of unresolved elements, "in much the same way as the 'grain' of a developed photographic plate is due to irregularities in the distribution of the silver particles," to cite M. Minnaert.

As a student of the planets, Nasmyth is almost forgotten, except for two persistent traces. He proposed in 1853 that Jupiter's activity indicates some internal source of heat, an idea that was confirmed a century later. And popular writings often cite his description of the close conjunction of Venus and Mercury in September, 1878. For several hours, he had the rare opportunity of watching the two planets in the same telescopic field of view, and he noted that Venus resembled "clean silver" while Mercury looked "as dull as lead or zinc." Even as an astronomical observer, Nasmyth was thinking as a foundryman.

29. Leon Foucault's heritage to telescope making

EVERY AMATEUR astronomer who pushes one glass disk across another to grind a telescope mirror knows the name of Leon Foucault (1819–1868), who invented the widely used knife-edge test for checking the figure of a mirror. This accurate and beautiful method, in its present improved form, is ample reason for lasting fame, but amateur telescope making owes much more to Foucault than that. His ideas guide nearly every step taken by the present-day hobbyist in constructing a Newtonian reflector.

Foucault was a brilliant experimental physicist who, like so many great 19th-century scientists, did much of his best work as an amateur. Because he could not stand the sight of blood, he gave up his medical studies in his native Paris to become a science journalist and an author of textbooks. Parisians of the 1850's regularly read his column in the newspaper *Journal des Débats* to stay abreast of current happenings in science. In his home, he set up a laboratory for experiments in optics, in collaboration with his friend Louis Fizeau (1819–1896), who also became a distinguished physicist. Both men were interested in the recently invented art of photography, and on April 2, 1845, they together took the first daguerreotype of the Sun. Although the sensitized silver surfaces used in that process were so slow that exposures of minutes were needed for landscapes and portraits, the solar disk was so bright that the main difficulty was to obtain a short enough exposure. But 1/60 second gave an image that showed two large sunspot groups and the solar limb darkening.

In 1849 Fizeau made the first crude experimental measurement of the velocity of light. He and Foucault then turned to the related problem of ascertaining whether light traveled more slowly in water than in air, a test of the wave theory of light. No sooner was an appropriate apparatus designed than the friends quarreled. Each man conducted the experiment by himself and independently reported his own results to the Paris Academy of Sciences, in papers read at the meeting of May 6, 1850.

Foucault's 1851 pendulum demonstration in the Pantheon, Paris. From A. Berget's *Le Ciel,* 1923.

Foucault became a famous man in the following year by his pendulum experiment to demonstrate the Earth's rotation. It was suggested to him by the chance observation that a steel rod, clamped in the chuck of a lathe, tended to vibrate in the same plane when the lathe was slowly turned. After trials with a seven-foot pendulum in his own cellar and with a 36-foot one in the Paris Observatory, Foucault carried out a dramatic public exhibition. As C. A. Young described it: "From the dome of the Pantheon in Paris he suspended a heavy iron ball about a foot in diameter by a wire more than 200 feet long. A circular rail some 12 feet across, with a little ridge of sand built upon it, was placed under the pendulum in such a way that a pin attached to the swinging ball would just scrape the sand and leave a mark at each vibration. The ball was drawn aside by a cotton cord and allowed to come absolutely to rest; then the cord was burned, and the pendulum set to swinging in a true plane; but this plane seemed to *deviate slowly towards the right,* cutting the sand in a new place at each swing and shifting at a rate which would carry it completely around in about 32 hours if the pendulum did not first come to rest. In fact, the floor of the Pantheon was seen turning under the plane of the pendulum's vibration."

This experiment, which today is shown at many planetariums and science museums, made a sensation in 1851. Foucault was awarded the Cross of the Legion of Honor and, at the instigation of Emperor Napoleon III, was appointed to the staff of the Paris Observatory as a physicist. It was there that his attention turned to the improvement of reflecting telescopes.

By the 1840's, the reflector had fallen into general disrepute among practical astronomers, who greatly preferred refractors to speculum-metal reflectors such as the Herschels had used. Only a few enthusiastic amateurs like Lord Rosse, William Lassell, and James Nasmyth were building large mirror telescopes. More typical was the action of G. B. Airy who, after he became Astronomer Royal in 1836, promptly dismantled the 15-inch Ramage reflector that stood on the Greenwich Observatory grounds.

In those days of primitive optical test methods, opticians seldom attempted aspheric surfaces. Rather, as many beginners do today, they made spherical mirrors of large focal ratio to reduce spherical aberration. Big reflectors with their heavy metal mirrors were too long and unwieldy to mount equatorially, so they were usually placed on clumsy altazimuth stands. Also, speculum metal tarnished quickly, and the only way to restore a bright surface was to repolish the mirror at the risk of impairing its figure. Sir Howard Grubb probably had just that in mind when he said, "Reflectors very seldom do good work except in the hands of their makers."

Foucault realized that these disadvantages could be avoided by making mirrors of glass and coating them with silver by the chemical process described by Justus Liebig in 1856. The same idea occurred to Carl August von Steinheil (1801–1870), who in March, 1856, announced his construction of a 4-inch silver-on-glass reflector. Honors must be shared with this able German astronomer and engineer, who founded a famous Munich optical firm. The great influence of Foucault is due to his having published lucid and detailed accounts of his methods. In fact, his informative 41-page "Memoir on the Construction of Silver-on-glass Telescopes," in Vol. 5 of the *Annales* of Paris Observatory (1859), is the true forerunner of the more modern *Amateur Telescope Making* series.

He began by studying the glass-working processes used in the Paris optical shop of Marc Secretan, a well-known commercial telescope maker. Foucault quickly saw that standard grinding and pol-

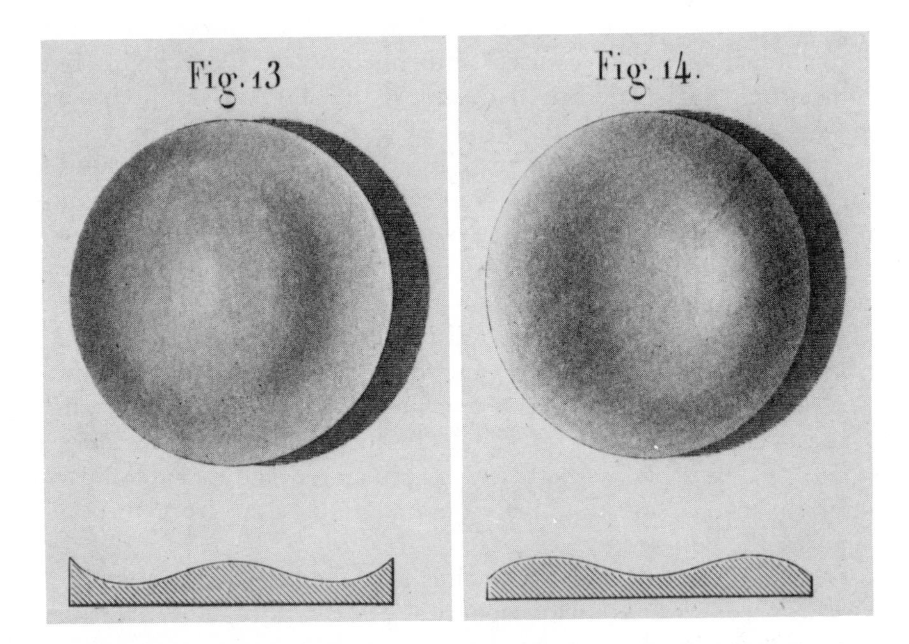

These drawings by Foucault show typical patterns produced by his knife-edge test. That at the left is for an undercorrected mirror (high center), the other for an overcorrected one (center too deep). These pictures are from his 1859 memoir in the Paris Observatory *Annales*. For a preliminary check on the sphericity of a mirror, he also employed a forerunner of the modern Ronchi test. Between the mirror and an illuminated pinhole at its center of curvature, he placed a fine rectangular grid (near the pinhole) and examined the image of this grid for the straightness of its lines.

ishing procedures seldom gave a good concave spherical surface, much less the required paraboloidal one. To solve this problem, he invented his famous knife-edge test. His 1859 description of it begins:

"Insert an illuminated pinhole near the center of curvature of the mirror in such a manner as not to block the returning rays. After these rays cross, they form a diverging beam, into which the eye is placed and is moved toward the focus until the whole mirror surface appears illuminated. Then, pass an opaque straightedge through the image to the point of complete disappearance. During this operation, the eye sees a progressive dimming of the mirror, all parts of whose surface are of uniform brightness up to the moment of extinction, in the case of a perfect sphere. In the contrary case, the extinction does not take place simultaneously at every point, but contrasting shaded and bright areas give the observer an impression in exaggerated relief of the prominences and depressions by which the figure differs from a sphere. This effect is the necessary result of the paths of the rays which converge more or less exactly to a common focus."

The French scientist explained how to interpret this pattern of light and darkness seen on the test mirror: "Generally, if the surface so tested has high and low areas distributed in any manner whatsoever, all of the slopes inclining toward the side where the knife-edge is appear black, and the oppositely inclining slopes appear bright. The aspect of such a mirror should be the same as that of a matte surface that presents (with greatly magnified relief) hills and hollows similarly placed, and illuminated obliquely by light from the side opposite the knife-edge."

Using his test Foucault could thus tell what kind of correction a mirror needed to be made spherical. Even more important, he showed how by repeated application of his test a mirror surface could be changed from a sphere to a paraboloid. Suppose the polished mirror is initially spherical, showing no fault under the knife-edge test with the illuminated pinhole placed at the mirror's center of curvature. If the optician now moves the pinhole closer to the mirror and tests again, the surface will appear mildly defective. But a slight figuring of the mirror, using a smaller polisher, corrects the surface until it again darkens evenly with the pinhole at its new location. This process is repeated several times, the operator at each step placing the pinhole still closer to the mirror and letting the image move farther out, then figuring the mirror some more and checking the result with the knife-edge. Each step produces an ellipsoidal surface that is closer to the desired paraboloid. The process is finished when the approximation is sufficiently close, and the mirror is then ready for silvering.

Foucault tells us that it took him about six hours of polishing to parabolize a 10-inch mirror in this manner, testing it every 10 or 15 minutes. The whole operation differs considerably from what amateurs practice today, when it is usual to keep the pinhole fixed and to use the knife-edge to measure the radii of curvature of different zones on the mirror.

In this artist's concept, Leon Foucault applies his test to a new mirror. A beam of light emerges from a pinhole in the lamp shade, strikes the mirror in the foreground, and is reflected back to Foucault's hand-held knife-edge. Sketch by Steven Simpson.

The first two mirrors that Foucault made were of 9.8-inch diameter, the second being an f/4 paraboloid. His third was a 13-inch f/6.8 mirror of such quality that with a power of 600 it neatly divided the close and difficult binary star Gamma[2] Andromedae, whose components were at that time separated by only about 0.58 second of arc.

Some specific shop practices of Foucault deserve mention. He abandoned the traditional use of a convex iron grinding tool, finding that especially for large mirrors it produced only an approximately spherical surface. Instead he used a second glass disk as a tool, fastened on top of a heavy pedestal, around which he walked and pushed the mirror blank to and fro, just as modern amateurs do. His abrasive for grinding was emery, since carborundum was not known until 1891. Polishing was done with rouge-impregnated paper laps attached to disks of wood or cork. After silvering, the mirror was given a brilliant finish by gentle burnishing with a rouge-sprinkled chamois skin, a process which may have been original with Foucault.

The mountings for his telescopes were made by F. W. Eichens. To avoid flexure, Foucault supported his mirrors in air cushions that could be inflated as required. This idea in refined form is incorporated in some large modern telescopes, such as the 140-inch of the European Southern Observatory at La Silla, Chile.

Because he was not a professional optician guarding trade secrets, Foucault published the full details of his methods in the *Comptes Rendus* of the Paris Academy of Sciences (1857) and, as already mentioned, in the *Annales* of Paris Observatory (1859). His ideas took quick root in England, where in 1859 two articles of his appeared in translation in the *Monthly Notices* of the Royal Astronomical Society. Many silver-on-glass reflectors were built from the 1860's on by such skilled British amateurs as G. H. With, George Calver, Warren De la Rue, and A. A. Common.

In 1879 Common built the 36-inch telescope that, after an undistinguished career in an English amateur's hands, was reerected at Lick Observatory in 1895 as the Crossley reflector (see Chapter 60). Here its remarkable power as a photographic instrument was demonstrated. Its success led in a few years to the construction of the 60-inch and 100-inch Mount Wilson reflectors. The line of descent from Foucault to George Ellery Hale is short and direct indeed.

The largest mirror Foucault ever made was one of 31.5 inches aperture, which in 1864 was installed at the Marseilles Observatory. This telescope made two contributions of much significance to modern astrophysics, both at the hands of Edouard Stephan. The first, in 1873–1874, was the earliest attempt to measure the apparent diameters of stars interferometrically. The other, in 1877, was his visual discovery of Stephan's Quintet. For over a century the physical interpretation of this tight grouping of galaxies in Pegasus has perplexed astronomers.

30. The Clarks and some of their refractors

THERE IS A FASCINATION about the name Clark among all telescope enthusiasts. Five successive times the Massachusetts firm of Alvan Clark and Sons made the world's largest telescope objective, from the 18½-inch Dearborn in 1862 to the 40-inch Yerkes in 1897. To the present day the reputation of the Clarks' lenses large and small is widespread, and the dream of many an amateur is to own a Clark telescope.

Before the publication in 1968 of Deborah Jean Warner's little book, *Alvan Clark & Sons: Artists in Optics* (issued by the Smithsonian Institution), it was not easy to find detailed information about Alvan Clark (1804–1887) and his sons George Bassett Clark (1827–1891) and Alvan Graham Clark (1832–1897). The outlines of the father's story are familiar; he was a Yankee inventor and portrait painter who became a self-taught optician. But all three were unassuming men who wrote little or nothing and whose firm's records no longer exist. Their story had to be pieced together from scattered sources.

It appears that Alvan Clark's attention was first directed to telescope making around 1844, when he ground some small speculum-metal mirrors. In 1846 he turned to lenses and by two years later had advanced sufficiently to sell a 5-inch refractor to a school in Newburyport, Massachusetts. Nevertheless, until 1860, while establishing his optical business in partnership with his sons, Alvan continued to paint portraits for a livelihood.

Two personal contacts seem to have been important in the Clarks' emergence from obscurity. One was Alvan's friendship with William Rutter Dawes, an English amateur who was an enthusiastic observer of double stars (see Chapter 69). Their association began in 1851 with a correspondence about some close pairs Alvan had discovered with his own telescopes, and it led to Dawes' purchase of a

Alvan Clark is flanked by his sons Alvan Graham (left) and George Bassett. Courtesy Mary Lea Shane Archives of Lick Observatory.

7½-inch Clark refractor in 1854 and other instruments in the next few years. The American went to England in 1859 to visit Dawes and met such influential British astronomers as John Herschel and Lord Rosse. At this time Clark telescopes were better known in Europe than in America.

The second contact was with F. A. P. Barnard, president of the University of Mississippi, probably during the 1856 meeting of the American Association for the Advancement of Science at Dudley Observatory in Albany, New York. Barnard in 1860 commissioned Alvan to build an 18½-inch refractor, but it could not be delivered to Mississippi because of the Civil War. Instead the instrument was

Sirius is one of the most famous double stars. The primary, about 1¾ times the size of our Sun, is the brightest star in the nighttime sky. Its companion, over 11,000 times fainter and only 1/50 the Sun's size, is the celebrated white dwarf Sirius B. The two stars move around each other in a 50-year orbit. This photograph by R. B. Minton shows the pair near maximum separation, 11.3 seconds of arc; it was taken February 23, 1972, with the 61-inch reflector of the Lunar and Planetary Laboratory.

erected at the old Dearborn Observatory of the Chicago Astronomical Society. This was the first of a long series of American orders for large Clark telescopes.

It was during the testing of this objective at the Clark plant that the companion of Sirius was first seen. Warner tells us that this discovery was made on January 31, 1862, with Alvan Graham Clark at the eyepiece and his father by his side. They were waiting for Sirius to appear from behind the wall of a building when a faint star came into view three seconds before dazzling Sirius itself. This object was the famous "dark companion" that had been predicted by Bessel and others, and which was much later recognized as a white dwarf star.

Warner is brief on the subject of Alvan Clark's optical methods, but these were clearly primitive. She writes, "One visitor to the Clark workshop [the Scottish astronomer Ralph Copeland] thought the appliances both few and rude compared with those used by European artisans; he attributed the Clarks' success to skillful manipulation and personal supervision rather than reliance on precise mechanisms." Similarly, when Alvan was awarded the Rumford medal in 1867 for his ability to figure near-perfect lenses, he was asked by the Rumford committee to write an account of his original methods. But he had first to invite the committee to his shop, for he knew so little of the way others had made lenses that he could not say which of his techniques were original!

The Clarks usually tested their lenses in a 230-foot horizontal tunnel under their factory, by placing a point source of light at the focus and examining whether or not the lens appeared uniformly illuminated. Alvan and his sons had great skill in local correction of lens surfaces, by which they sought the sharpest possible focus rather than a mathematically true curve.

Much of the credit for the later Clark telescopes must be attributed to Carl Lundin, who joined the firm in 1874 at the age of 23 and remained with it until he died in 1915. Apparently Lundin did the actual work on the 40-inch Yerkes objective. His accomplishments have never been properly recognized, because he had an "exceedingly retiring and modest disposition." Equally obscure is his son C. A. R. Lundin of the Warner and Swasey Co., who made the 82-inch McDonald Observatory mirror in the late 1930's.

Since the principal resource of the Clark firm was the personal skill of its leaders, its activities shrank quickly with the retirements of Alvan in the early 1880's and George Bassett in 1891, and again with the deaths of Alvan Graham in 1897 and the elder Lundin in 1915. Although the company continued under different names until the middle of this century, its later work was rather undistinguished.

For this reason, the catalogue of Clark-made instruments that forms the second half of Warner's book extends only to 1897. It is arranged alphabetically by owner's name. From the considerable number of large Clark telescopes that were later remounted by other firms, one gathers that the Cambridge firm was valued primarily for its optics. European observatories, for example, often ordered a Clark objective and a Repsold mount.

The emphasis in this catalogue is on smaller telescopes. The most famous of these is the 6-inch refractor made for S. W. Burnham in 1870, with which he discovered about 400 double stars at the start of his astronomical career (see Chapter 17). The remarkable quality of this objective is demonstrated by many discoveries of pairs as close as 0.5 second of arc. In 1881 or 1882 the 6-inch and its dome were transferred to the Students' Observatory of the University of Wisconsin. As of 1984, its mechanical parts were on display at Chicago's Adler Planetarium.

Alvan G. Clark (left) and Carl Lundin with the crown-glass element of the Yerkes Observatory 40-inch objective, completed in 1897. Courtesy Yerkes Observatory.

Another traveling-telescope episode involved the 12-inch Clark refractor supplied in 1868 to Wesleyan University, Middletown, Connecticut. Complete with equatorial mounting, finder, circles, driving clock, and micrometer, it cost $6,000 delivered. About 1920, this telescope was sold to Miami University, Oxford, Ohio, which in July, 1959, presented it to Leslie C. Peltier. The telescope and its dome were reerected in the famous comet hunter's backyard in Delphos, Ohio.

A third perambulating Clark refractor is the 4-inch made in 1893 and used by William Tyler Olcott, well known early in the 20th century for his amateur observing manuals. From him it went to Phoebe Haas of Philadelphia, who presented it to the American Association of Variable Star Observers. Thereafter, this fine telescope was loaned to Walter Scott Houston of Haddam, Connecticut, who for many years has made observations with it for his "Deep-Sky Wonders" column in *Sky & Telescope* magazine.

There are also tantalizing accounts of lost Clarks. For example, where is the 12-inch refractor at West Point, New York, erected in 1884 by the U.S. Military Academy and pictured on page 100 of Warner's book? "The observatory and large refractor were seldom used," she says. "The tube has been scrapped and the objective lost."

31. An episode in early astrophotography

FOR A VIVID IMPRESSION of how celestial photography has advanced since the middle of the last century, consider the improvement in the limiting magnitudes attainable. The first star photograph ever taken was a daguerreotype of Vega (magnitude 0) with the 15-inch Harvard Observatory refractor on the night of July 16, 1850. By 1857, with the same telescope, George P. Bond could record 6th-magnitude stars on wet-collodion emulsions. In 1887 at the Paris Observatory, the brothers Paul and Prosper Henry photographed 16th-magnitude stars in an 80-minute exposure with a 13-inch refractor and dry plates. By 1960, the 200-inch Palomar telescope was reaching magnitude 23.

The innovations of an able American amateur astronomer, Lewis Morris Rutherfurd (1816–1892), played an important part in this advance. He was a socially prominent New York City lawyer of independent means whose taste for astronomy had been aroused during his undergraduate days at Williams College. He was also a friend of the famous telescope maker Henry Fitz, whose optical methods he learned, and of the astronomer Benjamin Apthorp Gould.

Having given up the law for a scientific career, Rutherfurd in 1856 erected an observatory in the garden of his residence at 175 Second Avenue in Manhattan. It contained an 11¼-inch Fitz refractor whose objective Rutherfurd figured himself under the direction of Fitz.

Lewis Morris Rutherfurd (1816–1892), after whom the Rutherfurd Observatory of Columbia University was named, was not just an astrophysicist but a lawyer and yachtsman. He greatly advanced the art of celestial photography by inventing and using telescopes specifically designed for the purpose. He took many extremely detailed photographs of the solar spectrum, using diffraction gratings he had ruled himself; one grating had as many as 17,000 lines per inch. From *Contributions* from the Observatory of Columbia University, Nos. 1 and 2, 1906.

Next to the 20-foot dome were rooms for a transit instrument and for computing. "The transit," wrote Rutherfurd, "is 189 feet N.W. from Second Avenue, and 76.3 feet N.E. from Eleventh Street."

After the refractor had been furnished with a good driving clock, Rutherfurd in 1858 took up celestial photography in earnest. In the days of the wet-collodion process, each photographic plate had to be prepared just before use by coating a piece of glass with a solution of guncotton and various iodides and bromides. As soon as the plate dried it was sensitized by dipping it into a solution of silver salts. The plate then had to be placed in the camera immediately, while still wet, as it lost its sensitivity on drying. Thus Rutherfurd's exposures were ordinarily limited to about six minutes.

Photographic experiments with this first telescope were disappointing, except for some plates of the Sun and Moon taken at reduced aperture. The main problem was that photographic emulsions in that era were sensitive only to blue light. Thus, the refractors made for visual use at longer wavelengths did not give sharp photographic images. Even at the best photographic focus, 0.7 inch longer than the visual focus, stars were embedded in a violet blur. During 1859 Rutherfurd tried interposing various lenses between the objective and plate to correct this condition, but he obtained satisfactory results only close to the optical axis.

Rutherfurd then obtained a 13-inch Cassegrain reflector from Fitz, which he knew would be inherently achromatic (free from false color). This telescope was firmly strapped to the main tube. But after a three-month trial this instrument was abandoned. The tremors of the city, though imperceptible in the refractor, were greatly enhanced in the Cassegrain. Also, in the damp and corrosive New York air the optical surfaces had to be resilvered at least every 10 days.

Finally, Rutherfurd overcame his problems by constructing the first refracting telescope ever designed specifically for photography. With an aperture of 11¼ inches, the objective consisted of a crown lens combined with a flint element that was ground to produce a combined focal length about 10 percent shorter than that required to satisfy visual use. Novel test methods were needed, as Rutherfurd explained: "I had a most delicate task to produce the correction for figure, since the judgment of the eye was useless unless entirely protected from the influence of all but the actinic rays. A cell of glass inclosing a sufficient thickness of the cupro-sulphate of ammonia, held between the eye and the eye-piece, enabled me to work for coarse corrections upon α Lyrae and Sirius, but so darkened the expanded disk of a star in and out of focus that all the final corrections were made upon tests by photography, which gave permanent record of all the irregularities of surface to be combated."

Completed in December, 1864, the new telescope was a complete success. A three-minute exposure on the Praesepe star cluster revealed 23 stars, some as faint as magnitude 9, within an area 1° square. Lunar negatives of great sharpness could be obtained in two or three seconds. Of course, this photographically corrected objective glass was useless for visual work. Hence Rutherfurd in 1868 erected a 13-inch visual refractor that could be converted at will to a photographic instrument by adding a third component, of flint glass. This corrector lens could be attached in front of the visual objective with three set screws within a few minutes, shortening the focal

This pair of photographs by Rutherfurd was originally printed for use with stereo viewers that were common in his day. This version may be seen in stereo by placing the edge of a pocket mirror on the white line, with the mirror perpendicular to the page and its silvered side to the right. Put your nose close to the mirror, view the left image with your left eye and the *reflection* of the right image with your right eye. By slightly adjusting the mirror, the images will overlap and seem three dimensional.

length to 13 feet to give sharp star images at photographic wavelengths.

Now well equipped, the New York amateur began an extensive observing program – the photography of star clusters and of Milky Way fields in Cygnus and Cassiopeia, with the ultimate purpose of measuring star positions and the motions of stars across the sky. By 1877, when failing health curtailed his observing, he had accumulated well over 1,400 celestial photographs of which more than 650 were of star groups. For example, there were 54 plates of the Pleiades, 27 of the Perseus double cluster, 23 of Praesepe, and 58 of the field around the proper-motion star Mu Cassiopeiae. Moreover, Rutherfurd hired assistants to measure many of these photographs with a measuring engine of his own invention. When in 1890 the aging amateur presented his plate collection to Columbia College, he also gave 20 folio volumes of unreduced plate measures. The reduction and preparation of these results for publication was a major activity of Columbia College Observatory for the next two decades.

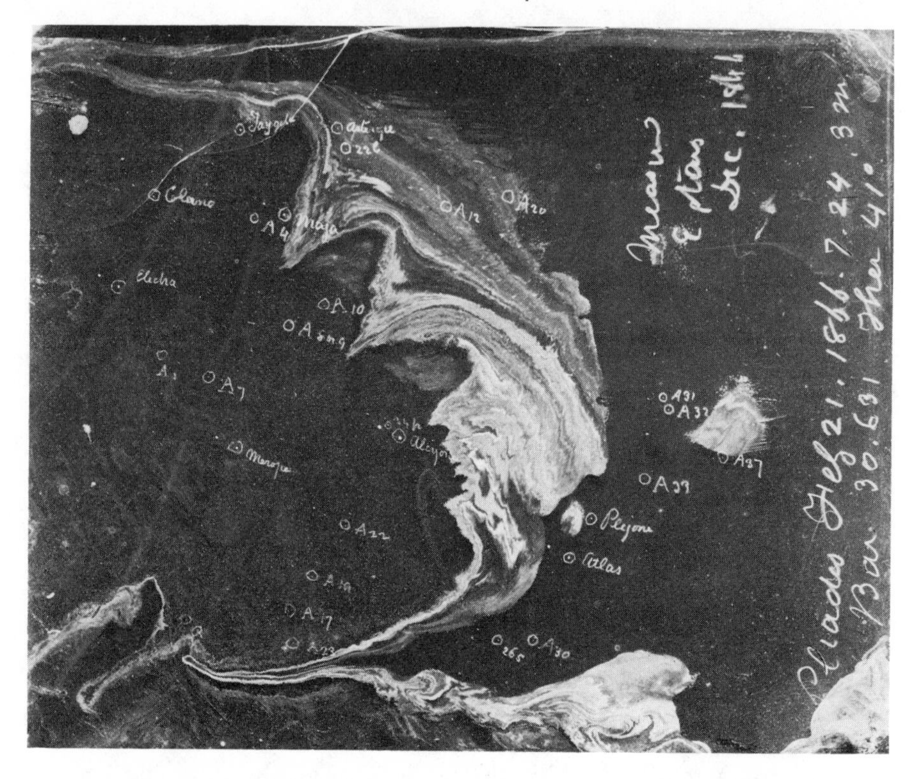

Lewis Rutherfurd took this photograph of the Pleiades star cluster on February 21, 1866, with his 11¼-inch refractor. Stars to about magnitude 8.0 can be found on the original – roughly 1,600 times fainter than the first star image to be recorded, Vega in 1850 by J. A. Whipple and W. C. Bond at Harvard Observatory. Rutherfurd circled the visible stars and identified the brighter ones with their names. The swirls are chemical stains on the plate. Along the right margin "7.24, 3m" may be interpreted as the time of mid-exposure and its duration, respectively. Actually, this is a double exposure of unequal length; two images of the brightest stars are visible. North is up, east to the left. Courtesy Columbia University and Kevin H. Prendergast.

The 13-inch refractor had already been donated to Columbia in 1883. However, its new location, only 250 feet from heavily traveled railroad tracks, was very unfavorable, particularly as New York's light pollution rapidly increased. In 1927 the telescope was sent on loan to what is now called the National Museum of History and Technology, in Washington.

This historic instrument was temporarily returned to Columbia's Rutherfurd Observatory in 1934 to repeat after six decades the original photographs of Praesepe and the Pleiades, in order to detect internal motions in these clusters. To duplicate the early conditions as closely as possible, Jan Schilt and John Titus used specially manufactured Eastman plates, designed to have the same fineness of grain, color sensitivity, and speed as the wet-collodion plates used by Rutherfurd.

In this way, extremely accurate relative proper motions could be derived for 37 Praesepe members.

32. Early photography and the Great Comet of 1882

IN 1950, DORRIT HOFFLEIT published a pamphlet, *Some Firsts in Astronomical Photography,* which contains much interesting information about the earliest efforts to take pictures of celestial objects. Hoffleit tells us that the first comet photograph ever made was one of Donati's comet in 1858 by the English amateur Usherwood. This must have been a noteworthy feat in those days when the exceedingly slow daguerreotype process was being replaced by the faster but inconvenient wet plate, with which only short exposures were feasible. Usherwood's achievement was probably made possible only by the great brightness of Comet Donati. With the invention of gelatin dry plates in the 1870's, long exposure times became common, and a spectacular advance in celestial photography began. The first application of the dry plate to astronomy was a spectrogram of the bright star Vega taken by William Huggins in 1876.

According to Agnes M. Clerke's *A Popular History of Astronomy During the Nineteenth Century,* Tebbutt's Comet (1881 III) was the first to be photographed really satisfactorily. In the following year, one of the most brilliant comets of modern times appeared. Many persons in the Southern Hemisphere independently discovered it during early September, 1882, as a naked-eye object in the morning twilight. Comet 1882 II brightened rapidly as it approached perihelion, its closest point to the Sun. On the 17th of that month it passed only 305,000 miles from the solar surface. In a few hours the comet moved across the Sun's disk, reached elongation, and traveled westward behind the Sun.

David Gill, director of the Cape Observatory in South Africa, described the splendor of the comet as it came above the horizon on the following morning. "The Sun rose a few minutes afterwards, but to my intense surprise the comet seemed in no way dimmed in brightness, but becoming instead whiter and sharper in form as it rose above the mists of the horizon." Later in the morning, "It was only necessary to shade the eye from direct sunlight with the hand at arm's-length to see the comet with its brilliant white nucleus, and dense white, sharply-bordered tail of quite ½° in length."

During the remainder of September and October, the comet continued as a spectacular feature of the morning sky. At Vizagapatam, India, one amateur astronomer described the tail as shaped like the tusk of an elephant. As seen with the naked eye, the whole tail took nearly an hour to rise above the sea horizon.

The best-known photographs of 1882 II are those taken by Gill that October and November. He used an ordinary camera with a portrait lens of 2½-inch aperture and 11-inch focus, attached to the declination counterweight of the Cape Observatory's 6-inch refractor. Guiding on the comet's nucleus, he obtained exposures, one as long as two hours 20 minutes, on six different nights. Prints of his pictures were sent to the Royal Astronomical Society in London.

Gill's photographs were of historic importance because of the unexpected number of stars they recorded. It was this circumstance that led him to undertake the *Cape Photographic Durchmusterung,* in collaboration with J. C. Kapteyn.

In Australia, R. L. J. Ellery of Melbourne Observatory was experimenting with many kinds of astronomical photography. He was taking plates of the Moon and the Orion nebula with the 48-inch reflector, and a systematic photographic record of sunspots was maintained. In his annual report for 1882–1883, Ellery says that some exposures were made on the Great Comet of 1882, "with more or less success." Also at Melbourne, an unnamed local photographer obtained a picture of the comet on October 10, 1882. A print was forwarded by Ellery to the Royal Astronomical Society, but it was

This photograph of the great September comet of 1882 is among the first ever made of this kind of celestial object. It was taken in Australia by an unidentified photographer, probably at Bombala, New South Wales. Courtesy Lawrence H. Aller.

noted, "The camera in this instance was not equatorially mounted, and the nucleus of the comet is therefore very much elongated."

It is to be expected that other now-forgotten pictures of 1882 II were taken. One of these, reproduced here, was rescued from a trash can at Mount Stromlo Observatory, Australia, by Lawrence H. Aller. It had been sent to the observatory by Mrs. G. W. Wright, Goulburn, New South Wales. The only information Wright could furnish is that the photograph shows the Great Comet of 1882 and that it came from Bombala, a small town in New South Wales. Unfortunately, the photograph has little or no scientific value because it has apparently been extensively retouched. The print shows many fine brush marks in the comet tail, whose edges seem to have been sharpened. Nevertheless, the picture is of some historic interest in recalling the impression made by a remarkable sight in the dawn sky of late 1882.

33. Five lives of a 60-inch reflector

ANYONE WHO TRIES to compile a detailed list of the world's large telescopes knows how some of them resist straightforward tabulation. For example, not a few telescopes change in size over the years. A case in point was the 40-inch Alvan Clark reflector erected at Lowell Observatory in 1909, which in 1925 became a 42-inch when a one-inch iron ring around the perimeter of the mirror was removed. Another problem is what happened to some once-celebrated instruments. No one seems to know the ultimate fate of the 16-inch refractor that Henry Fitz completed in 1863 for W. S. Van Duzee of Buffalo, New York. Other telescopes may change their identities and reappear in distant countries, or even divide into two, as if to taunt the cataloguer. The most curious case I have run across is that of the Common 5-foot reflector.

Among amateur telescope makers, Andrew Ainslie Common (1841–1903) has a probably unique distinction: The first mirror he ever ground was 60 inches in diameter. Like many other ATM's, this Englishman was mainly self-trained, but he had the advantage of being a well-to-do engineer who was able to retire at 49. Even earlier he appears to have had much leisure to pursue his hobbies in the workshop and observatory adjoining his home in the London suburb of Ealing.

Before attempting the 60-inch, Common had erected an 18-inch silver-on-glass reflector in 1877 and a 36-inch two years later. Both of these had mirrors made by George Calver, but Common designed and made the equatorial mountings.

The 36-inch deserves special comment, for this telescope later became the famous Crossley reflector at Lick Observatory (see Chapter 60). Also, it was used in Common's pioneering photography of nebulae. At that time dry plates were coming into use, and he was quick to exploit their advantages over the old-fashioned wet collodion plates. Some highly successful photographs of the Orion nebula were taken with the 36-inch in 1882-1883, winning Common

Andrew Ainslie Common (1841–1903), the British amateur astronomer who designed and built the Crossley telescope. He received the Royal Astronomical Society's gold medal in 1884, four years before this picture was taken, for his photographs of nebulae. Courtesy Mary Lea Shane Archives of Lick Observatory.

the gold medal of the Royal Astronomical Society the following year.

To make room for the 60-inch, Common sold the 36-inch in 1885 to Edward Crossley, an English amateur who had made some reputation as an observer of double stars and Jupiter. However, the acquisition seems to have been a white elephant. Crossley made little use of the 36-inch, and in 1895 he presented it to Lick Observatory in California. After James Keeler improved the mounting, the Crossley in 1898 became a very effective photographic telescope, achieving superb long-exposure photographs of galaxies and nebulae.

But let us return to 1886, when Common began work in earnest on his five-foot mirror. The disk was of plate glass, 61 inches in diameter but only four inches thick, cast with a central hole 10 inches across. It cost £130. A large grinding machine was completed that September.

The fork mounting that Common designed for the 60-inch reflec-

On January 30, 1883, A. A. Common photographed the Orion nebula with his 36-inch reflector. That 37-minute exposure was so striking that Agnes M. Clerke wrote: "Photography may thereby be said to have definitively assumed the office of historiographer to the nebulae; since this one impression embodies a mass of facts hardly to be compassed by months of labour with the pencil. . . ." The photograph reproduced here, taken February 28th of that year, is a one-hour exposure; it is the frontispiece of Clerke's *A Popular History of Astronomy*, 1887 edition.

tor was unusual. Its polar axis was a hollow wrought-iron cylinder eight feet in diameter that floated in a tank of water to relieve the friction. With an intended focal length of 28 feet 7 inches, the telescope would have required a dome about 60 feet in diameter. For compactness and economy, a shed with a roll-off roof was built instead.

The big telescope was ready for use in February, 1889, but it gave persistently elliptical star images even after the mirror had been refigured several times. Common at first attributed the poor image quality to the fact that the disk had been standing in an inclined position in his shop for four years before grinding commenced. Later he concluded that the hole in the middle was also to blame. During the pouring of the disk, so he reasoned, the molten glass must have flowed around the central core of the mold in two streams that produced an inhomogeneity where they met. To overcome this problem, Common in 1888 had ordered a second mirror blank from France, this time without a central perforation. It arrived two years later and was ground, polished, and figured in the short space of three months. Because of the unperforated mirror, the telescope was altered to work as a Newtonian instead of a Cassegrain.

Common continued to modify the 60-inch telescope. One night, while observing at the Newtonian focus, he narrowly escaped a bad fall from the high staging. Resolved never to run this risk again, he redesigned the instrument so that the observer worked at the lower end of the tube. For this purpose, the Cassegrain secondary was reinstalled, and a diagonal plane mirror was inserted just above the primary to bring the focus out to the side. (This "modified Cassegrain" concept was later used on the Mount Wilson 60-inch.)

Apart from taking some photographs of the Orion, Dumbbell, Pleiades, and other nebulae, Common made relatively little use of his big telescope. Evidently it was inconvenient to operate, and every year the city lights and smoke of London increased. Moreover, his interests were changing. He became active in the British Aluminium Co. and also developed a successful telescopic gunsight for the Royal Navy before his sudden death in 1903. His executors advertised the 60-inch telescope with its two primaries for sale.

I suspect that the success of the Crossley reflector at Lick Observatory encouraged Edward C. Pickering to consider acquiring the Common 60-inch for Harvard Observatory in Cambridge, Massachusetts. Ever since becoming director in 1877, Pickering had a special enthusiasm for visual measurements of the brightnesses of stars. During a quarter-century he and his assistants had made 1.5 million photometric settings on some 70,000 stars to as faint as magnitude 12, using 2- and 4-inch meridian photometers and later a 12-inch horizontal telescope. He was eager to extend this work to still fainter stars. Hence it must have been with keen satisfaction that he issued a public announcement on August 18, 1904. It said in part: "A reflecting telescope of sixty inches aperture was constructed by the late A. A. Common, and for several years has been idle. From its great

E. C. Pickering, fourth director of Harvard Observatory, during the early years of his administration (1876–1919). Courtesy Harvard Observatory.

aperture it should show extremely faint stars, and would be especially adapted to measuring their light. Some years ago, an attempt was made to purchase this telescope, but the means of the Observatory would not then permit. In 1902, the anonymous gift of $20,000 was received, and it has supplied several urgent needs of the Observatory. Representing these facts to Professor Turner of Oxford, during his recent visit to Harvard, he recognized the importance of utilizing so valuable an instrument, and that the nature of the observations and other conditions were favorable to securing valuable results. He therefore wrote to Mr. T. A. Common, with the result that this Observatory has purchased the telescope on such liberal terms that Mr. Common may fairly be regarded as having contributed a large portion of the cost.

"Steps are being taken for packing and transferring the instrument at once to Cambridge. It is hoped that in a few weeks the telescope may be received and mounted, and that observations to supply one of the great wants of Astronomy, a measure of the light of the very faint stars, can then begin. The work of many years has supplied this want for the brighter stars, and may now be extended to the faintest objects within the reach of human knowledge."

In the *Annual Report* of Harvard Observatory for the year ending September 30, 1904, Pickering added: "The instrument has arrived safely in Cambridge, and is to be mounted according to a plan proposed for the large reflector of the Yerkes Observatory. The position of the eyepiece will be fixed and always directed towards the South Pole, and the observer and recorder will work in a warm room. Attaching the photometer described above [the artificial-star and neutral-wedge device used on the 12-inch horizontal telescope] we expect to have not only the largest, but the most convenient instrument of its kind, with which accurate measures can be made of the light of all stars, from the brightest to the faintest known."

A small wooden building for the 60-inch reflector was erected on the Harvard Observatory grounds a little east of where the director's residence then stood. But the telescope didn't get to work as promptly as the sanguine Pickering had hoped. His 1906 *Annual Report* stated that the mounting was "still unfinished, but it is hoped that its completion will not be delayed much longer." In 1907 and in 1909, however, he speaks of continued delays in finishing the mounting.

The final abandonment of the project by Pickering is announced in the 1914 annual report. There he states that the 60-inch reflector did not form sufficiently good star images for visual photometry of faint stars, but other Harvard telescopes were giving excellent results on a large scale by photographic methods.

Following Pickering's death in 1919, Harlow Shapley came from Mount Wilson to Harvard and became director in 1921. He had been an active observer with the large Mount Wilson reflectors and unquestionably felt the need for a large Harvard telescope to continue his studies of star clusters. With characteristic energy, Shapley strove to get the 60-inch into working order. In 1922 he announced that the mirrors and mounting had been tested with satisfactory results, and that it was hoped to put the instrument in operation within a year.

The Common 60-inch telescope as re-erected in Cambridge, Massachusetts, is seen from the southwest in a photograph probably taken about 1906. Courtesy Harvard Observatory.

This task must have been largely the responsibility of Willard P. Gerrish, who for many years was the observatory's instrument designer. When the American Astronomical Society held its Christmas, 1922, meeting in Cambridge, Gerrish gave a paper, "The Adjustment of the 60-inch Harvard Reflector." The 60-inch could not be pointed at the north celestial pole because the observing house was in the way, and thus usual methods of effecting the polar alignment of the mounting were impossible. Hence Gerrish invented a simple new method: A 6-inch telescope with an eyepiece reticle was clamped temporarily to the main instrument in such a way as to afford an unobstructed view of Polaris. Details of the method, which has practical interest even today, can be found in *Popular Astronomy, 31*, 193, 1923.

Despite Shapley's and Gerrish's efforts, the Common telescope never got into effective use. The last mention of it in the Harvard annual reports was in 1922. In his *History and Work of the Harvard College Observatory 1838–1927,* S. I. Bailey is quite brief with what must have been an embarrassing subject. He does add: "Little use was made, therefore, of this telescope, although some investigations were undertaken for the determination of the total intensity of stellar radiations, a field for which its great light-gathering power seemed to render it well fitted." This passage evidently refers to attempts to measure the radiation of bright stars with a bolometer or similar device.

How startled Common would have been to hear that his 5-foot giant would be 40 years old before it became a success, and then as three telescopes rather than one!

In 1927 Harvard Observatory obtained funds for a modern 60-inch reflector for its Southern Hemisphere station, which had just been moved from Arequipa, Peru, to Bloemfontein, South Africa. One of the two Common mirrors in Harvard's possession was refigured, but a new mounting of standard two-pier design was built by the firm of J. W. Fecker. In the very capable hands of John S. Paraskevopoulos, the South African telescope began taking excellent photographs of southern galaxies, clusters, and nebulae. Eventually Harvard found its station in South Africa too expensive to operate alone, so in 1955 an arrangement was made with five other observatories in Ireland, Sweden, Belgium, and West Germany to share its facilities and costs. Under this new joint management, the 60-inch telescope continued its useful life.

The other Common mirror was also put to good use. When the Warner and Swasey reflector for Perkins Observatory, near Delaware, Ohio, was installed in 1925, it had a 61-inch disk of concrete in lieu of a mirror. Director Clifford Crump searched in vain for a glass works that would provide the mirror blank. The European masters had evidently been ruined by World War I, and no American firm would attempt the job. Finally, in 1927 the National Bureau of Standards was persuaded to undertake the task, and the following year cast a two-ton blank of optical glass, 72 inches in diameter and 12 inches thick. From that disk, probably the largest ever cast from optical glass, a 69-inch mirror was eventually made for Perkins Observatory. However, according to Ernest H. Cherrington, Jr., the big reflector at Delaware would have stood blind and useless for nearly seven years, had not Harvard director Harlow Shapley generously loaned Common's second 60-inch mirror. The Common mirror was replaced by the 69-inch one in 1931.

The fifth incarnation of the Common reflector was at Harvard's Oak Ridge (later Agassiz) station in Massachusetts. There the second Common mirror was installed in the new 61-inch reflector that was completed in 1933. Four years later, however, it was replaced by a 61-inch by Fecker of 310 inches focal length. Therefore this telescope, which continues to do good work to this day, no longer has any connection with Common.

Until 1963 that Common mirror remained on display in the dome of the 61-inch telescope at the Agassiz Station. Then Donald Menzel, Harvard Observatory's director since 1954, rented it to the firm of Varo Electronics in Garland, Texas, which purchased the mirror in 1969. Presumably it is still in Texas.

34. The lost legacy of Dr. Peate

An amateur telescope maker's mirror 62 inches in diameter – over a ton of clear green glass – is an eye-catching possession of the Smithsonian Institution's Museum of History and Technology in Washington, D.C. Upon the completion of this mirror in 1897, it was the largest ever made of glass for use in a telescope. Yet by 1955 this giant and its maker had become so nearly forgotten that neither is mentioned in H. C. King's comprehensive *The History of the Telescope*. The detailed story was rescued from oblivion in 1962 by F. W. Preston and William J. McGrath, Jr., in *Bulletin* 228 of the United States National Museum.

Now that his accomplishments in optics are becoming better known, it is clear that John Peate (1820–1903) narrowly missed astronomical fame. This unusual man was a bricklayer in Buffalo, New York, who in 1851 became a Methodist minister and for over half a century served the Erie Conference as a successful and popular preacher in Ohio, Pennsylvania, and New York. Just how or when Peate first became interested in astronomy is uncertain. Perhaps it was in 1859, when he visited several observatories during a European walking tour. He made his first telescope, a 3-inch refractor, about 1870. Later he constructed a 6-inch that accompanied him from one residence to another until it was mounted in his observatory at Greenville, Pennsylvania.

Preston's historical researches have traced 10 mirrors figured by Peate, and the possibility of others is mentioned. Some were of considerable size, but in most cases their fate is obscure. A 12-inch primary was made for a college in Harriman, Tennessee, but the college, observatory, and mirror have all disappeared. We are further told: "A 15-inch mirror in a reflector located at Allegheny College [Meadville, Pennsylvania] was probably made by Peate, although the College records do not show its origin, nor do they mention a 30.5-inch mirror which Peate was making for Allegheny College in 1891, according to an article in *The Scientific American.* Definitely Peate's was a 22-inch reflector found in about 1935, still in its packing case, at Thiel College, Greenville, Pennsylvania."

All this effort was fruitless, for none of the clergyman's mirrors seems ever to have been put to serious astronomical use. By presenting them to colleges that could not find funds to build and staff observatories, he effectively buried them.

The most striking instance of this began in 1894, when it was arranged that he should build the world's largest telescope for the newly founded American University in Washington, D. C. The choice of 62 inches for its aperture seems explained by the fact that A. A. Common in England had constructed his 60-inch reflector a few years before. (Lord Rosse's famous 72-inch at Parsonstown, Ireland, had been inactive since 1878.)

Making a mirror blank over five feet in diameter and seven inches thick was an unprecedented feat for American glassmakers of that day, but Peate placed a contract with the Standard Plate Glass Co. of Butler, Pennsylvania. The first attempt failed because a zinc sheet in the mold volatilized when the molten glass was poured over it and bubbled up through the melt. In a second attempt, casting was safely accomplished and, after several days of annealing, the covering of sand was removed and the glass examined. But the iron mold, contracting with tremendous force as it cooled, had fragmented the disk. This problem was solved by George Howard, a young engineer at Standard Plate who later became a famous inventor of glass-

Amateur John Peate's 62-inch, 2,500-pound mirror, here seen in its protective crating at the Smithsonian Institution. This picture, together with most of the information in the text, is from *Bulletin* 228 of the United States National Museum, "Holcomb, Fitz, and Peate: Three 19th Century American Telescope Makers."

making machinery. For the third and successful casting, the mold's iron bolts were replaced by wooden ones, which could break to relieve the strain as the mold cooled. Thus in May, 1895, Peate received his 62-inch blank, with a surprisingly modest bill for only $450.

The 75-year-old clergyman had erected a small optical shop in Greenville and equipped it with grinding and polishing machines.

Using a waffle-pattern tool four feet in diameter, Peate rough-ground his mirror with steel shot and emery as abrasives. Eventually the big job was completed, after some 750 hours of polishing and figuring.

In 1898 the finished mirror was sent to American University, to await some donor who would pay for a mounting and an observatory. No one ever appeared. After Peate's death five years later, the big glass disk slumbered nearly forgotten in its packing crate. Transferred in 1934 to the Smithsonian Institution, it was to spend another quarter century and more in storage.

How good a mirror is the 62-inch? Apart from Peate's own testing, it seems to have been critically examined only once. This was about 1925, when Perkins Observatory was preparing to erect a reflector some 60 inches in aperture and sent a committee to see if the American University mirror could be used. No optical tests of its figure were made, but the disk was found to be too thin and poorly annealed. Instead of trying to salvage the Peate mirror, Perkins Observatory arranged with the National Bureau of Standards for the pouring of a new blank from which a fine 69-inch paraboloid was made.

John Peate's story evokes the picture of an unusually enthusiastic and able amateur telescope maker who was entirely out of touch with astronomers and astronomy. Alvan Clark, Henry Fitz, and John Brashear brought their instruments to the attention of leading observers, and they made themselves known as optical craftsmen. Peate never did this, and his mirror making remained only a hobby. Robert P. Multhauf, a well-informed historian of American science, comments: "In a professional telescope-maker of the end of the 19th century, Peate's accomplishment would have been remarkable. In an amateur it is amazing." Certainly it is a feat worth recalling.

35. The great Paris telescope fiasco

A FRIEND ONCE TOLD ME how he went to the 1958 Brussels Fair and saw the display of Maksutov telescopes and other astronomical optics in the Soviet pavilion. It seems that no international

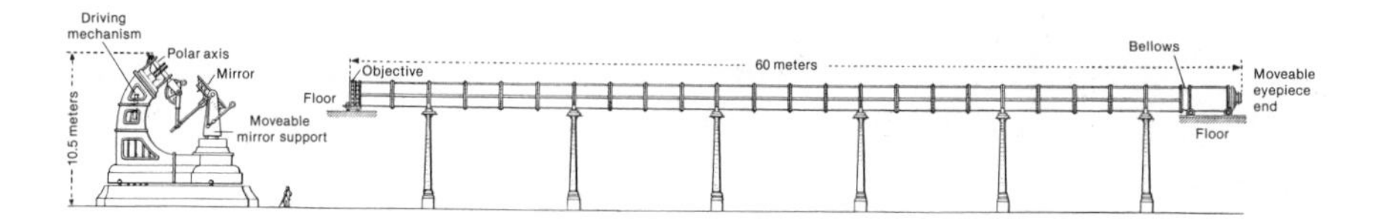

The main components of the great Paris refractor are labeled in this diagram adapted from *Le Ciel,* 1923.

exhibition is quite complete without astronomical instruments of novel design or unprecedented size. The tradition is an old one; it goes back at least as far as the 1893 Columbian Exhibition in Chicago, where wondering crowds gazed at the huge 40-inch refractor soon to be erected at Yerkes Observatory.

As fairs go, the Paris Universal Exhibition of 1900 must have been one of the most striking ever held. Some 51 million persons visited the immense fairgrounds that stretched along both sides of the river Seine, and on one gala evening 22,000 mayors of French municipalities sat down to dinner as the guests of President Emile Loubet. Since France at that time was a leading nation in the manufacture of precision optics, it was fitting that one of the French buildings at the exhibition should be a Palais de l'Optique, and it was natural that this should contain the largest telescope in the world.

That instrument was the project of a syndicate headed by Francois Deloncle. At first, a reflector of 120 inches aperture was considered, but facilities could not be found for casting a glass mirror blank of such unprecedented size. Finally, a 49.2-inch refractor of 187 feet focal length was decided upon. A telescope of this enormous length could not be mounted in the ordinary manner inside a dome. Instead, a fixed horizontal tube was used, into which starlight was reflected by a mirror 79 inches in diameter that could track the stars as they moved across the sky.

The firm of Mantois cast the 50-inch blanks for two refractor objectives, visual and photographic. And, after many unsuccessful trials, the Jeumont glassworks provided a satisfactory disk for the siderostat mirror. The figuring and polishing of the optical components were carried out by P. Gautier, the leading French maker of large telescopes, who also was responsible for the mounting.

The siderostat mirror was 6½ feet in diameter and one foot thick. From *Le Ciel,* 1923.

To permit focusing, the eyepiece assembly was carried on a four-wheeled carriage that moved along rails. With the lowest power, 500, the field of view was only three minutes of arc, but celestial objects could easily be located thanks to the accuracy of the mounting and its adjustments. The observer was connected by telephone to an assistant at the siderostat who could read the setting circles and operate the controls. A brief conversation – "more right ascension," "easy does it," and "back up" – would suffice to get the desired object centered in the eyepiece field. The weight-driven clockwork of the siderostat functioned so well that a star would stay in view with 500x for 45 minutes without the need for using the slow motions.

The throngs of visitors who lined up for a look through this great instrument were turned away at midnight, when E. Antoniadi of Juvisy Observatory would take his place at the eyepiece for observations of nebulae.

Set up at the Paris exhibition, the tube of the 49.2-inch refractor was 197 feet long. From *Le Ciel,* 1923.

Antoniadi's drawings of the Ring nebula in Lyra, NGC 7009 in Aquarius, and a few other planetaries seem to represent the only attempt to make any scientific use of the 49.2-inch refractor. As reproduced in the *Bulletin* of the Société Astronomique de France for 1900, they are rather disappointing, and probably do not show much (except for faint field stars) that could not have been detected visually with much smaller apertures. The meagerness of these results was partly due to the very unfavorable location of the telescope, amid the searchlights of the fairgrounds and the smoke of a great city. In addition, the quality of the seeing must have been seriously impaired by the use of a steel tube without any provision for adequately ventilating its interior.

When the Paris exhibition of 1900 closed, the heavy financial investment in the telescope had been only partially met by the admissions collected. Deloncle's syndicate tried unsuccessfully to sell the instrument to the French government, and it was finally broken up, the optical parts being stored at the Paris Observatory.

In addition to the dubious merit of having shown that the telescope of the future would be of a different type, the Paris giant enjoyed for about a year the clear title as the largest refractor in the world. Perhaps it should also be called the second largest telescope of that day, as Lord Rosse's historic 72-inch reflector had been inactive since 1878, while A. A. Common in England had erected his 60-inch reflector in 1889. The first large modern telescope, the 60-inch Mount Wilson reflector, was not completed until 1908. No refractor as big as the Paris 49-inch has ever been built again.

36. Beginnings of the Space Age

ASK ANYONE what was the first application of rocketry to astronomical observation. The answer you hear will probably involve the Lunik Moon probes of the 1950's or the Pioneer solar-system probes of the 1960's. And, perhaps, someone with a good memory will recall the study of the ultraviolet solar spectrum by V-2 rockets fired from White Sands, New Mexico, in the late 1940's. All of these replies happen to be wrong by more than a century.

If you are betting on the answer, a good referee is J. G. F. Bohnenberger's classic manual of practical astronomy; its second edition appeared in 1852. He is speaking of methods for determining geographical longitude that depend on comparing the local times at two places of some sharply defined event. Bohnenberger tells:

"Rockets are admirably suited for finding the difference of two meridians. To avoid using intermediary stations between two rather distant points, . . . one needs rockets that ascend 8,000 or 9,000 feet, exploding at this level, and which can be seen from a distance of about 120 miles. If these signals are launched from a mountain 5,000 or 6,000 feet high, they can be seen quite well at a distance of 140 miles, making it possible to determine directly the longitude difference of two places 250 to 275 miles apart. At peak altitude, the rocket automatically fires a charge of half a pound or one pound of gunpowder, forming a momentary brilliant flash of light that can be timed with great certainty."

Rockets were actually used in this way, and Bohnenberger reports in detail a determination on May 6, 1822, of the longitude difference between Vienna and Neustadt, Austria. Three rockets were launched that evening, from which it was deduced that Neustadt lay $33^{s}.02$ of longitude west of Vienna.

This early use of rockets can properly be regarded as a true forerunner of such contemporary achievements as astronomical geodesy. Perhaps the advance has been more in technology than in scientific imagination.

As early as 1837 W. Beer and J. H. Mädler made a study of the possible advantages of an astronomical observatory on the Moon; it is summarized on pages 22-24 of their selenographical treatise, *Der Mond*. Some of their thoughts have a modern ring, such as the convenience of being able to watch the development of sunspots uninterruptedly over a 350-hour interval, the length of daylight on the Moon. However, in 1837 it was not known that the lunar atmosphere is virtually nonexistent. Thus, the German astronomers supposed that faint stars would be invisible from the near side of the Moon, because of the glare of the Earth, even when the Sun was below the horizon. The advantages that they predict for astronomical work from the averted hemisphere in fact also apply to the side that we see.

The problem of practical means of communication with intelligent beings existing elsewhere in the universe attracted considerable attention in 1959 and 1960. Those years saw an analysis of possible techniques by G. Cocconi and P. Morrison of Cornell University and also Frank Drake's pioneering search for radio signals (Project Ozma). These ideas have a long prehistory; the only essential novelty of the modern proposals was the realization that interstellar distances are involved and that radio is the only feasible medium of communication. In the last century the problem seemed simpler, when there was hope that only interplanetary distances had to be bridged, which could be done with the aid of light. It is said that C. F. Gauss envisioned vast geometrical symbols, such as crops planted in regular patterns, that could be seen telescopically from other planets. Their inhabitants, thereby recognizing intelligent life on Earth, might be expected to respond in similar fashion. In Jules

Verne's novel of a trip to the Moon a related idea is mentioned. It is easy to smile at these naive suggestions, but the current proposals may be equally unrealistic in supposing that extraterrestrials have faithfully duplicated our own technology.

Even the idea of space communication by radio is not quite new, as readers of H. G. Wells' *The First Men in the Moon* know. This novel appeared originally as a magazine serial in 1901. The year before, a great deal of publicity resulted from Nikola Tesla's ill-founded announcement that he had received radio signals presumably originating from the planet Mars. The newspapers magnified the story by linking it with Percival Lowell's press release a few days earlier that he had observed a bright "projection" at the edge of Mars. Enthusiastic editors failed to realize that the Arizona astronomer was referring to a Martian cloud, and they assumed some artificial attempt at signaling by supposed inhabitants of the red planet. Also in 1900, the Paris Academy of Sciences accepted a bequest of 100,000 francs to be awarded as a prize for the first person to communicate with another planet. The stir occasioned by this combination of events lingered long in the popular mind, and even by the time of Project Ozma, a fair number of people must have remembered it.

PHENOMENA OF THE EARTH, MOON, AND PLANETS

37. Thoughts about twilight

EVERY ENTHUSIASTIC OBSERVER knows the pleasant anticipation brought on by a cloudless, gradually darkening sky. If his or her telescope has been readied and other preparations attended to, there is time to watch the stars come into view, one after another, and to see the constellations take form.

While the Sun sinks below the western horizon, in the east rises the grayish blue shadow of the Earth. Just above it is a purplish arc, merging gently into the deep blue of the sky. As the Earth's shadow mounts higher into the eastern sky, its initially well-marked upper boundary grows less distinct, usually becoming unrecognizable by the time the sun is 12 degrees below the horizon. Simultaneously, the purple border has been fading away. Meanwhile, in the west, the bright yellows and reds have vanished, and a reddish purple glow appears, covering much of the western sky. As the darkness deepens, this glow fades and contracts; its final disappearance on the western horizon marks the end of astronomical twilight and the beginning of true night.

The length of time between sunset and the close of twilight varies with the time of year and with geographical latitude. At the latitude of Philadelphia, Pennsylvania, astronomical twilight lasts two hours in late June. But this contracts to only 90 minutes at the times of the March and September equinoxes, when the path of the setting sun crosses the horizon most steeply. At the Earth's equator, each year contains about 4,400 hours of day, 850 of twilight, and 3,500 of night. At the North and South Poles the corresponding numbers are about 4,450, 2,400, and 1,900 hours.

Predictions of astronomical twilight are based on the rule that evening twilight ends and morning twilight begins when the center of the Sun's disk is 18 degrees below the horizon. Of all the varied observational data used by those who calculate astronomical ephemerides, this 18-degree figure is the hoariest, as it dates back to Claudius Ptolemy – about A.D. 130!

Many early astronomers made their own determinations of this number. Johannes Kepler (1618), Pierre Gassendi (1647), and J. H. Lambert (1760) all agreed with Ptolemy in assigning a solar depres-

The Earth's shadow. Gareth Wynn-Williams of the University of Hawaii photographed the shadow of the Mauna Kea volcano on the morning of September 16, 1978. Seen near the apex of the mountain's shadow is the nearly full Moon, somewhat flattened by atmospheric refraction, shortly before it set.

sion of close to 18 degrees for the end of astronomical twilight. But the Portuguese scientist Nonius, in his book of 1542 on twilight, preferred 16 degrees, and the German Christoph Rothmann cited 24 degrees in a letter he wrote to Tycho Brahe in 1588.

Much more trustworthy was the result announced in 1864 by J. F. J. Schmidt, the famous selenographer who was director of Athens Observatory. The city of Athens was beginning to illuminate its streets by gas, and the mayor asked Schmidt for information on when the lights should be turned on and off. From a series of 170 timings when the last vestiges of twilight color vanished on the western horizon, Schmidt found 15.9 degrees for the average depression of the Sun at the end of astronomical twilight.

For many practical purposes, it is more convenient to consider civil twilight, the interval after sunset or before sunrise during which everyday activities are possible without artificial light. By convention, civil twilight is defined as ending in the evening or beginning in the morning when the center of the sun's disk is six degrees below the horizon. An approximate statement of this rule is embodied in the motor-vehicle laws of many states, which require that automobile headlights be turned on half an hour after sunset.

In the past, several other definitions of the end of civil twilight were widely used. At sunset, looking back and forth between the bright western sky and the darker eastern heavens, it is possible to locate roughly the imaginary line where the illumination changes most rapidly. The 18th-century German astronomer Lambert timed when this line, the so-called crepuscular arc, passed directly overhead. By this means he found that civil twilight ended when the solar depression was 6.4 degrees. The end of civil twilight has also been described as the moment when it is no longer possible to read a printed page outdoors. Using a copy of *Sky & Telescope* for the experiment, I found that this time seemed definite to within a few minutes; it corresponded to the Sun being 7.2 degrees below the horizon.

Closely akin to the problem of the duration of twilight is the question of the visibility of stars and planets during dusk. One of the most complete studies undertaken was by Schmidt, who reported his results in an 1864 paper. To decide whether a particular star is visible in a bright, blank sky, it is necessary to know just where to look. Otherwise, you will record not the time of first visibility but merely the time when you happen to scan the exact spot. Hence, Schmidt first brought a star inside the field of view of his telescope, and then sighted along the tube with the unaided eye from time to time until the star was seen. In this way, he established that the zero-magnitude stars Vega, Capella, and Arcturus could first be detected when the sun was still ¾ degree above the horizon. Similarly, 2nd- and 3rd-magnitude stars were first visible when the solar depression was 4.3 and 5.1 degrees, respectively.

The observations become simpler for fainter objects, for once the stars down to magnitude 3 are visible to the naked eye, dimmer ones can be located by their positions amid familiar groupings. Thus, Schmidt was able to recognize 4th-, 5th-, and 6th-magnitude stars when the Sun was 6.8, 8.9, and 11.6 degrees, respectively, beneath the horizon.

Very different figures can be found in older books. According to both Ptolemy and Kepler, 1st-magnitude stars could not be seen until the Sun was 12 degrees down, and J. F. Wurm in 1782 found that a solar depression of 6½ degrees was needed. The fact that Schmidt could see stars like Vega before the Sun had set suggests that his forerunners had not taken special precautions to be looking in the exact directions of the stars.

For the amateur who wants to make visibility tests, it is simpler to do the observations during morning rather than evening twilight, and to time when particular stars disappear in the growing dawn.

A century or two ago, it was thought that detailed observations of twilight might provide valuable information about the then inaccessible high atmosphere. Nowadays artificial satellites and other devices routinely probe the upper air. Nevertheless, the varied and beautiful phenomena of twilight remain of interest for their own sake.

38. Darkness at noon

IN THE MORE OUT-OF-THE-WAY PARTS of rural Massachusetts, where a family is apt to have lived in the same community for many generations, there is a surprising continuity of oral tradition.

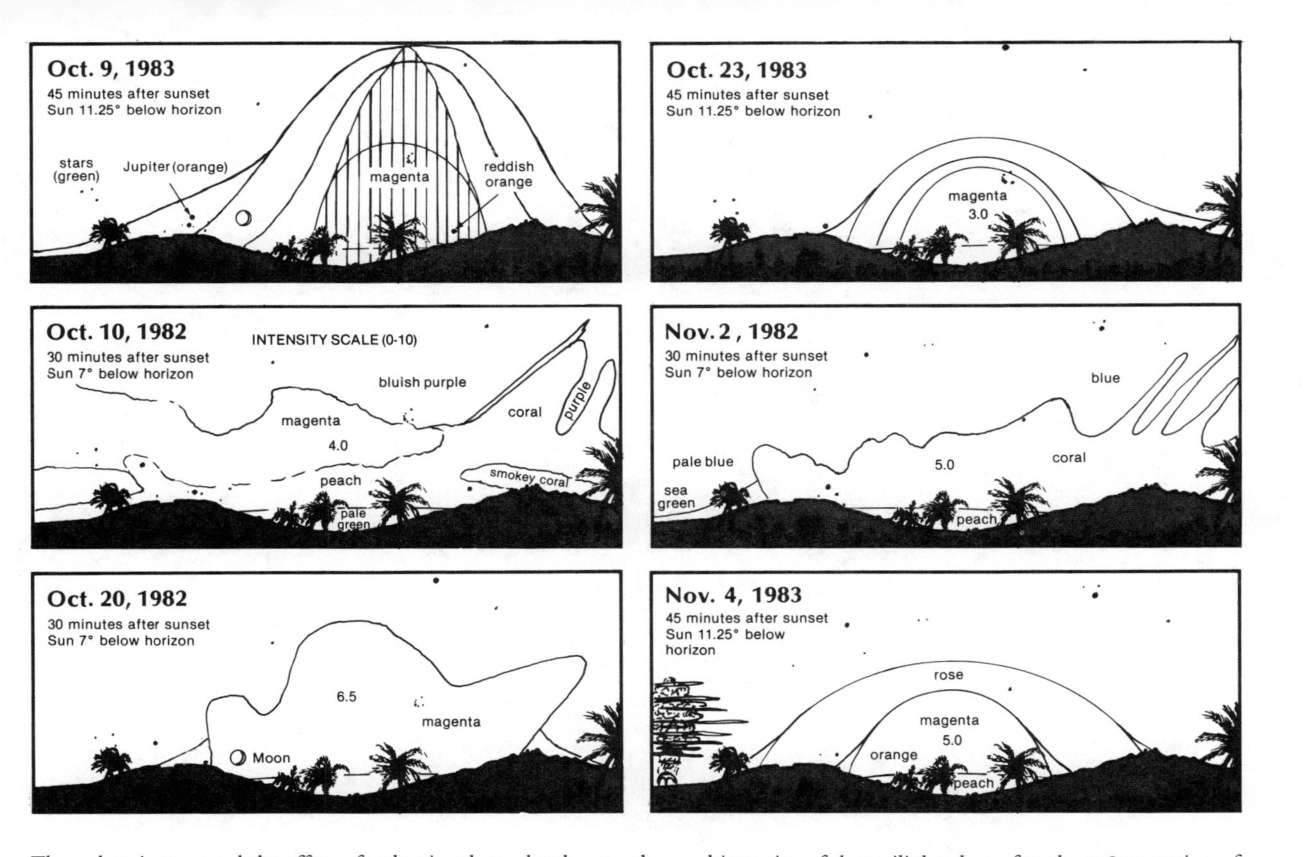

These drawings record the effect of volcanic ash on the shape, color, and intensity of the twilight glow after the 1982 eruption of Mexico's El Chichon volcano. The observations were made from Hawaii's "Big Island" 45 minutes after sunset. At that time the sky had darkened considerably, enhancing the outline of the skyglow due to the ash. The glow disappeared 10 minutes later. The ash also changed the normal colors of stars and planets. From the Hawaiian diaries of Stephen J. O'Meara.

You will hear old-timers tell verifiable stories of Indian attacks on their town 250 years ago, or of their ancestors' doings in the Revolutionary War. One widespread lingering memory of this kind concerns a remarkable phenomenon in the late 18th century that still deserves scientific notice.

This was the famous dark day of Friday, May 19, 1780, that extended over nearly all of New England. One of the many eyewitness reports still extant is a letter by Jeremy Belknap of Boston. He tells that the forenoon had been cloudy, and about 10 or 11 o'clock the clouds assumed a strange yellowish hue, which tinted all the landscape. An hour later the light began to fail, and by 1 o'clock the darkness was so great that candles were lighted and kept burning all afternoon. The atmosphere was not simply dark, said Dr. Belknap, but seemed full of "the smell of a malt-house or a coal-kiln."

Compare that description with the diary of Phineas Sprague of Melrose, Massachusetts: "Friday, May the 19th 1780. – This day was the most Remarkable day that ever my eyes beheld. . . . About ten oclock it began to Rain and grew vere dark and at 12 it was almost as dark as Nite so that wee was obliged to lite our candels and Eate our dinner by candel lite at Noon day. But between 1 and 2 oclock it grew lite again but in the Evening the cloud caim over us again. The moon was about the full [but] it was the darkest Nite that ever was seen by us in the World."

Some more specific facts were recorded in the manuscript journal of Bishop Edward Bass, who presumably was in the neighborhood of Newbury, Massachusetts. The unusual darkness, he wrote, set in about 10:40 a.m. and lasted all day, though varying: the obscurity was deepest about 12 to 1 o'clock. "Afterwards there was a larger glint at the horizon, which made it somewhat lighter. It was, however, at the lightest, darker I think than a moonlight night. The sky had a strange yellowish, and at times reddish appearance. The night was the darkest I remember to have seen, till about midnight, when a slight breeze sprung up from the north or northwest, after which it soon began to grow light. At Falmouth, Casco Bay, it was not dark at all. Upon Piscataqua River, Berwick, Dover, and so forth, it was very rainy . . . but not uncommonly dark as I am told by a person who traveled there that day. I hear of the darkness at Danbury, in Connecticut. It did not extend to North River."

We read in one contemporary account, in the *Memoirs of the Amer-*

ican Academy: "Candles were lighted up in the houses; the birds having sung their evening songs, disappeared and became silent; the fowls retired to their roosts. The cocks were crowing all around as at break of day; objects could not be distinguished but at a very slight distance, and everything bore the gloom and appearance of night."

This alarming day produced a deep moral effect on the pious inhabitants of New England, and there was a widespread fear that the end of the world was at hand. At Salem, Massachusetts, Nathaniel Whitaker's congregation met together at church and heard him preach a sermon that maintained the darkness was a divine rebuke to the people for their sins. In various other places, sermons were preached on such texts as Isaiah *13:10;* Ezekiel, *32:7, 8;* Joel *2:31;* and Revelation *6:12.*

The Connecticut legislature was in session at Hartford that day as the darkness gathered. The journal of the state House of Representatives reads, "None could see to read or write in the House, or even at a window, or distinguish persons at a small distance, or perceive any distinction of dress in the circle of attendants. Therefore, at 11 o'clock adjourned the House till 2 o'clock afternoon." In a neighboring room, the governor's council was also in session, and a motion to adjourn was proposed. But Col. Abraham Davenport objected with firm dignity: "Either the day of judgment is at hand or it is not. If it is not, there is no cause for adjournment. If it is, I wish to be found in the line of my duty. I wish candles to be brought." This widely remembered episode became the subject of a poem by John Greenleaf Whittier.

What is the scientific explanation of the dark day of May 19, 1780? If there is any modern literature, I have not yet seen it. But the better contemporary accounts agree that the obscurity was at least partly due to smoke from great forest fires in northern New Hampshire. At the time that region was rapidly being opened to settlement, and it was the custom to clear away the forests by burning.

It is possible that somewhere a blue Sun was viewed just afterward? On September 26, 1950, a blue Sun and a blue Moon were seen at Edinburgh Observatory, where with commendable enterprise a spectrographic study of the phenomenon was made with the 36-inch reflector. From this observation it appeared that the blue was due to scattering by oil globules and soot particles, which had spread eastward across the Atlantic at about 40,000 feet from big forest fires in Alberta on September 23rd.

39. Some notes about earthshine

ASTRONOMY, like natural history, has the attractive feature that many phenomena studied by the expert are known and enjoyed by every layman. The beautiful ashen light of the Moon's night side is an example. This sometimes very conspicuous glow was vividly described by J. F. J. Schmidt, the German-born astronomer who resided in Greece from 1859 to 1884 as director of Athens Observatory (see Chapter 50). Although a trained professional, all his life he had an amateur's enthusiasm for skywatching.

On the very clear evening of January 8, 1867, he was at the observatory, situated on the historic Hill of the Nymphs west of the Acropolis. As the Sun set at 5 p.m., the two-day-old crescent Moon hung in the southwest over the waters of the Saronic Gulf. Attracted by the unusual brightness of the ashen light, Schmidt watched in his 6-inch refractor during the next hour and a half.

In the darkening twilight, the night side of the Moon was spread out like a map, with many craters and the outlines of the lunar seas clearly visible. Schmidt could recognize the long rays of Tycho, bright spots like the one near Hell, Pico and other mountains, and such small craters as Bessel, Censorinus, and Pytheas. Going outdoors, the astronomer stood where the sunlit crescent was just hidden by the dome and noted that the Moon's dark side looked brighter than the most prominent portions of the Milky Way or the zodiacal light. Later, Schmidt watched the Moon set behind the mountains on the distant island of Salamis. Even after its bright sickle was gone, the earthlit part was conspicuous to the naked eye, though it was less than two degrees above the true horizon.

This particular report is part of a long series of observational records on earthshine by Schmidt, of which a portion was printed in his 1878 book about the Moon. He had been quite interested in the phenomenon as early as the 1840's, when he was a young ama-

The brightly earthlit Moon was only 29 hours old as it set behind trees a mile away in this picture taken on Mount Pinos, California, January 17, 1961. Courtesy Alan McClure.

teur using very small telescopes. His notes from those days contain color descriptions (bluish, yellowish gray, reddish brown) and repeated mentions of the ashen light being recognized one day after first quarter. Other visual observers have seen this secondary light as much as three days after first quarter. But in 1949 Audouin Dollfus managed to detect it only 38 hours before full Moon, using a coronagraph at Pic du Midi Observatory.

In ancient times, earthshine was curiously explained by the Greek astronomer Posidonius as due to a partial transparency of the lunar globe, which allowed some sunlight to pass as through a cloud. This view, reasserted by Vitello in the 13th century, lingered long enough to be rebutted by Galileo. In 1542, the German Erasmus Reinhold advanced the opinion that the Moon's secondary light was some kind of phosphorescence of its surface, an interesting conjecture in view of 20th-century evidence that luminescence occurs on the Moon.

Recent astronomy books, if they allude to earthshine at all, are apt to attribute its modern explanation to Kepler in 1604. This is only part of the truth. It was his teacher Michael Mästlin who correctly stated in his rare 1596 book on eclipses that the ashen light is sunlight that has been reflected first by the Earth and then by the Moon before reaching our eyes. In fact, the far-ranging genius of Leonardo da Vinci (1452–1519) had given the same explanation a century before, but it was forgotten until his manuscripts were published in 1797.

The long account of earthshine given by Galileo in his famous *Sidereus Nuncius* (1610) is a fine example of deduction from simple observations. The great Italian scientist begins: "Here I wish to assign the cause of another lunar phenomenon well worthy of notice. I observed this not just recently, but many years ago, and pointed it out to some of my friends and pupils, explaining it to them and giving its true cause. Yet since it is rendered more evident and easier to observe with the aid of the telescope, I think it not unsuitable for introduction in this place, especially as it shows more clearly the connection between the moon and the earth."

Galileo mentions that the brightest earthlight at any time is along the lunar disk's edge that is farthest from the Sun – a fact easily verified by the reader. Galileo gives various hints to the observer, advising him to interpose a chimney or roof to mask the bright crescent, and to allow the Sun to sink as far below the horizon as feasible. "Moreover," writes Galileo, "it is found that this secondary light of the moon (so to speak) is greater according as the moon is closer to the sun. It diminishes more and more as the moon recedes from that body until, after the first quarter and before the last, it is seen very weakly and uncertainly even when observed in the darkest sky."

With swift, sure arguments, he disposes of earlier theories of ashen light and concludes with Kepler that the Earth's light is its cause. He points out that when the Moon is new for us, the Earth is fully illuminated as seen from our satellite. At any time, the phases of Earth and Moon as seen from the other are complementary. Thus the fading and brightening of the earthshine each month correspond precisely to the changing phases of the Earth. Galileo urges that the phenomenon of earthshine is good evidence to support the Copernican theory, since it rebuts critics who exclude the Earth from the planets on the grounds that it is "devoid of light."

The modern scientific interest in the ashen light was foreshadowed in a paper that D. F. Arago read before the Paris Academy of Sciences on August 5, 1833. He pointed out that the brightness of this light should be enhanced by widespread terrestrial cloudiness, which would make the Earth a better reflector of sunlight. He prophesied: "When we have better photometric instruments at our command, we may be able to read in the moon the record of the average clearness of our atmosphere."

Although quite a few astronomers made measurements of earthshine, it was not until a century later that the "better photometric instrument" really materialized. It was the visual double-image photometer of André Danjon (1890–1967), who like Arago was a director of Paris Observatory. Danjon realized that the main source of error in measuring the intensity of earthshine is the intervening luminous veil of terrestrial atmosphere. He neatly eliminated this problem by inventing the double-image photometer, in which the observer sees two Moon images, the dark limb of the first just touching the bright limb of the second. The observer then reduces the intensity of the second image (with a calibrated photometric wedge, for example) until the two adjacent areas appear of equal brightness. Such a differential measurement is unaffected by either atmospheric extinction or superimposed light.

Danjon could observe with this photometer until three days after first-quarter Moon, in effect determining the magnitude difference between the Sun and the Earth for various phases of the latter. In his 1954 summary, Danjon reported that the visual magnitude of the full Earth as seen from one astronomical unit distance is -3.92. But there were seasonal changes in the reflecting power of our planet, some 20 percent between October and July. During the International Geophysical Year (1957–1958), a worldwide chain of Danjon photometers was operated to monitor such changes. If we recall that NASA did not launch the Tiros 1 spacecraft until 1960, it is evident that the first weather satellite was our Moon itself!

40. Observing very thin lunar crescents

SOMETIMES an astronomically useful word becomes obsolete prematurely, like the adjective *combust.* This is defined in large dictionaries as referring to a planet that is invisible to the naked eye because of the Sun's light. Thus, we might speak of the Moon as being combust from the time the waning crescent is last seen in the morning sky until the waxing crescent first becomes visible in the west after sunset. Authentic cases of naked-eye sightings of a Moon younger than 22 hours are not very common.

Every so often, variants of the following two stories crop up in the popular literature. Neither stands close inspection. Johannes Kepler is sometimes cited as having seen the old Moon in the morning and the new Moon that same evening. However, Kepler actually was just quoting a statement by his teacher Michael Mästlin, who seems to have given neither observation nor other authority for his belief. This matter has been discussed critically by such Kepler experts as Edward Rosen and Ludwig Günther. The other story was written over a century ago by William Henry Smyth in his *Cycle of Celestial Objects:* "The late excellent Mr. W. Frend used to relate a remarkable circumstance: early one morning, a lady of his acquaintance noticed the wire-like crescent of the moon, then approaching nearly to her conjunction; and the day after, in the evening, she observed the opposite crescent in the west, soon after sunset." This is mere hearsay from a book with a poor reputation for accuracy. Nevertheless, William Frend (1757–1841) was a well-known English author, who from 1804 to 1822 published a popular astronomical magazine entitled *Evening Amusements, or the Beauty of the Heavens Displayed.*

Among valid young-Moon sightings are those of Terence W. Quigley of Green Bay, Wisconsin, who described himself as a compulsive observer. This amateur's interest in hairline crescent Moons began on March 5, 1954, when he saw the lunar crescent only 20 hours 58 minutes after new. He could not match this personal record until April 6, 1970, when during a search for Mercury with 7 x 35 binoculars he chanced upon the Moon when it was only 20 hours 41 minutes old. About five minutes later the sliver of light was comfortably visible to his unaided eye.

The youngest crescent ever photographed may well be that captured by William D. Pence of Charleston, Illinois. He happened to sight the Moon in the 8 x 50 finder of his 6-inch reflector and, by knowing just where to look, could see it without optical aid on April 25, 1971, when its age was 21 hours 13 minutes. "The sky was very transparent," he noted, "but owing to twilight, earthshine was not visible to the naked eye or in the telescope. At 32x the crescent appeared broken into several pieces . . . I took the picture [seen on page 202] immediately afterward with my 6-inch reflector."

A. D. Thackeray, while director of the Radcliffe Observatory, Pretoria, South Africa, wrote that on April 5, 1962, during a binocular search for Comet 1962c, he saw the 20½-hour Moon with the naked eye and without difficulty. "I have often regretted that my house on the observatory grounds does not command an unobstructed view of the western horizon; otherwise I feel sure that I would have recorded many other planned views of a very young moon," he wrote. "The problem has a practical significance, because at time of Ramadan members of the Moslem community in Pretoria constantly telephone the observatory to ascertain whether the new moon will be visible or not. I have always understood from my predecessor, the late Dr. Harold Knox-Shaw (who dealt with the same problem in his young days at Helwan, Egypt), that the record

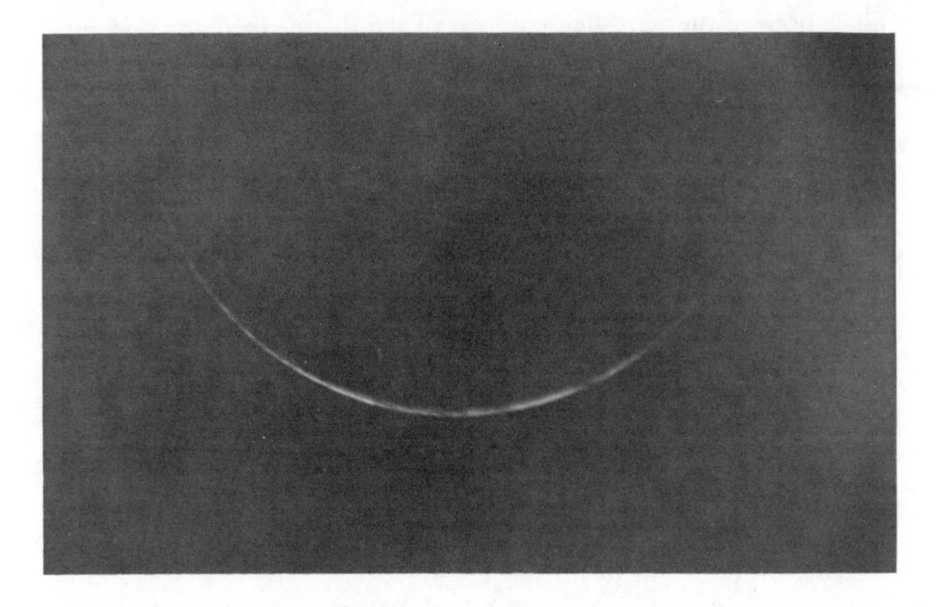

The Moon only 21¼ hours after new, photographed by William D. Pence at Charleston, Illinois, in the evening twilight on April 25, 1971. Note that the crescent extends appreciably less than a semicircle.

was 17½ hours. In the case of my own record, the visibility was so easy (despite the proximity of a large city) that I have no doubt this record could be beaten by more than an hour, given good enough conditions."

The modern record for sighting a young Moon seems to have been established by a member of the China Lake Astronomical Society in California. Realizing that March 15, 1972, would provide the opportunity to see a Moon less than 15 hours old, James H. McMahon prepared a detailed diagram showing the location of the Moon relative to Mercury and the sunset point. He reported, "Robert Moran drove in his jeep to the highest peak in the Rademacher Hills, eight miles due south of China Lake and 3,698 feet above sea level. . . . Moran was plagued by small clouds and some low haze. Nevertheless, using 10 x 50 binoculars and a copy of the diagram, he caught sight of the crescent at 6:28 p.m., when the moon's age was 14 hours 53 minutes. With difficulty, but positively, he continued to view the crescent for three minutes. Because the moon was hard to see, he made no attempt at a naked-eye sighting, moving just his eyes to read the time from his wristwatch. The orientation of the crescent was as expected, and its length from cusp to cusp was estimated at only 60 degrees. At the rate the crescent was approaching the horizon, it probably set 1½ minutes or less after it was last seen by him through the haze."

The topic of young Moons had a lifetime fascination for the English amateur astronomer Charles T. Whitmell (1849–1919). By profession a school district administrator, he was a skillful computer who delighted in the solution of curious and unusual astronomical problems. Whitmell wrote many articles about them in popular journals, and he published the following account of a 14½-hour Moon that is widely regarded as the naked-eye record: "From Scarborough, in Yorkshire, about 8^h p.m. (G.M.T.) on Tuesday, 2nd May 1916, the Moon was observed by Lizzie King and Nellie Collinson, two maids in the service of Mrs. Ackroyd, of 43, Westbourne Grove. I have been in correspondence with this lady, and wish to thank her for convincing evidence most courteously given. On the same evening, about 8^h 15^m, the crescent was also observed with the naked eye by Mrs. Willimott and her daughter, residing at Heighington in county Durham. Mrs. Willimott has also kindly replied to inquiries. We have thus the evidence of four persons. As might have been expected, atmospheric conditions were perfect . . . It seemed to me exceedingly probable that a mistake of a day might have been made with regard to the observations at Scarborough and Heighington. But this was certainly not the case. Among other circumstances which fixed the date as the 2nd May was the occurrence on the same night of a Zeppelin raid over Yorkshire. . . . So far as I am aware, this Moon, 14½ hours old, is the youngest yet observed in England."

One would like to know what the eyewitnesses themselves said, but the original evidence satisfied Mr. Whitmell, who seems to have been a competent judge. He also reports the case of a Mr. Hoare of Faversham, Kent, who saw a 14¾-hour crescent on July 22, 1895, and of D. W. Horner, Tunbridge Wells, Kent, who viewed a 16-hour-old Moon on February 10, 1910. In each instance, he communicated with the observer and made a detailed investigation of the circumstances, as far as they could be reconstructed. Whitmell's own personal record was 22 hours, on April 1, 1919.

Good vision, a suitable climate, and years of perseverance are all that is needed for anyone to see some very young crescents. Not every new Moon offers a favorable occasion, for the moment of conjunction with the Sun must fall within certain limits of local time. For example, if at the observer's site new Moon occurs in the late morning, at sunset that day the crescent will be a hopelessly difficult eight hours or so, whereas on the next evening it will be a commonplace 32 hours old. If you are trying for a record, pick a month when the moment of new Moon falls sometime between midnight and sunrise.

The season of the year is also important. In north temperate latitudes, around the March equinox is a favorable time for viewing a young evening crescent. At such a time the ecliptic crosses the horizon most steeply. This situation causes the Sun to sink rapidly, darkening the sky, while placing the waxing crescent quite far above the horizon. Similarly, around the September equinox is favorable for observing the morning crescent. Several of the dates cited earlier are near such times. Finally, it is advantageous to have the Moon at perigee (closest to the Earth) on the date of an attempted observation, for its orbital motion is then fastest, thus increasing the angular distance of the crescent from the Sun. Indeed, this was the case for the observations of Quigley and of Thackeray.

It is well known that many calendars, both ancient and modern, begin the month with the first appearance of the Moon. Historians, therefore, would like to have some simple criterion for deciding whether or not the crescent could have been seen from, say, Babylon on a specific date. One answer to this question was given in 1910 by J. K. Fotheringham, an English specialist in the history of astronomy and ancient chronology. His material was a collection of 76 naked-eye observations of the visibility or nonvisibility of the crescent in the years 1859 to 1880, mainly by J. F. J. Schmidt at Athens. For each observation he calculated the true altitude of the Moon at the moment of sunset and the difference between the azimuths of the Sun and Moon at that time. Fotheringham's data are plotted here. He noted that a clear dividing line could be drawn between the positive and negative observations, which he approximated by the curve. If for a given date at Babylon the calculated altitude of the Moon at sunset fell under the curve, the crescent probably would not have been seen, according to him.

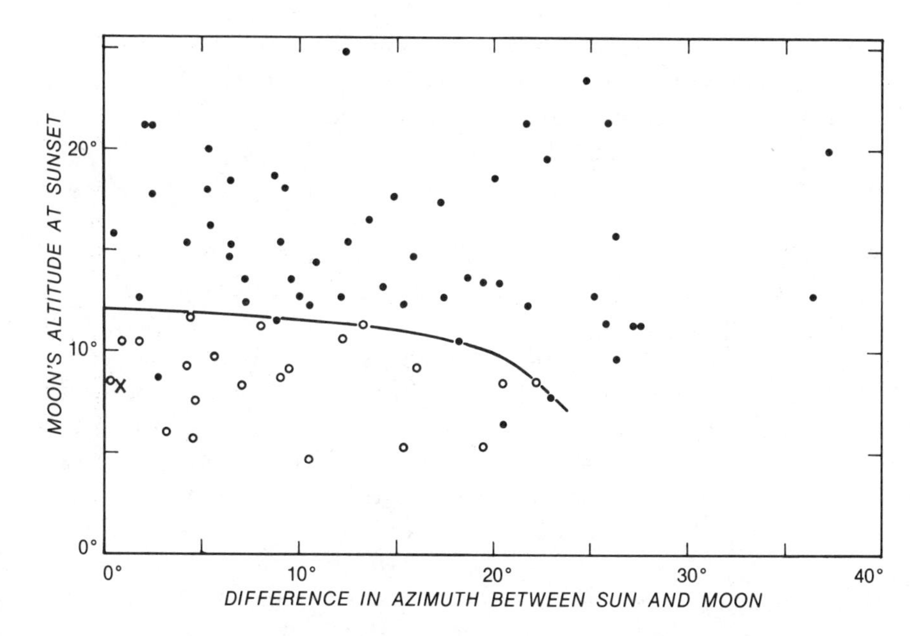

In this diagram by J. K. Fotheringham, each dot represents a sighting of the crescent Moon, and each open circle is an observation when the Moon was looked for but not seen. He drew the curve as the approximate division between the successful and unsuccessful cases. A cross has been added for the 14½-hour Moon said to have been seen from England on May 2, 1916. A few of Fotheringham's points lie outside the chart as redrawn here.

Somewhat similar criteria were developed by E. W. Maunder of Greenwich Observatory and by the *Indian Ephemeris and Nautical Almanac*. Both are listed in this table, along with Fotheringham's.

Criteria for crescent visibility

Moon-Sun Azimuth Difference	Moon's Altitude		
	Foth.	Maunder	Ind. Eph.
0°	12°.0	11°.0	10°.4
5°	11°.9	10°.5	10°.0
10°	11°.4	9°.5	9°.3
15°	11°.0	8°.0	8°.0
20°	10°.0	6°.0	6°.2
23°	7°.7		

What about the record-breaking observations made in 1916 at Scarborough and Heighington? At both places the Moon's altitude at sunset was 8.3 degrees and the Moon stood nearly vertically over the below-horizon Sun. These observations correspond to the cross on Fotheringham's diagram. By all the three criteria the Moon should not have been seen! Are the prediction rules at fault? Or was Whitmell mistaken after all in accepting these observations as genuine? Perhaps the best guide to the extreme limit of visual detection is supplied by the distinguished French astronomer André Danjon.

In 1931, when director of Strasbourg Observatory, he was engaged in determining how the light of the Moon varied with lunar phase. That August 13th, he saw the Moon when it was only 16 hours 12 minutes before new. To his surprise, in a 3-inch refractor at 25 power the crescent extended less than a quarter of a circle, being estimated as 75 to 80 degrees from cusp to cusp. Other observations, as well as an examination of printed records, showed that this shortness of the crescent was a general phenomenon, which became less marked as the angular distance of the Moon from the Sun increased.

There is a simple explanation for the shortening, which Danjon pointed out in 1932. When the moon is a slender crescent, the shadowed sides of lunar mountains are turned toward us and tend to mask the sunlit areas beyond them. Thus, the illuminated crescent looks narrower and shorter than if our satellite were a perfect sphere.

This effect is shown by Danjon in the diagram opposite. He explains as follows: "Let us represent the moon . . . by its projection on a plane passing through its center and those of the earth and sun. Light coming from the direction SO illuminates the left half of the globe, limited by the terminator BD. Since the earth is in the direction OE, the hemisphere turned toward us is bounded by the great circle that projects as AC. On a smooth sphere, the zone AOB would appear sunlit, forming a 180-degree-long crescent with one cusp at O, the other at the diametrically opposite point of the sphere.

"But the moon is not smooth, and the mechanism described above displaces the cusp from O to Q. The lunar surface in the little triangle OPQ remains invisible. We call PQ the *deficiency arc,* and evaluate it as follows. If a is the angular distance of the moon from the sun (taking account of lunar parallax), 2ω the length of the cres-

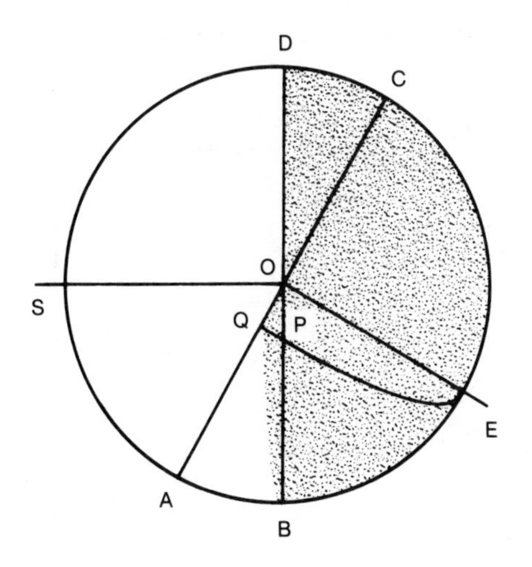

This diagram by André Danjon is explained in the text. From *L'Astronomie,* February, 1932.

cent (which would be 180 degrees on a smooth sphere), the deficiency arc α is given by the formula

$$\sin \alpha = \sin a \cos \omega."$$

From here and there, Danjon collected more than 50 measurements and estimates of crescent length, and from each he calculated the deficiency arc, using this formula. The result, taken from his paper of 1936, is the graph on page 208. It shows that when the Moon is more than 40 degrees from the Sun, the arc is negative; more sunlit area is visible than a smooth surface would afford, and the bright cusps extend for slightly more than a semicircle. But for elongations of less than 40 degrees, the deficiency arc rises steadily. Finally, when the Moon is seven degrees from the Sun, the deficiency is also seven degrees. That is, *there is no longer any sunlit crescent visible.* For an elongation of seven degrees or less, the entire crescent is contained in the shadowed region and disappears from our view.

Thus, the limit of visibility of a young Moon is set by its angular distance from the Sun, not by its age. For a particular age, the angular distance can have quite a wide range of values, depending on the Moon's latitude (distance above or below the ecliptic) and whether it is near perigee or apogee. A calculation for the case of the

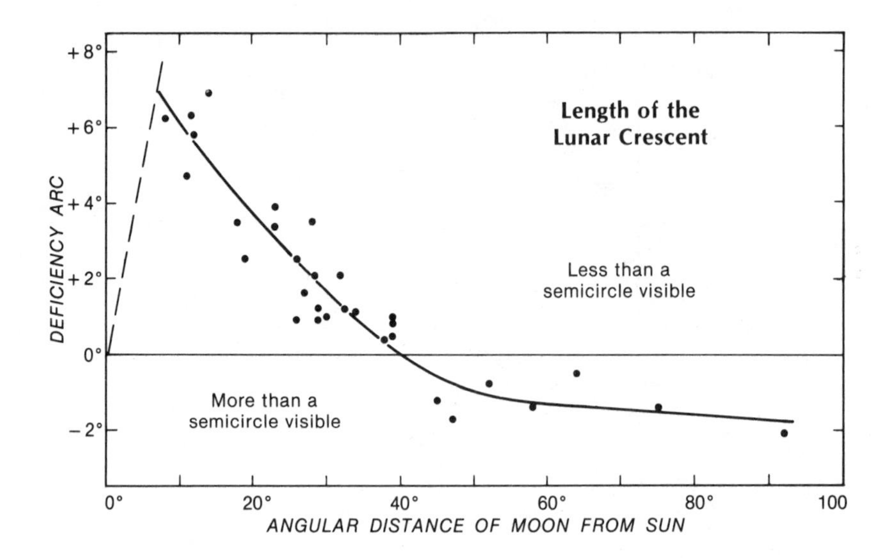

The solid curve in this diagram by Danjon shows the amount of the Moon's deficiency arc (contraction of sunlit crescent caused by lunar mountain shadows) at different angular distances of the Moon from the Sun. The dashed line indicates where the horizontal and vertical scales are numerically equal. The curve and line meet when the angular distance between the Moon and Sun is seven degrees; when the Moon is closer to the Sun, no sunlit crescent remains.

14½-hour Moon shows that it was 7.6 degrees from the Sun and therefore visible by Danjon's criterion.

Another consequence was noted by Danjon. At new, the Moon cannot pass more than 5½ degrees north or south of the Sun, which is less than the seven-degree Danjon limit. Hence the crescent must vanish for an interval of time during *every* lunation.

Under conditions when there can be no visible crescent, is it possible to observe the lunar disk by its earthshine? A succinct answer was given in a 1967 report by three rocket astronomers at the E. O. Hulburt Center for Space Research – M. J. Koomen, R. T. Seal, Jr., and R. Tousey. It reads: "From the earth's surface the moon has not been observed at elongations less than 7° from the sun (Danjon). Here the crescent vanishes and the twilight sky masks the earthshine. Photographs of the earthlit moon at 2° elongation were obtained on 12 November 1966. This was accomplished with an externally occulted Lyot white-light coronagraph of small size. Two of these instruments were flown side by side in a solar pointing control mounted on an Aerobee-150 rocket, launched from the White Sands Missile Range at 16:38 UT and reaching a 175-km peak altitude.

"The photographs resemble closely those of the full moon. Contrasts are approximately the same, leading to the conclusion that there was no appreciable quantity of cis-lunar dust, which would have manifested itself by strong forward scattering of sunlight. An irregular glint was present on the limb of the moon nearest the sun, and was apparently caused by grazing reflection from the smooth area of Mare Marginis. This glint may be considered a vestige of the new moon crescent."

The possibility of detecting the earthlit features of the lunar disk should be rather better at zero elongation than at an elongation of one or two degrees. For one thing, the Sun is totally eclipsed, and the earthshine is brightest. Even so, the seemingly black Moon is viewed through a heavy veil of scattered light from the corona and the sky. Visual attempts to see lunar features during a solar eclipse are probably meaningless because of contrast effects, but the vague outlines of the seas can be traced in a photograph taken during totality on March 7, 1970, by W. T. Peters at Cameron, Mexico. This picture is the only one of its kind that I have seen. In a sense, it is the ultimate in young-Moon photographs.

41. The first observed occultation of one planet by another

ON THE NIGHT OF JANUARY 9, 1591, young Johannes Kepler, then a student at the German university of Tübingen, went outdoors with his teacher Mästlin to observe a close conjunction of Jupiter and Mars. The two planets rose shortly after midnight, but the watchers saw no trace of Jupiter. It appeared to them that the giant planet was hidden behind Mars, and the reddish color convinced them that the latter was indeed the nearer of the two planets.

Taken at face value, this seems to be an observation of one of the rarest of all planetary phenomena – the occultation of one planet by another.

However, this pretelescopic observation may merely concern a close approach or *appulse* of the two planets, with their separation less than the resolving power of the eye. It is well known that the bright planets appear to the naked eye considerably larger than they are in actuality. Tycho Brahe estimated the diameter of Mars as 100 seconds of arc when it is one astronomical unit (92,960,000 miles) from the Earth – a value 10 times greater than the planet's actual diameter. Naked-eye testimony is not enough, unless backed by precise calculations of the planets' positions, a verification that is lacking in the above case of Mars and Jupiter.

Only a solitary case is on record of the occultation of one planet by another having been observed telescopically. This was on May 28, 1737, when Venus passed in front of Mercury. In the evening twilight at Greenwich Observatory, John Bevis watched the event low in the west with a nonachromatic refractor of 24-foot focal length. He saw the two planets only about three minutes of arc from each other in the field of view. They were drawing together rapidly when clouds blocked the view. Eight minutes later Venus could be seen again at its full brilliance. But Mercury was invisible, concealed behind the disk of Venus. Once more the clouds blotted out the spectacle and prevented further observations.

Bevis' unique record was critically analyzed more than a century later by the great French astronomer, Urbain Le Verrier, who had begun his scientific career as an organic chemist but turned to celestial mechanics. Already director of the Paris Observatory and world-famed for his independent prediction of the planet Neptune, he had just completed his new and greatly improved tables of the motions of the inner planets. As one of the tests of these tables, he computed from them the circumstances of the 1737 occultation of Mercury by Venus.

Le Verrier calculated that Mercury was indeed behind Venus when seen by Bevis through the parted clouds. But the innermost planet was only partly occulted; 2.27 seconds of arc of its diameter had already emerged. However, Mercury was near elongation at the time, about "quarter-moon" phase, and the portion of its disk that protruded beyond Venus' limb was unilluminated, so Bevis could

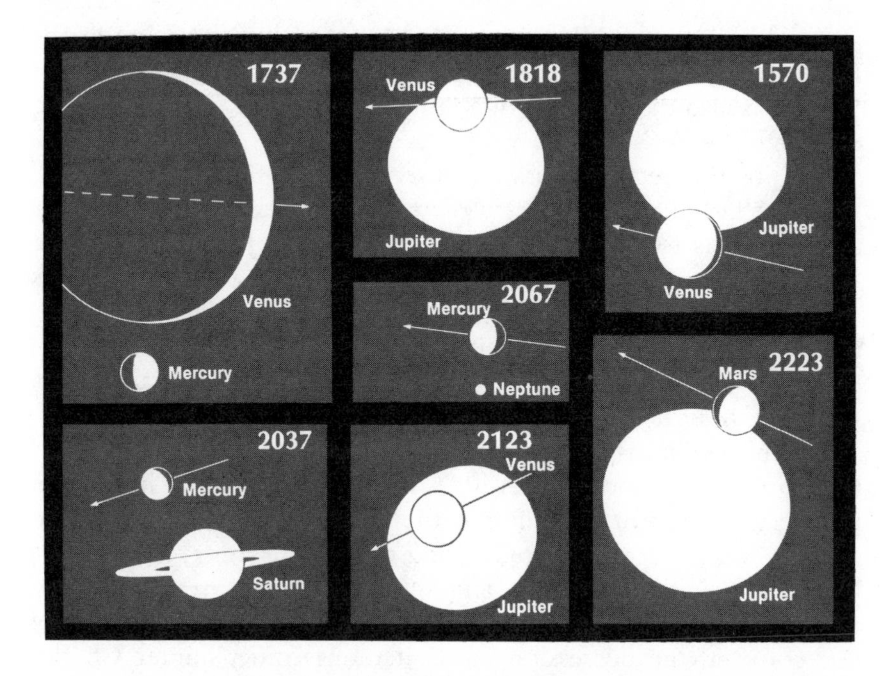

Six mutual occultations of planets and the near miss of 2037 are illustrated here, on the basis of Steven C. Albers' detailed calculations. All drawings are to the same scale, with north up, and show the geocentric circumstances (as would be seen by an observer with the planets directly overhead). In each case, the long arrow shows the track of the nearer planet relative to the farther one, with the occulting planet's position shown at the time of least distance of centers. In the Mercury-Venus event of 1737 and the Mercury-Neptune one in 2067, no occultation occurs geocentrically, but in both cases Mercury occults the other planet for more northerly observers. In addition to the Mercury-Saturn near miss of September 15, 2037, pictured here, Albers notes three others. On July 20, 1705, Mercury's disk cleared Jupiter's by only about one second of arc for the Far East. On August 23, 2100, Mars goes past Neptune with 1.6 seconds clearance, and on July 16, 2173, Mercury and Mars miss by one second.

not have seen it. Le Verrier commented with pride about the accuracy of his calculations: "The precision with which the principal phases were observed in the occultation of Mercury by Venus on May 28, 1737, is a remarkable confirmation of the exactness of our tables of the sun, Mercury, and Venus."

Although he observed at Greenwich, John Bevis was an amateur astronomer. Born in 1695 at Old Sarum, in Wiltshire, England, he became a physician with wide scientific interests. Besides the astronomical observations that comprised most of his 27 papers in the *Philosophical Transactions,* he made pioneer electrical experiments. Bevis was also the inventor of a "satellite slide rule," for predicting eclipses of Jupiter's satellites.

In 1738 he erected a private observatory at Stoke Newington, where he began an ambitious project, the preparation of a great star atlas that was to contain many more stars than Johann Bayer's of 1607 and to have greater exactness. To obtain star positions for this work he made so many observations of when stars crossed the meridian – often 160 in a single night – that his material was complete by 1745. The final product was his *Uranometria Brittanica,* an atlas in 50 sheets, to be accompanied by a catalogue of the star positions.

But this enthusiastic amateur became the victim of circumstances that robbed the world of his life's work and consigned him to an undeserved obscurity. Although the atlas was actually printed in 1750, it was never distributed owing to the bankruptcy of his publisher. The edition was lost, and surviving copies are exceedingly rare. J.-J. Lalande, writing in 1792, seems to have known of only one copy, in the possession of the famous comet hunter Charles Messier. The star catalogue was never published, and there is no trace even of its manuscript. This double loss deprived the astronomical world of a useful tool that would have given its author lasting fame. At that time the lack of adequate star maps was a serious handicap to observers. For example, it greatly retarded the growth of variable star astronomy.

John Bevis died at London on November 6, 1771. While he is nearly forgotten today, he deserves notice for his observation of a remarkable planetary phenomenon. Venus will not again occult Mercury until December 3, 2133. Even this event will hardly be observable, since the two planets will be a scant four degrees from the Sun.

42. Edmond Halley at St. Helena

In March, 1675, the first Astronomer Royal, John Flamsteed, received a letter from an enthusiastic 18-year-old amateur astronomer. Edmond Halley, it appeared from this letter, was an Oxford undergraduate born in London, who two years earlier had resolved to devote his life to astronomy. Three months later Flamsteed wrote glowingly to his friend Richard Townely: "I have met also with an ingenious youth versed in calculations and almost all parts of mathematics, tho yet scarce 19 years of age; Mr. Edmond Halley, whose assistance I hope to have often since he lives commonly in the city and we agree exceedingly well in our thoughts concerning the most convenient and useful methods."

This young man was already an active observer who had acquired skill in measuring star positions with a quadrant of two-foot radius. At his father's home in London he viewed the total lunar eclipse of July 7, 1675, with a nonachromatic refractor 24 feet long. And while at Oxford he observed a remarkable sunspot during July and August of the following year. Thus, while Halley was still in his teens he was becoming known for his businesslike astronomical reports published in the *Philosophical Transactions* of the Royal Society.

Very early his thoughts were directed toward the compilation of an extensive star catalogue, in order to provide more precise reference points for tracking the motions of the planets. But he knew that such catalogues were already being prepared by Johannes Hevelius at Danzig and by Flamsteed at Greenwich, while G. D. Cassini at Paris was expected to undertake a third. Therefore he decided to observe the far-southern stars that were invisible from Europe and never before measured by a well-equipped astronomer.

Halley considered possible observing sites in the tropics. Early in 1676 his college friend Boucher wrote to him from Jamaica that the climate was excellent for astronomical observations. Despite poor eyesight, said Boucher, he could count nine Pleiades stars with his unaided eye. Halley consulted scientific friends in England, who recommended Rio de Janeiro and the Cape of Good Hope as favor-

Edmond Halley (1656–1724), as pictured in an engraving in A. v. Schweiger-Lerchenfeld's *Atlas der Himmelskünde,* 1898.

able localities. But these places were then occupied by the Portuguese and Dutch, respectively, and Halley did not want to spend the time needed to master a foreign language.

His choice therefore fell on the small South Atlantic island of St. Helena, at that time the southernmost land under England's flag. Lying 1,200 miles off the African coast, it was discovered by the Portuguese in 1502, and since 1651 it had been garrisoned by the East India Company as a waystation for its shipping. The youthful Halley acted promptly and with worldly wisdom. Beginning in July, 1676, he and his Oxford friends began writing letters to influential persons. The matter reached King Charles II, who at once requested the East India Company to furnish Halley and his companion Mr. Clerke with free passage to St. Helena and all necessary aid. This was quickly arranged, and Halley's father agreed to contribute £300 a year for the duration of the expedition. In November Halley and Clerke sailed from England aboard the East Indiaman *Unity,* and after a voyage of three months they arrived at Jamestown, the port of St. Helena.

From this small settlement, a narrow valley ran up into the rugged interior of the island toward the central mountain, Diana's Peak.

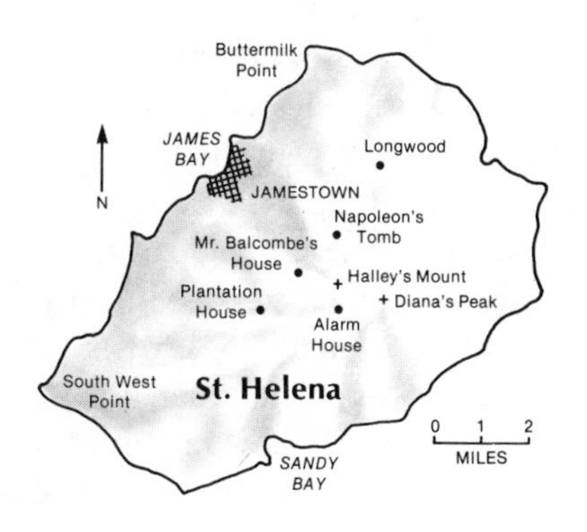

This chart of St. Helena is redrawn from an early 19th-century map reproduced in Gilbert Martineau's book *Napoleon's St. Helena,* 1968. Halley's Mount is still named after him. When Emperor Napoleon came as a prisoner to the island in 1815, he was quartered for a few months at Mr. Balcombe's House.

From its 2,700-foot summit, one could see an unbroken circle of sea; two miles north was the bleak plateau where later was built Longwood Old House, residence of the exiled Napoleon Bonaparte from 1816 until his death in 1821.

Halley set up his instruments on another nearby eminence, known to this day as Halley's Mount. For his catalogue observations he used a "great sextant" of 5½-foot radius, fitted with telescopic sights, to measure the angular distances between stars. He also brought from England his 24-foot refractor and several smaller ones, a pendulum clock, and his two-foot quadrant for time sights.

The year that the young astronomer spent on the remote island was very busy but often frustrating. Cloudy weather was much more prevalent than Halley had anticipated, and often the mountains were enveloped in fog. It was necessary to use every moment of clear nighttime sky. He got along badly with Governor Gregory Field, who is said to have been discourteous and arrogant. Also, living conditions must have been primitive in the isolated interior. But in spite of these obstacles, Halley succeeded in compiling a catalogue of 341 stars (six of them in the constellation Lupus from observations aboard ship). Bad weather prevented any measurements of stars in Piscis Austrinus (except Fomalhaut) and in Indus. As a compliment to his royal patron, Halley arranged 12 stars of Argo into a new constellation, Robur Carolinum or Charles' Oak.

The original edition of Halley's catalogue, published just after his return to England, is exceedingly rare, but Francis Baily's 1843 revision of it is accessible today in any large observatory library. The 341 stars are arranged by constellation, with celestial longitudes and latitudes given to a nominal accuracy of half a minute of arc. Star brightnesses are given to the nearest whole magnitude and occasionally to the half magnitude. Rough though these data are, they provide some of the earliest available information about the luster of the far-southern stars. Three objects are listed as nebulae. One is merely a group of stars in Scorpius, but the "nebula between the tail of Scorpius and the bow of Sagittarius" is the star cluster now known as Messier 7, while that "in the horse's back" is the fine globular cluster Omega Centauri.

Although Halley's chief effort was to measure stars for this catalogue, he also observed the lunar eclipse of May 17, 1677, the eclipse of the Sun that May 31st, and the transit of Mercury on November 7th. It was this last event that first suggested to Halley the possibility of using transits of Venus to measure the Sun's distance. He also seems to have independently discovered that a pendulum beating seconds is shorter near the Earth's equator than in European latitudes, a fact noted by Jean Richer at Cayenne in 1672–1673. Thus, when Halley returned to England in 1678, he was laden with scientific fruits and already a famous man, who was quickly rewarded with election to the Royal Society and an honorary degree from Oxford. With St. Helena as a stepping-stone, he was fully launched on his career by his 22nd birthday.

Two centuries after Halley, Sir David Gill (later director of the Cape Observatory) visited St. Helena for a week in July, 1877. On a tour of the interior, he and Mrs. Gill climbed Halley's Mount to search for any traces of the old observatory. Mrs. Gill described this excursion:

"We did not know whether any record of this work remained in stone and lime, and it was a pleasant surprise to find, on the spot that an astronomer's eye at once picked out as the most favourable, a bit of low wall, duly oriented, and overrun with wild pepper *(Cluytia pulchella)*. This had been the Observatory, without doubt; and near to it is a quarry from which the stones for its erection had evidently been taken. So charmed was my husband with this interesting record of the work of 200 years ago, that his investigations and surmises regarding it left us short time to linger in the little

David Gill (1843–1914) photographed about 1896. Courtesy Mary Lea Shane Archives of Lick Observatory.

hollow [the site of Napoleon's tomb] lying near the foot of Halley's Mount."

Gill's identification of Halley's observing site seems to have been accepted generally by later writers (apart from occasional confusion of Halley's Mount with Diana's Peak). But a correspondent living on St. Helena, Arthur H. Mawson, tells me that the story is not quite right.

The low stone walls are still there on Halley's Mount, though the wild pepper trees mentioned by Mrs. Gill are gone. Mawson, who speaks with authority as a mining engineer, says the stones are not from the quarry but probably from a nearby road cutting. There is even some reason to doubt that these ruins are connected with Halley at all. A modern historian of the island, G. C. Kitchins, says, "Halley moved his instruments about all over the mountain and had no permanent observatory." Mawson suggests that the old walls may be remains of a shelter erected for artillerymen after Halley's visit, and he cites Kitchins' book: "In 1682 Alarm guns were placed on Halley's Mount, as the settlers in Sandy Bay could not hear those at Alarm House."

43. Father Hell's reputation

A LONG TRAIN OF CONSEQUENCES followed from a famous paper read by Edmond Halley in 1716 before the Royal Society of London. It was entitled "A New Method of Determining the Parallax of the Sun, or his Distance from the Earth." This paper explained how the Sun's distance could be determined with vastly improved accuracy, by timing the beginning and ending of a transit of Venus across the solar disk as seen from widely separated places over the world.

This phenomenon had last occurred in 1639, when no useful observations of it were obtained. Consequently, astronomers awaited with keenest interest the next transits of Venus, on June 6, 1761, and June 3, 1769. Elaborate preparations were made and numerous expeditions set out for places as remote as Tahiti, Siberia, and California, in an international scientific effort unprecedented at that time.

Of the many stories of these expeditions, one of the most curious concerns Maximilian Hell's 1769 observations in Lapland and their aftermath. This Jesuit astronomer was born in 1720 at Schemnitz, Hungary, and became one of the best-known astronomical figures of his generation. His order sent him to found a new observatory at Tyrnau in 1752, and three years later he moved to Vienna to take charge of the recently established university observatory. There he gained fame for his observations and writings, and especially for publishing 37 annual volumes of the Vienna astronomical ephemeris. Surely this useful career should have entitled him to a respectable if obscure niche in astronomical history.

He was long denied this entitlement, however, owing to a letter handed him on September 5, 1767, by the Danish ambassador at Vienna. It was an invitation from King Christian VII of Denmark and Norway to travel at royal expense to Vardo, in the extreme northeast corner of his domains, to observe the transit of 1769. This

Maximilian Hell, S. J. (1720–1792). From a contemporary print, courtesy Owen Gingerich.

island station north of the Arctic Circle was of great importance, as observations of the transit made there could be combined with others from the South Pacific to give a solar parallax determination of considerable weight.

Father Hell left Vienna on April 28, 1768, in the company of his assistant, Johann Sajnovics, S. J. Their travel diaries, published many years later, give some curious details of their journey. One stop was at Znaim in Austria, where at the observatory they were shown an iron octant said to have belonged to Tycho Brahe. In Saxony, Sajnovics noted, the beer was so bitter that it could scarcely be drunk. The two traveled through Leipzig to Lübeck, where they were met by an agent of the Danish king and escorted to Copenhagen. Christian VII received them cordially.

The last leg of their travels was by ship to Vardo, where they arrived on October 11th to find a foot of snow on the ground. Erection of a small observatory began two days later. On the 18th their clock was set up, and the next day the first time-sights on the Sun were taken. Hell's diary records trouble with the native helpers at the observatory, who would report for work at 9:30, take lunch from 11 to 2, and leave at 3:30 when it started to get dark. The remainder of the winter and spring were spent by the astronomers in geodetic surveys, observations of auroras, hunting, and enjoying the hospitality of the islanders.

On the date of the transit, June 3, 1769, the Sun did not set in the high northerly latitude of Vardo. The entrance (ingress) of Venus onto the solar disk was timed shortly after 9:30 p.m., through a brief break in the clouds, and the captain of the astronomers' ship fired off his nine guns and ran up his flag. The overcast remained solid until just before Venus left the Sun about 3:30 a.m., when successful observations of egress were made and celebrated with another salvo.

On the next day the jubilant astronomers checked their clock correction by Sun sights and wrote letters to Vienna and Copenhagen, which were sent off by a courier armed with a flintlock. Father Hell and Sajnovics returned by sea to Copenhagen, arriving in late October, and spent the winter there.

Meanwhile, in Paris, the famous French astronomer Lalande was collecting all available timings of the transit in order to calculate the solar parallax. He wrote for Hell's observations, which the Vienna astronomer declined to furnish in advance of publication. This aroused suspicions, and Lalande hinted that Hell was waiting for other observations so that he could manipulate his own to fit them. There were even rumors that Hell had not seen the transit at all and was forging his records in Copenhagen. Even though the publication of the Vardo observations by the Danish Academy of Sciences quelled these stories, some lingering doubts remained for many years.

In 1835, C. L. Littrow, director of Vienna Observatory, discovered among its archives Hell's original journals and published them in a book that attracted wide attention. Littrow took a highly critical view of Hell's veracity, and he denounced the Vardo observations as proven forgeries. The evidence was clear, according to Littrow: The original ink figures for the contact times had been scraped out of the paper and others inserted; in many places in the manuscript, changes had been made with ink of a slightly different color.

The reputation of Father Hell remained in eclipse for half a century, until the famous American astronomer Simon Newcomb happened to visit Vienna Observatory in 1883. Being especially interested in transits of Venus, he asked to be shown Hell's journal.

Studying the manuscript with great care, he began to find curious discrepancies with Littrow's description of it. For one thing, many of the alleged alterations had been made before the ink had dried; the writer on making a mistake had rubbed it out with his finger. To settle whether the original contact times had actually been erased, Newcomb examined the paper under a magnifying glass by grazing light. Nothing had been erased, but evidently ink had not flowed freely from Hell's pen, so he had written his figures over again for clarity.

There remained only Littrow's allegation that changes had been made with ink of a different color. To Newcomb, the manuscript had merely been retouched with the same kind of ink, which looked a little darker on drying. On inquiry, he found that Littrow's color vision was so defective that he could not distinguish the orange tint of Aldebaran from that of the whitest star. In Newcomb's words, "No further research was necessary. For half a century the astronomical world had based an impression on the innocent but mistaken evidence of a color-blind man respecting the tints of ink in a manuscript."

44. Father Perry's expedition to Kerguelen Island

AIRLINES AND ARTIFICIAL SATELLITES support the comfortable belief that we live in an increasingly compact and familiar world. Actually, some parts of the Earth's surface were better known in the 19th century than today. One good example is Kerguelen Island, a French possession nearly as large as Delaware, lying in a lonely part of the Indian Ocean well south of the shipping routes between South Africa and Australia.

Back in the 1820's and 1830's, hundreds of American whaling and sealing ships visited there, until the indiscriminate slaughter of sea animals made the trade unprofitable. Kerguelen Island regained brief public attention in 1874 when American, British, and German astronomical expeditions went there to observe the transit of Venus across the disk of the Sun. Today the storm-swept rocky land with its magnificent mountain scenery, glaciers, grassy meadows, fjords, and islets has only a handful of inhabitants and slumbers nearly forgotten.

This remote island was discovered in 1772 by a French naval officer, Y. J. de Kerguelen-Trémarec (1734–1797), during a voyage in search of a supposed new continent southeast of Africa. He sighted a coastline in latitude 49° 40′ south, longitude about 70° east, along which he sailed for 40 miles without landing until he was blown away from it by a gale. On returning to France, he published this remarkable account: "The lands which I have had the happiness to discover appear to form the central mass of the Antarctic continent. The latitude in which it lies promises all the crops of the Mother Country, from which the islands are too remote to derive fresh supplies. No doubt wood, minerals, diamonds, rubies, precious stones, and marble will be found. . . . If men of a different species are not discovered, at least there will be people in a state of nature living in their primitive manner, ignorant alike of offense or remorse, knowing nothing of the artifices of civilized society. In short, South France will furnish marvelous physical and moral spectacles."

This dream was deflated by Capt. James Cook's voyage of 1776. The great English explorer circumnavigated Kerguelen's South France, proving it a group of one large and many small islands. Rather unkindly, he rechristened it Desolation Island, a name that has since yielded to Kerguelen Island.

This distant outpost first attracted astronomical attention in 1857, when Sir George Biddell Airy, the Astronomer Royal, discussed early plans for observing the transit of Venus on December 8, 1874. Timings from stations distributed as widely as possible over the Earth's surface would be essential. For this reason, it was highly desirable to occupy posts in far southerly latitudes in the Eastern Hemisphere.

The great international program to observe the transit resulted from the efforts of Airy in England, Otto Struve in Russia, and Arthur Auwers in Germany. Agnes Clerke describes it in her *History of Astronomy During the Nineteenth Century:* "Every country which had a reputation to keep or to gain for scientific zeal was forward to co-operate in the great cosmopolitan enterprise of the transit. France and Germany each sent out six expeditions; twenty-six stations were in Russian, twelve in English, eight in American, three in Italian, one (equipped with special care) in Dutch occupation. In all, at a cost of nearly a quarter of a million [pounds], some fourscore distinct posts of observation were provided; amongst them inhospitable, and all but inaccessible rocks in the bleak Southern Ocean, as St. Paul's and Campbell Islands, swept by hurricanes, and fitted only for the habitation of seabirds, where the daring votaries of science, in the wise prevision of a long leaguer by the elements, were supplied with stores for many months, or even a whole year. Siberia and the Sandwich Islands were thickly beset with observers; parties of three nationalities encamped within the mists of Kerguelen Island. . . ."

The scientific leader of the British expedition to Kerguelen was Rev. Stephen Joseph Perry, S. J. (1833–1889), who since 1860 had been director of Stonyhurst Observatory near Manchester. He had already made a name by his sunspot studies, his geodetic and magnetic work, and his novel spectroscopic observations of the solar corona at the eclipse of December 22, 1870, in Spain. This hardworking, very able astronomer was a quiet-spoken and modest man, who seems to have made a deep personal impression on contemporary scientists.

His expedition gathered in July, 1874, at Cape Town, where the British government placed two small warships, HMS *Volage* and HMS *Supply,* at its service. In accordance with the instructions drawn up by Airy, the naval lieutenants assigned to the transit of Venus party practiced astronomical observations at the Cape Observatory. And because the expedition was to make photographic observations in addition to visual timings of contacts, four assistants sought to master the still-novel arts of wet-plate and dry-plate photography. During these preparations, Father Perry made the acquaintance of the American astronomers who had stopped at the Cape on their way to Kerguelen.

The British expedition sailed on September 18th and on October 8th sighted the northeast end of the island. Passing through Royal Sound into the broad landlocked expanse of Morbihan Bay, the ships found a sheltered anchorage at Island Harbor, nestled among four islands in mid-bay.

All of Kerguelen, wrote Father Perry in his journal, "as far as the

Places mentioned in this chapter are: 1, Royal Sound; 2, Morbihan Bay; 3, Island Harbor; 4, Observatory Bay (British site); 5, Molloy Point (American site); 6, Swain's Haulover (British site); 7, Swain's Bay; 8, Thumb Peak (British site); 9, Betsy Cove (German site). Most of these names were changed after France annexed Kerguelen Island in 1893. North is upward.

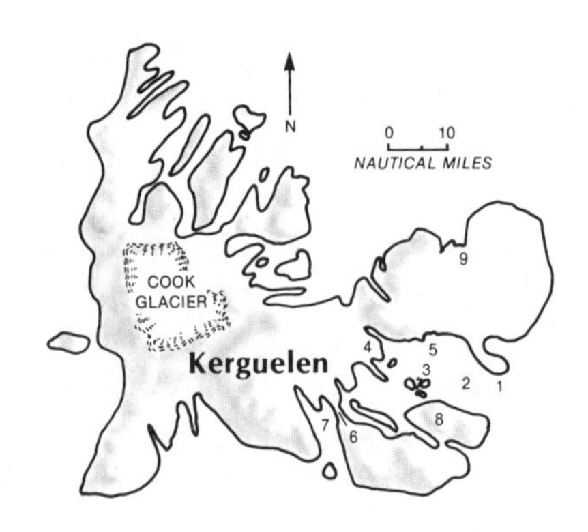

eye could reach, was completely buried in snow. This was the end of spring for the southern hemisphere, so we had pleasant prospects of rambles in snow-shoes over rugged hills and half-frozen marshes and bogs."

An excellent observing site was discovered near the west end of Morbihan Bay. At this inlet, promptly christened Observatory Bay, the British party built a pier and unloaded 600 crates of equipment from the *Supply*. Prefabricated huts for dwellings and instrument shelters were soon erected on the mossy plain. By the beginning of November the equipment of the little observatory was in full working order. There was a transit instrument for time determinations and an altazimuth instrument for finding geographical latitude from star altitudes. On nights when the Moon was visible, lunar altitudes and azimuths were measured with the latter instrument to give additional determinations of longitude. The Moon's meridian passage was observed on every possible occasion for the same purpose.

About the beginning of November Father Perry sailed in the *Supply* to the northern shore of Morbihan Bay to visit the American station at Molloy Point, where Comdr. G. P. Ryan, U.S.N., was in command. Perry inspected the American's transit, refractor, and solar camera, around which many hens and roosters were running. More important, he made arrangements for determining the longitude difference between the stations by means of signals simultaneously visible from the American and English camps.

Perry's big worry was the weather, for in December at Kerguelen average cloudiness runs about 75 percent. Therefore he sent telescopes and observers to two secondary sites to lessen the risk of the entire expedition being clouded out. Four men with two 4-inch refractors were carried by the *Supply* to Swain's Haulover, the narrow neck of land separating Morbihan Bay from Swain's Bay. ("Haulover" was the sealers' name for an isthmus over which they could drag their boats.) Another 4-inch refractor and two men were also placed at Thumb Peak, located near the southern shore of Royal Sound.

December 8th dawned clear, but soon fleecy clouds and haze appeared. As half-past six drew near, everybody was at his post, all preparations complete. At Observatory Bay a long series of multiple-exposure photographs of the Sun was begun with the solar camera. Father Perry himself used the 6-inch refractor to see Venus as a black notch on the Sun's limb, and he began a series of measurements of the size of the notch with a double-image micrometer. But a large cloud came over the Sun before ingress was complete, and when it lifted 20 minutes later Venus was well advanced on its path across the solar disk. The murk returned, and at times became so thick that the 6-inch showed nothing of the planet. With dismay, it was realized that the photographic program was largely ruined. Egress was due about 11 a.m. The sky cleared shortly before then, and the teams using the 6-inch and 4-inch refractors were able to get good timings of third and fourth contacts. Something had been saved!

The question was now how the two secondary stations had fared. That evening the *Supply* returned to Observatory Bay, bringing the two men from Thumb Peak. They reported excellent weather and timings of all four contacts, though it had been snowing the night before. Three days of storms intervened before it was possible to send the *Supply* to Swain's Haulover to pick up its team, which had been completely successful at ingress but clouded out at egress.

Although the transit was over, Father Perry's expedition could not leave Kerguelen. Airy's instructions required that over a hundred additional observations of the Moon be made to provide a more accurate longitude for the primary station. But so cloudy was the weather that this endeavor took 12 weeks, forcing the expedition to go on half rations. This time was also used to revisit Swain's Haulover and Thumb Peak for more geodetic work.

The German expedition to Kerguelen Island, as sketched by L. Weinek. From left to right, the conical shelters protected a visual refractor, a heliometer (device for measuring small angles), and a photographic refractor. The shed adjacent to the latter contained a camera obscura. From A. v. Schweiger-Lerchenfeld's *Atlas der Himmelskünde,* 1898.

Meanwhile, the German expedition at Betsy Cove, on the north coast of Kerguelen, was similarly winding up its affairs. In late December, the *Volage* sailed there carrying eight chronometers, in order to determine the longitude difference between Observatory Bay and Betsy Cove by comparing their local times. Father Perry was surprised to find the German camp located in the middle of a small cemetery, where sealers had been buried many years before. The 17-man team there enjoyed excellent weather on the morning of the transit, carrying out a full program of contact timings, photography, and heliometer measurements.

By late February, as the British at Observatory Bay neared the end of the long program of lunar observations, they began to dismantle their equipment and load it aboard ship. For safety against total loss, the scientific records were copied, one set to return on the *Volage,* the other on the *Supply*. Finally, at 5:30 a.m. on February 27th the last required transit of the Moon was observed, and three hours later both ships were under way.

45. The next transit of Venus

THE NEXT TRANSIT OF VENUS will take place on June 8, 2004. In passage across the face of the Sun Venus is a striking sight, a black disk large enough to be detected even with the suitably protected naked eye. For watchers in 2004 there will be the awesome thought that not a single human being remains alive who observed the last transit of Venus, in December, 1882.

In fact, this phenomenon has been seen only six times since the invention of the telescope. The first predicted occurrences were on December 7, 1631, and December 4, 1639; both were only imperfectly observed. The next transits, on June 6, 1761, and June 3–4, 1769, were widely observed from many parts of the world. By that time it was realized that timings of the beginning and ending of a transit from widely separated geographical locations afforded an accurate method of measuring the Sun's distance. This goal was also the motivation for many expeditions on the occasion of the transits of December 9, 1874, and December 6, 1882. This method of finding the Sun's distance has become hopelessly obsolete, being replaced by radio ranging to spacecraft and planets. Professional astronomers will probably feel much less urgency about the events in 2004 and 2012, but the appeal to amateurs will not be lessened.

The dates of the six historic transits show clearly the pattern of recurrence, at intervals of 8, 121½, 8, and 105½ years. The December transits occur when Venus is near the ascending node of its orbit (that is, crossing the ecliptic plane northward), while the June ones take place near the descending node.

Is it possible that some precisely dated ancient Chinese record of

On June 8, 2004, this will be the path of Venus across the Sun's disk for an observer with the Sun directly overhead. The planet's apparent diameter will be fully three percent that of the Sun.

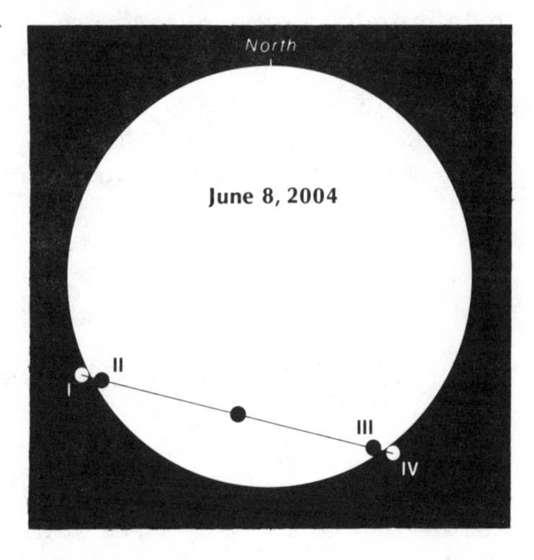

"a dark mark on the Sun" may refer to Venus in transit, instead of a naked-eye sunspot or sunspot group? To facilitate the search for such cases, the Belgian astronomer Jean Meeus published the dates and approximate circumstances of all transits between 2971 B.C. and A.D. 1526. In that same article (*Journal* of the British Astronomical Association, *68*, 98, 1958) Meeus also gives some geocentric predictions for the event of June 8, 2004. As seen from the center of the Earth, the event begins at 5:15 Universal time, when the disks of Venus and the Sun are externally tangent (contact I). Contact II is the moment of internal tangency, 19.4 minutes later. The least distance of centers (0.661 solar radius or 10.4 minutes of arc) comes at 8:21 Universal time. As the planet works its way westward across the solar disk it again reaches internal tangency (contact III), and 19.4 minutes later the transit ends with contact IV at 11:28 Universal time.

However, the contact times for an observer on the Earth's surface will differ by some minutes from the geocentric values. I have made some rough calculations of local circumstances.

North America is not particularly favored. If you live in the eastern part of the United States or Canada, when the Sun rises on Tuesday morning, June 8, 2004, the transit will already be nearly over. At New York City and at Montreal, the Sun will be only about 18° up when contact IV happens at 6:27 a.m. Eastern standard time. Farther south and west even less will be seen; Chicagoans and

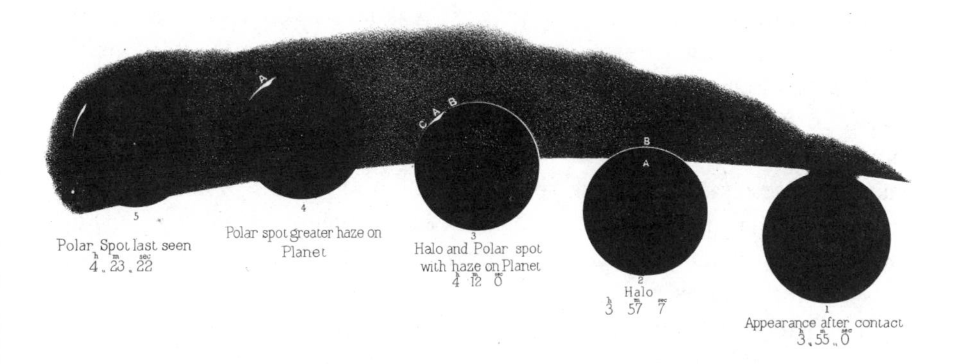

These observations of Venus at egress in 1874 were made by H. C. Russell, the director of Sydney Observatory, with an 11.4-inch refractor stopped down to five inches. The order of the sketches is from right to left. From *Observations of the Transit of Venus,* 1892.

Miamians will only be able to glimpse the last dark notch in the Sun's limb.

Europe will see the whole transit. At London, for example, the morning Sun will be 11° up when ingress starts and nearly on the meridian when the show ends 6¼ hours later. As we travel farther east around the globe, the transit slips later into the day. At Moscow, first contact comes at 8:13 a.m. and last contact at 2:33 p.m.; at Bombay, 10:42 a.m. and 5:14 p.m. But at Tokyo, where the transit does not start until 2:08 p.m., the Sun and Venus will set before the event is complete. Sydney, Australia, enjoys only ingress before the Sun sets, while New Zealand, Hawaii, and the western parts of the United States and South America miss it altogether.

A considerable number of interesting optical phenomena may be visible to observers on June 8, 2004. For example, at ingress, when Venus has partially entered upon the Sun's disk, a circle of light around the planet should be conspicuous, to judge from many reports in 1874 and 1882. This is a well-known effect of Venus' atmosphere, observed as early as the 1761 transit by the Russian scientist M. V. Lomonosov. Just after internal tangency the dark body of Venus may remain connected to the Sun's limb by a famous "black drop" – a dark ligament that may persist for many seconds before it ruptures and vanishes. Although the black drop features in many reports from 1761 and 1769, it was less often seen at the next two transits. Apparently this phenomenon is associated with small telescope apertures and low magnifications. The better equipment used in 1874 and 1882 sometimes showed instead a dusky, hazy appearance between Venus and the Sun's limb, causing an uncertainty of several seconds in timing contacts II and III. There are also some anomalous records, such as Venus during transit appearing gray with a black center, as reported by L. A. Vessey at Woodford, New South Wales, in 1874.

The Meeus article mentioned earlier contains a memorable quotation from William Harkness of the U. S. Naval Observatory, writing in 1882: "There will be no other [transit of Venus] till the twenty-first century of our era has dawned upon the earth, and the June flowers are blooming in 2004. When the last transit occurred the intellectual world was awakening from the slumber of ages, and that wondrous scientific activity which has led to our present advanced knowledge was just beginning. What will be the state of science when the next transit season arrives God only knows."

STUDIES AND STUDENTS OF THE MOON

46. Lunar studies before the telescope

WILLIAM H. PICKERING once discussed the question of how much detail on the face of the Moon can be distinguished by the unaided eye. He made up a list of 12 test features in order of increasing difficulty, the first being the crater Copernicus with its bright surroundings. Reasonably good eyes should be able to see No. 7, Mare Vaporum, as a very dark spot, while really fine vision can reach No. 10, the faintly shaded area near Sacrobosco. Persons with exceptional visual acuity might distinguish the 11th feature, a dark patch at the edge of Mare Imbrium, just across the Apennine Mountains from Mare Vaporum. But Pickering thought that his No. 12, the Riphaeus Mountains, might be beyond even the keenest naked-eye vision.

Careful experiments of this kind reveal a surprising amount of lunar surface features. The best results, I have found, are obtained by viewing the waning gibbous Moon during the latter part of morning twilight. Much less is visible by night, when glare hampers, or in full daylight, when contrasts are diluted.

Thus seen against the deep blue sky of a cool dawn, the pale lunar disk is richly dappled with recognizable markings. This sight suggests the question: How much can be deduced about the Moon's physical nature by a sharp-witted and keen-eyed observer without optical instruments? It is not easy for us today to answer this question directly, because we are too familiar with telescopic views and modern descriptions that cannot be fully erased from memory. Our best look into the problem comes from two remarkable books written about the Moon in the 1st and 11th centuries A.D. These reveal that selenography had made good progress long before the telescope was introduced.

The author of the first was a Romanized Greek, Plutarch of Chaeronea, who lived from about A.D. 46 to 120. Besides his famous *Parallel Lives,* which incidentally is rich in astronomical allusions, he wrote over 60 lesser works on varied topics in ethics, history, literature, and science. One of these is *On the Face of the Disk of the Moon.* Plutarch's book is not a technical scientific treatise but a literary work intended for educated laypersons. In the form of a dialogue, it

To test the sharpness of your vision, find the highest numbered of these 12 regions you can distinguish with the unaided eye. W. H. Pickering's scale runs: 1, bright surroundings of Copernicus; 2, Mare Nectaris; 3, Mare Humorum; 4, bright surroundings of Kepler; 5, region of Gassendi; 6, Plinius region; 7, Mare Vaporum; 8, Lubiniezky region; 9, Sinus Medii; 10, faint shading near Sacrobosco; 11, dark spot at foot of Apennines; and 12, Riphaeus Mountains. North is at the top.

presents a wide variety of opinions then held about the Moon, with arguments for and against each viewpoint. One speaker, Lamprias, is evidently the channel for Plutarch's own ideas, some of which have a very modern ring.

The opening pages summarize and refute several older theories concerning the Moon's spots. One notion was that the markings are merely illusions, another was the idea of Clearchus that the Moon was a mirror in which terrestrial lands and seas were visible. Also dismissed was the Stoic philosophers' doctrine that it was a mixture of the classical elements fire and air. Instead, maintained Plutarch through his spokesman Lamprias, the moon is another Earth. Reason demonstrated that it is a solid, opaque globe illuminated by the Sun. He quoted Aristarchus as saying that the Moon's diameter is between 19/60 and 43/108 of the Earth's, and he cited an unidentified author that it is distant 56 earth radii.

The roughness of the lunar surface was deduced from the absence of a specular reflection of the Sun, as from a polished glass ball; instead, an entire hemisphere is visibly sunlit at full phase. Plutarch went further and asserted that the Moon is covered with deep valleys and mountains. Topographical features inappreciable in themselves could be recognized by their long shadows. He mentioned as a specific analogy 6,350-foot Mount Athos in northern Greece, which near sunset casts a shadow that reaches 50 miles across the Aegean Sea to the island of Lemnos. According to J. H. Mädler, this very correct concept of the Moon's mountainous nature may have been suggested by the lunar Apennines. Near the time of first or last quarter, this range produces a conspicuous irregularity in the terminator for a naked-eye observer.

Plutarch's dialogue about the Moon was first printed in 1509, and one edition of it was prepared by no less an astronomer than Johannes Kepler. Probably this widely read classic had a good deal of influence on thinking about our satellite during the dawn of modern astronomy.

The second pretelescopic book on selenography came from the pen of one of the greatest Arab astronomers, al-Haitham, better known to us as Alhazen. Born at Basra in Iraq about A.D. 965, he moved to Egypt as a young man and died at Cairo in 1039. He was a very prolific author, famed for his commentaries on Aristotle and Galen, and for many books on mathematics, optics, and astronomy. His book *On the Nature of the Spots Seen in the Moon* was known only by a 13th-century mention until a manuscript was discovered early this century in the municipal library of Alexandria. The text was first published in 1925, in a German translation by Carl Schoy.

Alhazen began by emphasizing how careful observations of lunar spots show they are permanent features, not changing in shape, position, or size. He rejected the theory that the large dark areas are shadows of elevated parts, because such shadows would vary in extent. His conclusion was that the light and dark regions are composed of different materials, reflecting light in unlike amounts.

This conjecture had been mentioned in passing by Plutarch. The Arab astronomer discussed it in much detail, on the basis of his own optical theories. These sound rather strange today. One tenet of Alhazen was that the eye views an object by extending "rays of vision" out to feel it. His reasoning led him to believe that the Moon's darker parts were of denser rock.

Although Alhazen's tract was unavailable in Europe, its concept of visible differences in the composition of the lunar surface had a wide currency during the century before the invention of the telescope. The idea is said to be mentioned in the books of the German Erasmus Reinhold (1542) and the Italians Geronimo Cardano (1550), Giovanni Benedetti (1581), and Francesco Barozzi (1585). Three of these men are commemorated by the crater names Reinhold, Cardanus, and Barocius.

47. The "long night" of selenography

LOOKING BACK over the history of lunar studies since the invention of the telescope, we can see a repeated pattern. Four times in 400 years, a greatly improved observing method has attracted a new generation of scientists to lunar problems, and a vigorous but temporary flowering of selenographic observations and ideas has resulted.

The first of these occasions was the use of the newly invented telescope in studying the Moon – by Galileo, Kepler, Langrenus, Hevelius, Grimaldi, and their contemporaries from 1610 to 1650. In this period the Moon became recognized as another world to explore; it was also when maps were drawn. After a long lull the second period of great activity began early in the 19th century, when

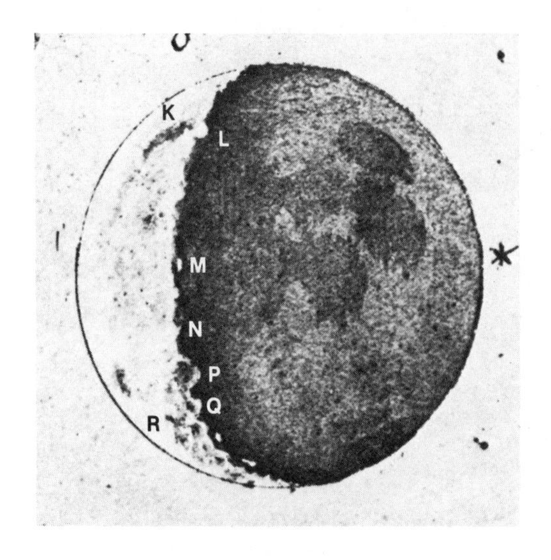

A drawing by Galileo of the waning Moon, dated by Ewen A. Whitaker as made January 19, 1610, at about 6 o'clock in the morning. The star Theta Librae (right) had just emerged from occultation. This illustration was supplied by Whitaker, who added letters to mark identifiable features near the terminator (the boundary between lunar day and night): *K,* shadow of the Montes Jura; *L,* Promontorium Laplace; *M,* T. Mayer and Gamma; *N,* Montes Riphaeus; *P,* Palus Epidemiarum; *Q,* Capuanus P and low area to the southwest; *R,* shadow of Hainzel. This picture appeared with an article by Whitaker in *Journal for the History of Astronomy, 9,* 1978.

achromatic refractors and filar micrometers were employed in lunar observing by W. G. Lohrmann, J. H. Mädler, and J. F. J. Schmidt, initiating scientific cartography of the Moon. The third revolution started in the 1890's, when large telescopes at Lick and Paris observatories began systematic lunar photography. The fourth revolution was, of course, closeup observations by spacecraft and manned exploration.

In this bird's-eye view, we see recurring peaks of scientific activity that usually last for several decades but eventually tail off. These intervals of selenographic productivity are coming at shorter and shorter intervals with increasing overlap.

One rather puzzling feature of this historical panorama is the "long night" of selenography that followed the early 17th-century triumphs. Why did lunar observing lose so much of its initial popularity and drop out of the forefront of astronomical research?

Actually, a good deal of selenographical activity was going on during this dark period, but it failed to produce lasting results. The work of Christiaan Huygens (1629–1695) exemplifies this thesis. One of Holland's great men of science, he was an active telescopic observer and is particularly remembered for discovering Saturn's satellite Titan in 1655 and for recognizing the true form of Saturn's

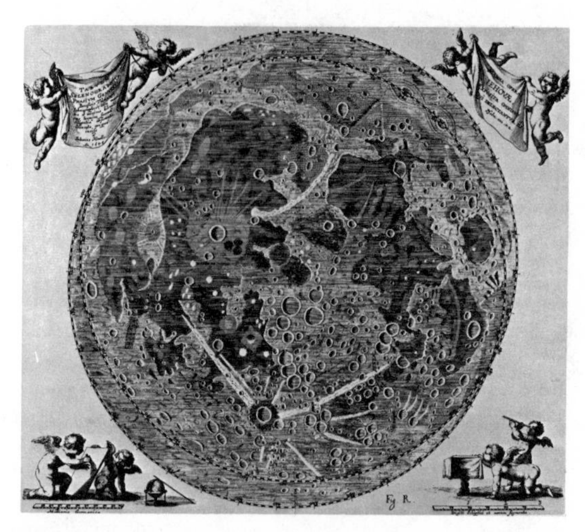

Hevelius drew this Moon map from observations with telescopes magnifying 30 to 40 times, and published it in his 563-page folio *Selenographia,* Danzig, 1647. North is upward, and the two overlapping circles show the lunar disk at extreme northerly and southerly librations. The map emphasizes the bright areas surrounding craters and the bright rays typical of full Moon. Hevelius thought that the rays were chains of hills. In his book, Hevelius presented a similar map with names for surface features; he called the crater Copernicus "Aetna," Tycho "Sinai," and Plato "Lacus Niger Major." Today only half a dozen of his names remain in use, among them "Apennines" and "Alps." Courtesy Tommy Brooks Memorial Library, Harvard Observatory.

ring system in 1659 (see Chapter 24). His lunar studies, however, remained generally unknown until 1925, when they were published for the first time in Vol. 15 of his collected works. It is now clear that Huygens was the discoverer of three famous lunar features that were long attributed to J. H. Schröter (1745–1816). On May 30 and 31, 1686, Huygens used his 123-foot-long nonachromatic refractor to view and draw the Straight Wall, the well-known cliff in Mare Nubium. That month he also recorded Schröter's Valley, and about the same time he saw the great Hyginus rille in a refractor of 60-foot length.

The lunar observations of Robert Hooke (1635–1703) have attracted much more attention, for he described them in a chapter of his widely read *Micrographia* (1665). This brilliant scientist served the Royal Society of London for nearly half a century as its curator of experiments. So versatile was he that he originated much but per-

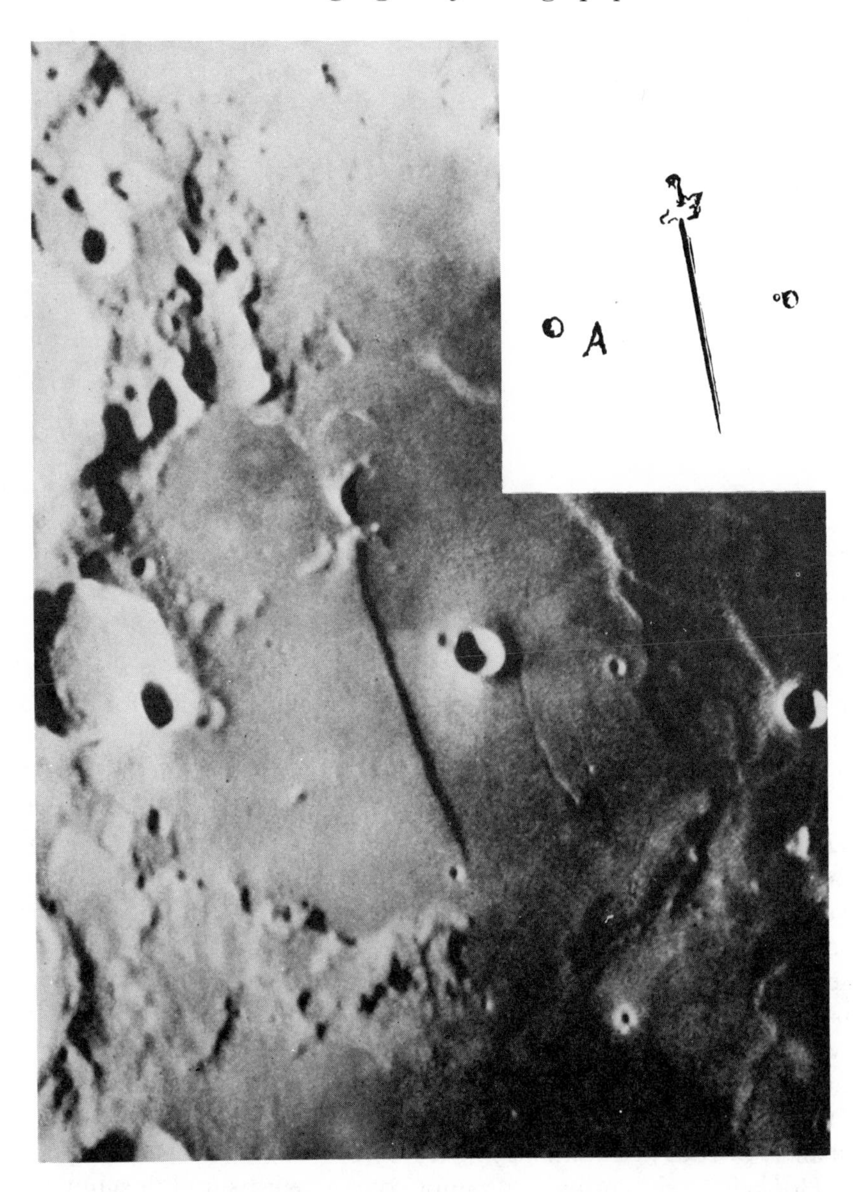

Compare Christiaan Huygens' discovery drawing of the famous lunar Straight Wall with a photograph taken with the 24-inch refractor at Pic du Midi Observatory. The prominent crater to the right (west) of the wall is Birt. The Huygens depiction is from *Complete Works of Christiaan Huygens, 15,* 1925.

Hooke's 1664 drawing of the ring plain Hipparchus is reproduced from his *Micrographia*. North is upward. The large crater above center is now called Horrocks, and the larger of the pair at bottom is Halley. Courtesy Ewen A. Whitaker.

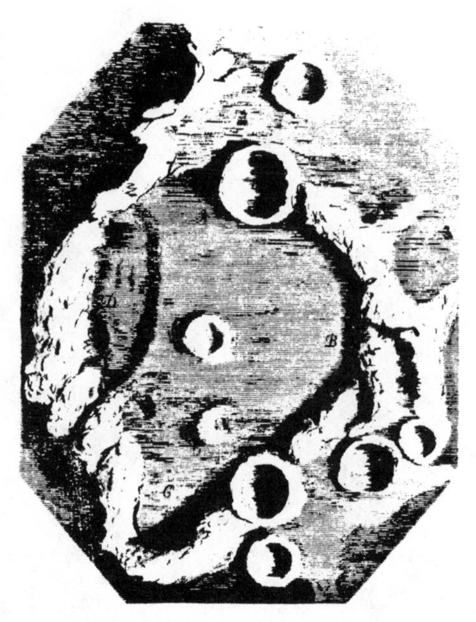

fected little, and the Moon captured his roving attention for only a short time.

Reproduced here from Hooke's *Micrographia* is his drawing of the crater Hipparchus, made with the aid of "a thirty foot Glass, in *October* 1664, just before the Moon was half inlightned . . . It appears a very spacious Vale, incompassed with a ridge of Hills, not very high in comparison of many other in the Moon, nor yet very steep. The Vale itself ABCD, is much the figure of a Pear, and from several appearances of it, seems to be some very fruitful place, that is, to have its surface all covered over with some kinds of vegetable substances; for in all positions of the light on it, it seems to give a much fainter reflection than the more barren tops of the incompassing Hills, and those are much fainter than divers other cragged, chalky, or rocky Mountains of the Moon."

Hooke's lively curiosity was not content with simply examining the Moon's craters, for he had to try various laboratory experiments to duplicate them. Dropping bullets into soft pipe clay gave realistic rimmed craters, but Hooke was dissatisfied with this idea as applied to the Moon, "for it would be difficult to imagine whence those [impacting] bodies should come." His second experiment consisted

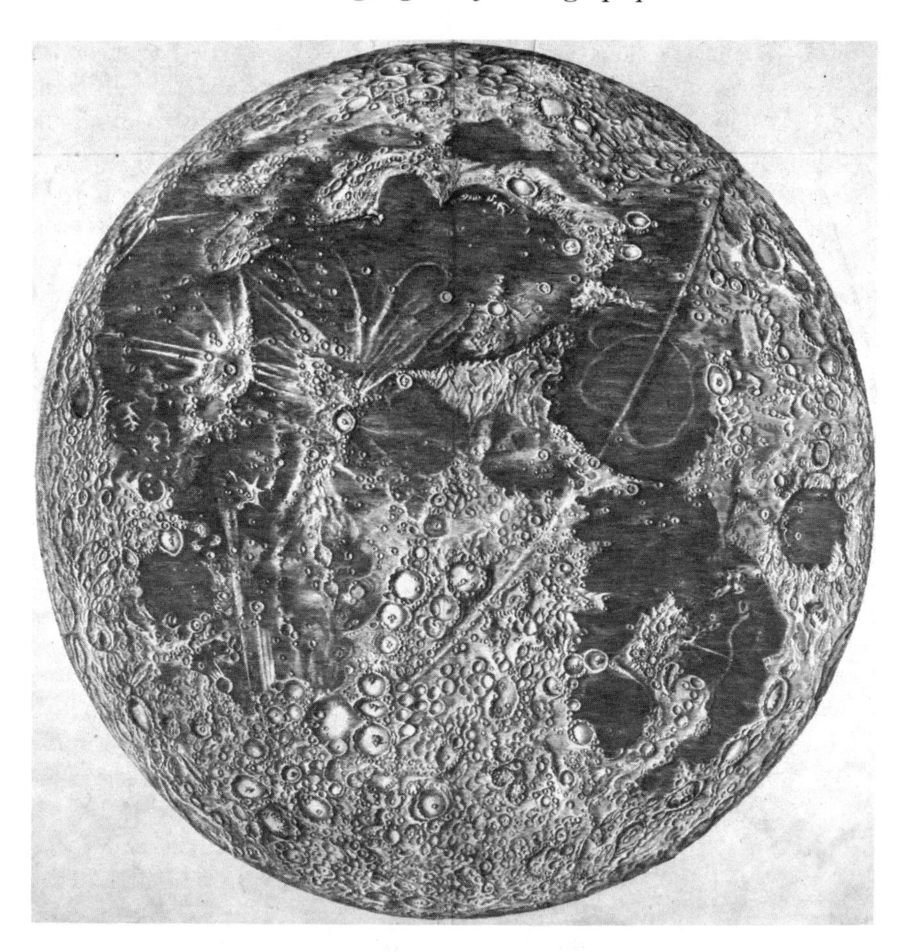

G. D. Cassini's Moon map of 1670 was a major advance over its predecessors. Even today it continues to have a quite modern appearance. Courtesy Ewen A. Whitaker.

of letting a pot of boiling alabaster cool gently, so that many small pits resulted. Viewing these pits by oblique candlelight in a dark room, he was delighted by their similarity to lunar craters.

On a much larger scale than these casual observations was the work of Gian Domenico Cassini (1625–1712) at Paris Observatory. In mapping the Moon, this famous astronomer was assisted by the civil engineer S. Leclerc (1637–1714) and a now-forgotten artist named Patigny. Cassini made many drawings of the Moon in its different phases, using black and white chalk on a blue background, and these were combined to make a full-Moon chart 12 feet in diameter. This work appears to be lost, except in the form of small-scale engravings of a century later.

One vestige of the Paris astronomer's selenographic labors is the name "Cassini's bright spot," which is sometimes used for the area called Hell Q on modern maps (see Chapter 52). It is conspicuous at full Moon to anyone using a small telescope to examine the rays running from Tycho north toward Arzachel. W. Beer and J. H. Mädler, who rediscovered it in the 1830's, quote what Lalande's edition of Cassini's chart tells about it:

"On October 21, 1671, M. Cassini noticed near Gauricus (a small feature located below Tycho) a kind of whitish cloud, of which some remnants remained on the 25th. On November 12th of that year, the same cloud reappeared in the same place.

"October 18, 1673. New large crater between Pitatus and Walter in exactly the spot where the whitish cloud was seen in 1671."

As Beer and Mädler correctly pointed out, the "white cloud" is a surface feature that can be seen every month for about 10 days around the date of full Moon. On October 18, 1673, when this region was close to the terminator, the spot would have been invisible. Since Cassini made no positional measurements, and since the craters Hell and Hell B lie fairly near the bright spot, one of them is undoubtedly his "new crater," according to Beer and Mädler.

Another Paris astronomer of the same period, Philippe de La Hire (1640-1718), spent a decade preparing a general map of the Moon. In his own words, La Hire worked with a telescope in one hand and his chalk in the other. This map is said to have been in existence as late as 1800 but is now lost.

Lost, too, are the more than 300 lunar drawings made by Maria Clara Müller (1676–1707), a skillful observer who was the daughter of the Nürnberg artist-astronomer G. C. Eimmart and the wife of the mathematician J. H. Müller. The latter, incidentally, wrote in 1710 a memoir with the title "Does the Moon Have an Atmosphere?"

Looking over this gallery of shadowy figures, we can see why none of them started a new era in lunar studies. One severe handicap was the unwieldiness of the immensely long nonachromatic refractors of the late 17th century. One glance at Huygens' aerial telescope (page 126) will suggest the exasperating labor of locating a sky object and of keeping it inside the field of view. It was feasible to use long telescopes for isolated observations, such as timing an eclipse of Jupiter's first satellite, but the labor of protracted work at the eyepiece must have been a powerful reason for taking no more than casual looks at the Moon.

Another reason for the backwardness of selenography was the small number and the isolation of its devotees, which tended toward the dispersal and loss of observing records. We should not forget the difficulties in reproducing lunar drawings in those days. Possibly one reason for Hevelius' reputation as a selenographer is that he was also an artist who could make his own engravings and a rich man who could publish his own books. Still another cause of the low ebb of descriptive selenography about 1700 was that it did not yield a product needed by other branches of astronomy.

But in the middle of the 18th century, selenography acquired a dollar-and-cents importance. Overseas commerce needed accurate maps, and these in turn required longitude determinations of distant places. One old-fashioned and rather inaccurate way was to clock the beginning of a lunar eclipse as seen from, say, Rome and Bombay; the difference between the two *local* times would be the longitude difference. A considerable improvement in this technique was pointed out by J. H. Lambert (1728–1777) and Tobias Mayer (1723–1762). Their idea was for longitude observers to time the entrances of individual craters into the Earth's shadow. If the observers at Rome and Bombay each timed the same long list of craters, a much more reliable result would be had. This plan required more accurate and complete lunar maps than were available. For this reason, although the method had been suggested a century before by Hevelius, it did not come into general use until better Moon maps were drawn for the specific purpose.

Apparently, the first use of Hevelius' crater method was at the lunar eclipse of November 1, 1724, for a determination of the longitude difference between Lisbon and Paris. At Lisbon, G. B. Carbone made crater timings with the aid of Riccioli's lunar map, but I do not know who made the corresponding observations in Paris. Carbone was a priest who later went to China as a missionary, and who published some observations of Jupiter's satellites. Perhaps this man has an overlooked importance in the history of lunar studies, as one of the first to show a practical need for better Moon maps.

48. Roger Boscovich and the Moon's atmosphere

EVER SINCE the first telescopic studies of the Moon nearly 400 years ago, astronomers have searched for phenomena that would indicate the presence of an atmosphere surrounding our satellite. Successively more and more sensitive tests failed, but a lingering rear guard of believers surrendered only when the Apollo landings put the airlessness of the Moon beyond any doubt. The history of the search for a lunar atmosphere seems very incompletely covered in books on the Moon. In particular, the important role played by Roger Boscovich (1711–1787) has gone largely unnoticed, and his book on the subject is mentioned seldom if at all.

This Jesuit astronomer was a Slovene (from present-day Yugoslavia) and not an Italian as usually stated. Born at Dubrovnik (Ragusa), he went to Rome in 1740 to become professor of mathematics at the Collegio Romano. His scientific work was extremely varied and influential, including the invention of astronomical instruments, studies of eclipses and transits of Venus, methods of computing comet orbits, hydraulic engineering, important contributions to pure mathematics, and a penetrating study of the nature of gravity. Besides this, he at one time served as the ambassador of his native Ragusa, then an independent republic, to the court of France.

His interest in the lunar atmosphere problem seems to have been stimulated by the annular eclipse of the Sun on July 25, 1748, which was widely observed in Europe. At Berlin, the famous Swiss mathematician Leonhard Euler watched the course of the eclipse by projecting the solar image on a white screen fastened behind the eyepiece of his telescope. On this screen he had drawn a circle just large enough to contain the Sun's disk. But when the eclipsed Sun

Roger Boscovich, S.J., was accomplished in so many fields that he is difficult to classify. He was mathematician, astronomer, philosopher, engineer, poet, and diplomat. His friends included King Louis XVI of France, and he was a member of the Paris Academy of Sciences and of the Royal Society in London. This sketch by Steve Simpson is based on a portrait that appeared in *l'Astronomie,* November, 1937.

had become a crescent, its cusps apparently extended beyond the circle. This effect, Euler noted, would result from the refraction of sunlight by air on the Moon. He critically analyzed this observation in a memoir read before the Berlin Academy of Sciences in 1748, concluding that the lunar atmosphere was dense enough to produce a horizontal refraction of 20 seconds of arc. (The terrestrial horizontal refraction is 34 minutes; we see the Sun on the horizon when it is actually this angular distance below it.)

Euler's paper led Boscovich to a comprehensive study, published at Rome in 1753 as a 76-page book, *De Lunae Atmosphera.* This is a systematic exploration of the phenomena that should be observable if the Moon possessed an appreciable atmosphere. As far as the optical theory of his day permitted, Boscovich made specific, quantitative predictions, which he compared with the observations available. The whole discussion is crisp and businesslike, and so modern is Boscovich's viewpoint that his book is worth reading even today.

If the Moon had a gaseous envelope, the Jesuit astronomer pointed out, a star or planet undergoing occultation would suffer very distinctive changes in appearance. As a star neared the Moon's limb, it would be displaced outward more and more by refraction, the shift at the moment of disappearance being twice the horizontal refraction: 40 seconds of arc, if Euler had been right. This effect could be tested by repeatedly measuring the angular distance of the star from the Moon's limb. Moreover, the star's light would become increasingly dimmed and reddened. A planetary disk about to be covered by the Moon would appear flattened, just as does the setting sun.

In addition, because blue light is refracted more strongly than red, a star's image very close to the Moon's edge would be spread out into a spectrum, with its red end toward the Moon. If the horizontal refraction by a lunar atmosphere were as great as 20 seconds, the spectrum ought to be nearly two seconds of arc long, according to Boscovich. This he considered too small to be recognized in the telescopes of his time. In the case of a planet, however, the effect would be easier to see, for the planet's edge nearest the Moon would have a red border. Something of this sort had indeed been recorded by Delisle and by Louville in France, when Venus was occulted in 1715. But they used nonachromatic refractors of inferior quality, and Boscovich dismissed their findings, particularly as such colored fringes had been looked for but not seen at other occultations of planets.

At solar eclipses a dense lunar atmosphere could betray its presence in several ways. The cusps of the solar crescent would be prolonged, and the irregularities of the Moon's edge smoothed out. And could the corona seen at total eclipses be interpreted as an extensive sunlit lunar atmosphere? "No," answered Boscovich. On pages 57 to 60 of his book he summarized the available reports on the corona, beginning with Kepler's of 1605. The corona must be an appendage of the Sun rather than the Moon, he decided. For one thing, at the total eclipse of 1715 it had unequal extent in different directions, unlike an atmospheric shell. Also, in 1724 Maraldi had noted that the corona was concentric with the Sun rather than the Moon.

Some of the other tests for air on the Moon marshaled by Boscovich were: The phases of the Moon should differ slightly from prediction; diffuse and gray mountain shadows instead of sharp and black; soft and blurred surface detail near the limb. Boscovich attributed to Huygens the idea that lack of lunar clouds denies an atmosphere, but regarded this as an insensitive test.

Partly with the aid of his own observations, Boscovich concluded that none of these signs of a lunar atmosphere had been observed beyond question. Euler was clearly wrong; the Moon's air, if any, had to be considerably less dense than he had announced.

Though Boscovich's book is little known now, it must have been widely read in its day. The year after it was published a second edition appeared at Leipzig and a third at Vienna in 1776. Some years later, when J. H. Schröter and Franz Gruithuisen wrote in support of a lunar atmosphere, their views met little acceptance. On the other hand, the skepticism in W. Beer and J. H. Mädler's book of 1837 was generally adopted by astronomers. These reactions may, in part, have been due to the continuing influence of Boscovich's treatise.

49. The first true mapper of the Moon

IF YOU HAD BEEN an amateur astronomer about 1800 and wished to study the Moon's surface, you would have been seriously handicapped by the lack of adequate maps. The best at that time was by the Göttingen astronomer Tobias Mayer, published in 1775, and on such a small scale that the lunar diameter was only 7½ inches. It was the only Moon map of the time that did not depend on mere eye estimates for crater positions.

Correcting this lack was the self-appointed task of a 25-year-old German amateur astronomer, Wilhelm Gotthelf Lohrmann (1796–1840). A native of Dresden, he was occupied in the geodetic survey of the kingdom of Saxony. This gave him a familiarity with practical cartography of the greatest value in his lunar project. His aim was

W. G. Lohrmann (1796–1840), after a portrait by Roessler. He, Mädler, and Schmidt were the leading German selenographers of the 19th century. Because of his retiring nature, his life is less well known than those of the other two, comments a modern German astronomer, Paul Ahnert, who adds: "To me he has always seemed a particularly sympathetic and attractive personality, and I believe his portrait bears this out."

"to represent the moon's surface features, and their gradations of light and shade, with all possible accuracy, and to carry out both measurements and drawings by scientifically sound methods." Both C. F. Gauss and J. F. Encke, the leading German astronomers of the day, encouraged the undertaking, and the latter furnished the mathematical formulas for the determination of crater coordinates.

After some experimental observations, in 1822 Lohrmann erected for the purpose an observatory on the top floor of his house in the Dresden suburb of Pirna. There was no dome, but instead a roof of planks that could be removed individually. Lohrmann's principal telescope was a refractor of 6-foot focal length, with a 4.8-inch achromatic objective by Fraunhofer; it was fitted with a filar micrometer. The instrument was equatorially mounted on a stone pier and had slow motions but no driving clock. There was also a 3.3-inch Fraunhofer refractor, and a small transit instrument for time determinations.

So equipped, Lohrmann enthusiastically began detailed charting of the Moon, on a scale of 38.4 inches to its diameter. To supply accurate reference points, he established from micrometric measurements the selenographic latitudes and longitudes of 79 craters. The pencil sketches made at the telescope were afterwards carefully combined into 25 ink drawings, which together covered the Moon's entire visible face. To describe the relative brightnesses of lunar formations, Lohrmann followed J. H. Schröter in adopting a scale of 0 to 10, where 0 represented black shadows, 5 light gray, and 10 the brightest regions, like the crater Aristarchus. These grades of brightness were carefully represented on the section maps. In drawing crater walls, Lohrmann indicated relief by making the shading darker the steeper the slope.

The first four of the 25 section maps were published in 1824 in Lohrmann's book, *Topography of the Visible Surface of the Moon, First Installment.* Its 130 pages contain a historical account of lunar studies, a full explanation of the author's working methods, and detailed descriptions of the regions charted.

This was the first modern treatise on selenography. Perhaps because of the book's rarity today, few modern astronomers have mentioned how closely its plan was followed, even in minor details, by the better-known *The Moon* of W. Beer and J. H. Mädler (1837). The merit of the latter book lies much more in its completeness than in its originality.

Even Beer and Mädler's skepticism about life and change on the Moon is foreshadowed. Lohrmann's attitude is shown by a footnote added to his description of Sinus Aestuum: "In this region, at longitude 8° E, latitude 6° N, Mr. Gruithuisen believes he has seen a city, a fortress, and other artificial works. He hopes soon to recognize the lunar inhabitants themselves, if they parade en masse through their forest glades, and he tells much in his selenographical writings of hot springs, minerals, animals, and plants. But these famed discoveries and the elaborate hypotheses based on them have no place in a straightforward book on lunar topography."

When publishing the first installment of his atlas in 1824, Lohrmann stated that he expected to finish the whole project in six years, and then republish all 25 sections together as a general map. However, it was only after many interruptions, partly due to his failing eyesight, that the charts were completed in 1836. Unfortunately, the copper plates for printing them had not been made when Lohrmann died at Dresden on February 20, 1840. He had, however, published

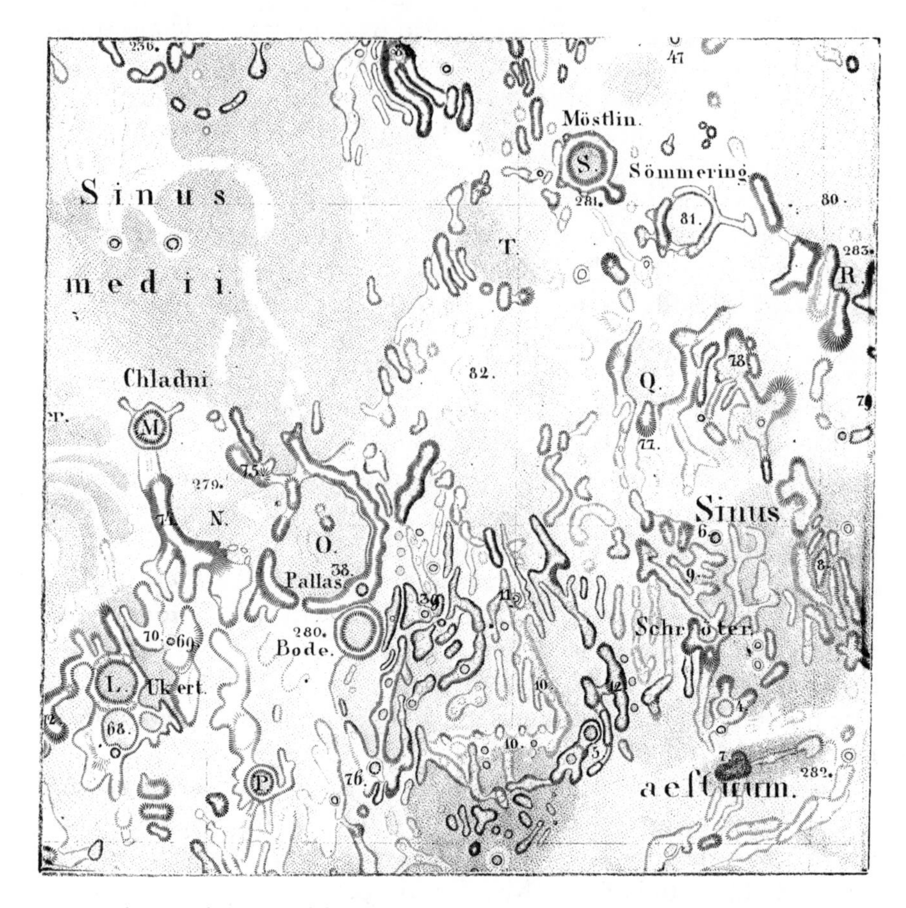

The central portion of the Moon's disk, reproduced at full scale from the 1878 edition of Lohrmann's atlas with labels and names added by J. F. J. Schmidt. This section covers about 15° of selenographic longitude and latitude. Look near numbers 6 and 9 for the group of hills and ridges that form Gruithuisen's "lunar city."

in 1838 a small general map, 15 inches in diameter, that was noteworthy for its rich detail.

Publication of the remaining sections of Lohrmann's atlas was undertaken in 1851 by the distinguished selenographer J. F. J. Schmidt (see Chapter 50). However, his prodigious energy as an observer apparently did not extend to business affairs, and it was not until 1878 that the other 21 section maps appeared. The names of two lunar craters record this protracted effort: Barth. after the Leipzig publisher of 1878, and Opelt, after the father and son who completed the reductions of Lohrmann's measures of crater coordinates.

Lohrmann's work far surpasses anything previously attempted, in scope, accuracy, and detail – his atlas contains 7,178 craters and 99 rilles. The long delay in publication kept most of it from the world until after the more extensive lunar maps of Mädler and Schmidt had appeared. Nevertheless, they followed in the paths he had opened, and thus Lohrmann ought to be recognized as the first modern selenographer.

50. Julius Schmidt: An incredible visual observer

ANYONE INTERESTED in the history of visual observing will soon come across the name of J. F. Julius Schmidt (1825–1884). Although he was director of Athens Observatory for a quarter century and a trained professional, amateur astronomers might well claim him as one of their own. Schmidt is famed for his great lunar atlas, his measurements of lunar mountain heights, and his discoveries of rilles and other detail.

Also, his energy as an observer has perhaps never been equaled. For example, variable stars were a favorite subject of his for four decades. During the last two years of his life Schmidt made no fewer than 85,000 observations of them. His record books contain 9,800 brightness estimates of the prototype Cepheid variable Delta Cephei and 5,800 of the eclipsing binary Beta Lyrae. Many of the more familiar naked-eye variable stars were discovered by him, including Eta Geminorum, Rho Persei, u Herculis, and Delta Librae.

As a planetary observer Schmidt showed equal enthusiasm. His great series of over 780 drawings of Jupiter provided the material for a later German's doctoral dissertation. At a single apparition of this planet, Schmidt timed 98 instances when the Great Red Spot crossed Jupiter's central meridian (that is, when it was equally distant from the planet's east and west limbs). His sunspot observations were so

J. F. Julius Schmidt, from a photograph taken at Athens on February 16, 1882.

numerous as to be reported in book form. Meteors, the zodiacal light, nebulae, lunar eclipses, and comets were also observed on the same extensive scale. But the special love of this unwearying man was the Moon.

As a 14-year-old boy in his native town of Eutin in north Germany, Schmidt bought at an auction a copy of Schröter's famous book on the Moon. His imagination was fired by the many pictures of lunar craters and mountains, and his one desire was to explore these wonders for himself. A small refractor with lenses ground by his father became Schmidt's first telescope, and with this instrument he began making drawings of the Moon. Having very acute vision and great skill in sketching, he realized that his efforts would become of scientific value as soon as he could use more powerful instruments.

This opportunity arrived in July, 1841, when Schmidt moved to Hamburg and was allowed regular use of the telescopes at the observatory there. In 1845 he got a position at the private observatory of J. F. Benzenberg at Bilk, near Düsseldorf. This experience proved to be a disappointment; the large refractor was reserved by Benzenberg for his own searchings for an intramercurial planet, and when not in use by him was kept under lock and key. For Schmidt there was only a small telescope, with an objective too defective for any serious lunar work.

A few months of this was enough, and Schmidt went to Bonn Observatory to spend eight years as F. W. A. Argelander's assistant (see Chapter 80). Here he found opportunity to continue the lunar drawings, in the intervals not filled by routine meridian-circle work, or by the zone observations for the great *Bonner Durchmusterung* star atlas and catalogue. On several visits to Berlin he had the use of the Royal Observatory's 9-inch refractor – the same instrument with which J. G. Galle first recognized the planet Neptune. Schmidt tells how on May 17, 1853, this telescope showed him no fewer than 78 craters on the floor of the lunar ring plain Clavius.

About this time, Schmidt was invited on Argelander's recommendation to take charge of the private observatory of E. von Unkrechtsberg, at Olmütz in Moravia. Here there was freedom from routine, and a Fraunhofer refractor of 5-foot focal length equipped with a good filar micrometer. This equipment was at once put to use for the measurement of lunar mountain heights from the lengths of their shadows. In the next three years Schmidt made almost 3,000 height determinations. During his stay at Olmütz, he wrote *Der Mond,* a 164-page treatise on selenography published at Leipzig in 1856. It is perhaps the first discussion of the lunar surface by anyone well versed in geology.

Schmidt's last move was to Greece, where he assumed the directorship of Athens Observatory in 1858. (The previous director, G. K. Bouris, once aroused an angry clamor by proposing that famous buildings of the ancient city be demolished to give the observatory a clear southern horizon.) When he arrived on December 2nd, he found the observatory in great disrepair and all of the instruments out of commission. But with characteristic energy he began work by taking a barometer reading the very day he stepped ashore

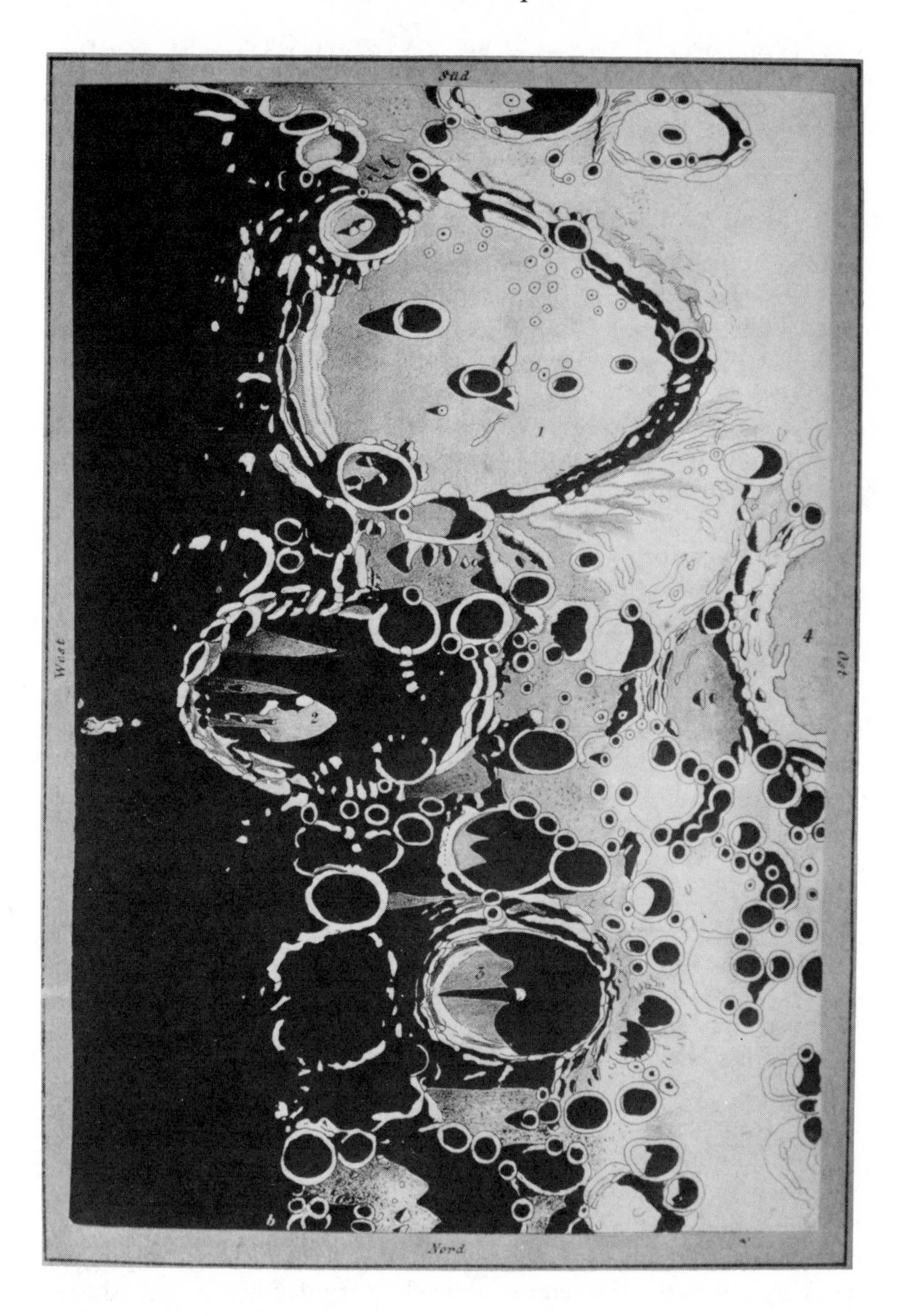
Sud
West
Ost
Nord
1
2
3
4

at Piraeus. The first telescope to be restored to working order was the 6-inch refractor that for the next quarter-century would serve for Schmidt's lunar studies. Later the small meridian circle was put back in use, but apparently only for determining time.

Under the favorable Greek sky, the German astronomer resumed his detail drawings of the Moon with renewed vigor. By 1865 their number was so great that he began to compile from them a general topographic map of the Moon, which was corrected and added to by new telescopic observations. In its final form, this map consists of 25 sections, which in combination show a lunar disk six feet in diameter. Some of the sections are based on 150 drawings, gathered over three decades. In all, 32,856 craters are represented; Schmidt estimated that a complete survey with the 6.2-inch Athens refractor at 600x would reveal 100,000 craters. This chart greatly surpassed in completeness and accuracy anything previously attempted.

The atlas was published at Berlin in 1878 as *Charte der Gebirge des Mondes (Map of the Mountains of the Moon).* Simultaneously, a companion text appeared under an identical title, except for the addition of the word *Erläuterungsband (Explanatory Volume).* This is still an invaluable historical source, containing, for instance, the full details of all Schmidt's height measurements, catalogues of rilles and ray systems, and a wealth of descriptive notes. (E. Neison's often-quoted *The Moon* came out in 1876, too early to incorporate Schmidt's work.)

Of special interest is the section on Linné, a feature in Mare Serenitatis that for decades was cited as evidence for probable change on the Moon (see Chapter 53). Up to the year 1843 Linné was generally recorded as a small, deep crater; in October, 1866, Schmidt reobserved it as a white spot. On pages 155 to 163 of his book, Schmidt reports individually his 200 observations of Linné from 1841 to 1874 – basic data for any serious attempt to solve the mys-

---

The neighborhood of Tycho under a setting Sun is shown in the frontispiece to *Der Mond,* Leipzig, 1856. Young Schmidt made this drawing on September 17, 1843, at the Hamburg Observatory. The very large ring near the top (south) is Clavius; Tycho, half-filled with shadow and with a central peak, is near bottom center. Schmidt says he did not have time to insert all of the fine detail then visible.

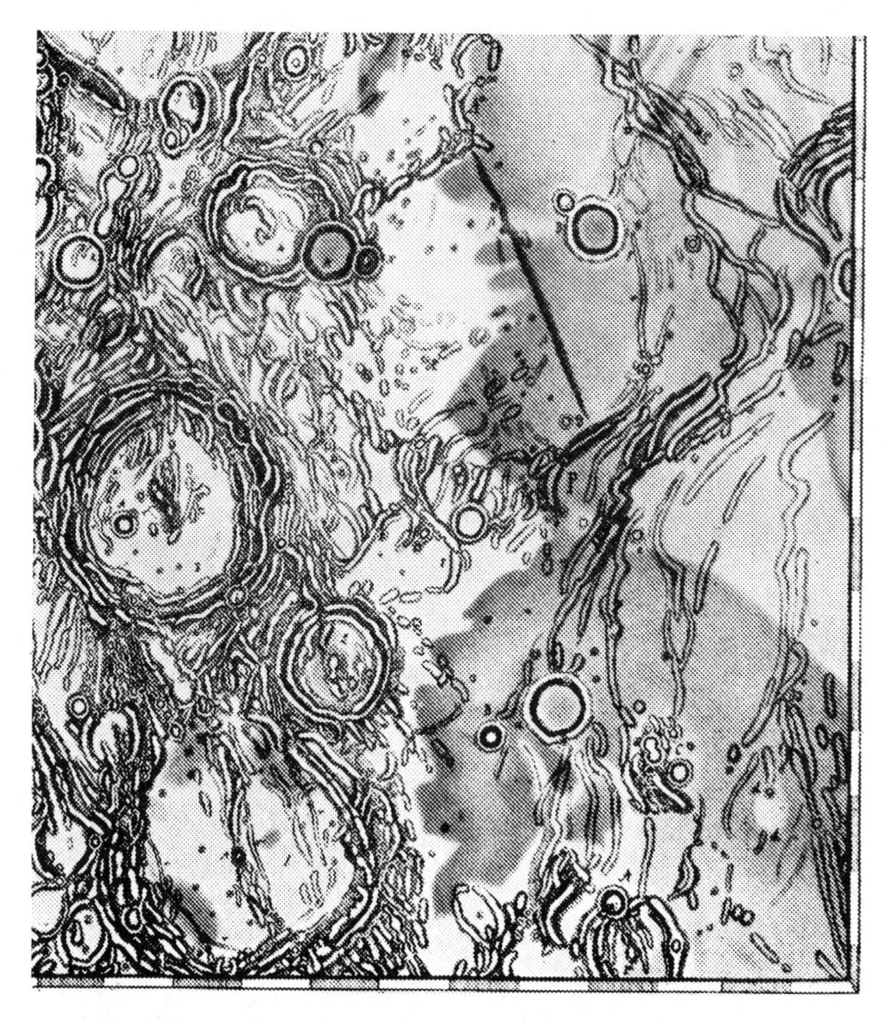

Part of Sheet 8 of Schmidt's lunar atlas, reproduced at about three-fifths original scale, with the Moon's central meridian running along the western (left) edge. The Straight Wall is in upper center. South is up. Courtesy Ewen A. Whitaker.

tery. He remarks that several other lunar features, such as Werner D and Lassell D (Alpetragius d), seem exactly analogous to Linné, each of them appearing as a bright patch containing a minute crater pit. In the above portion of his lunar atlas, Lassell D is the light area "dd" in the lower right corner.

Schmidt's atlas and companion volume of text incorporate his lunar observations only to the year 1874, but his selenographic studies continued until his sudden death on February 7, 1884. During this decade, according to his official Athens Observatory reports, he made 2,536 more lunar drawings, a larger number than he had used for the atlas. This additional work on the Moon seems never to have been published in any form.

Once, while using the library stacks at Harvard Observatory, I chanced upon another unpublished Schmidt manuscript. It was a 56-page, typewritten copy made in 1924 of his *Observations of Lunar Eclipses 1842–1879*. The text, in German, had been transcribed by Willard J. Fisher, who from 1922 until his death in 1934 at age 67 was a research associate at Harvard. Schmidt's book contains his extensive notes on 28 eclipses, beginning with the partial one of January 26, 1842, which he viewed as a 16-year-old. For some of these eclipses, Schmidt had published brief accounts in the *Astronomische Nachrichten,* but his manuscript supplies a good deal more information. The rest of the manuscript is simply raw observations, but the material is valuable for its very detailed notes on colors, long descriptions of various eclipse phenomena, and hundreds of timings of when craters entered or exited the Earth's umbral shadow. Although the work shows no application of photometry or photography, it has the special merit of being the comparative record of 38 years of lunar eclipses seen by the same expert eye.

The picture of Schmidt suggested by the eclipse records is that of a very determined but rather unimaginative man. He seems to have aimed at making great numbers of somewhat crude observations, rather than devising new and more accurate methods. During his directorship at Athens there was no expansion of the rather limited astronomical facilities. Schmidt appears to have worked unaided at the observatory, for in this part of the eclipse manuscript other names are mentioned very infrequently. For example, Demetrius Kokkides, who was to succeed Schmidt as director and become professor of astronomy at the University of Athens, appears as a collaborator only at the eclipse of November 4, 1873, when he made naked-eye timings of contacts.

One wishes for access to Schmidt's letters and a diary for a closer look at the personality of this unusual man. Clearly, his ability to make observations had far outrun his ability to analyze them. One impression was preserved by the Oxford astronomer Charles Pritchard, who in the spring of 1883 stopped in Greece on his return from an astronomical expedition to Egypt. "At Athens," he told the Royal Astronomical Society, "I had the satisfaction of a long interview with that eminent veteran in astronomical research, Dr. Schmidt. To me it seemed, and it still seems, a thousand pities that the invaluable results of many years of his labors still remain unpublished."

Less than a year later, Schmidt died in his sleep. The evening before, he had attended a social gathering at the German embassy in Athens. King George I and Queen Olga of Greece came to the observatory for his funeral. Noteworthy is the appreciation of Schmidt by the Dutch astronomer A. Nijland in an 1898 address entitled "The Rights of Small Observatories:" "Whoever has no instruments at all should take the shining example of Julius Schmidt, one-time director of the observatory at Athens, dilapidated but situated in an excellent climate. This astronomer of great talents and wonderful stamina has shown the whole world how in the hands of a skilled and ardent observer, even seemingly insignificant subjects gain special interest."

51. J. N. Krieger: The Moon half-won

THERE IS AN OFTEN QUOTED COMPLAINT by W. H. Pickering about lunar terminology: As a general rule, the larger a crater, the smaller the contribution to selenography by the person after whom it was named. Great observers like J. H. Mädler and J. F. J. Schmidt have been honored with minor formations, while the huge ring plain Plato is called after a philosopher whose addition to our knowledge of the Moon's surface was probably nil.

The earthly fame or obscurity of a selenographer also is sometimes

The Bavarian amateur J. N. Krieger (1865–1902) was one of the finest lunar observers of all time. This portrait is reproduced from the frontispiece of his 1912 *Mond-Atlas*.

an untrustworthy measure of his merits. For example, there are major figures from central and eastern Europe whose work is almost unknown by English-speaking amateurs, mainly because their writings are hard to find or use inconvenient languages.

A case in point is the German observer, Johann Nepomuk Krieger (1865–1902), whose beautiful lunar drawings were of an excellence that remains perhaps unmatched to this day. The brief allusions to Krieger in some recent books fail to convey the importance of his Moon atlas. The portion of it he managed to complete before his early death is still a storehouse of information.

Krieger's career was like that of many other amateurs, but with greatly enhanced lights and shadows. His intense urge to become an astronomer originated from a desire to escape the limited daily life of the mountain hamlet of Unterwiesenbach in Bavaria, where he was born on February 4, 1865. There his father ran a small brewery and an inn, in which the boy started work at the age of 15 after a year in agricultural school.

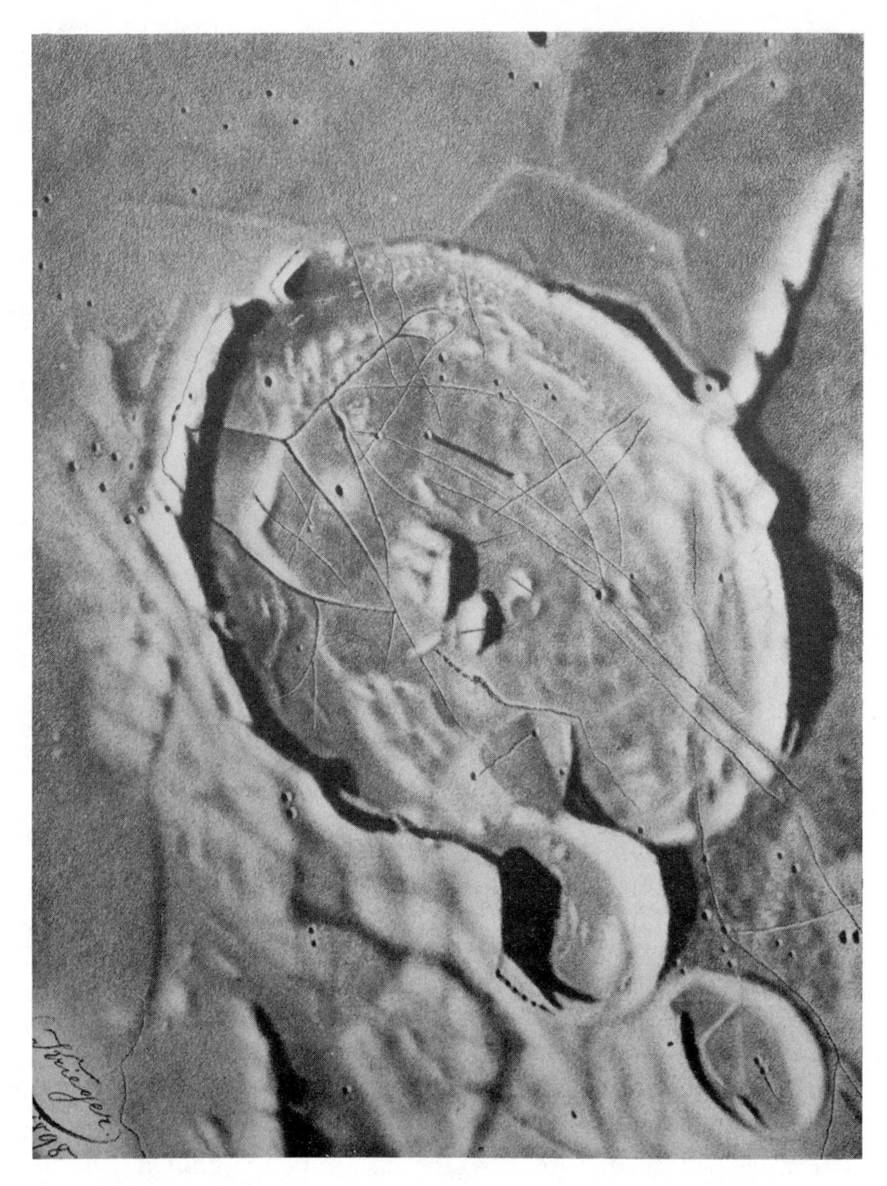

The lunar crater Gassendi, 69 miles in diameter, has fascinated generation after generation of amateur observers because of the delicate maze of hills, ridges, rilles, and pits on its broad floor. Krieger spent parts of 15 nights making this faithful depiction, here reproduced from Plate 46 of his 1912 atlas. South is up, west to the right.

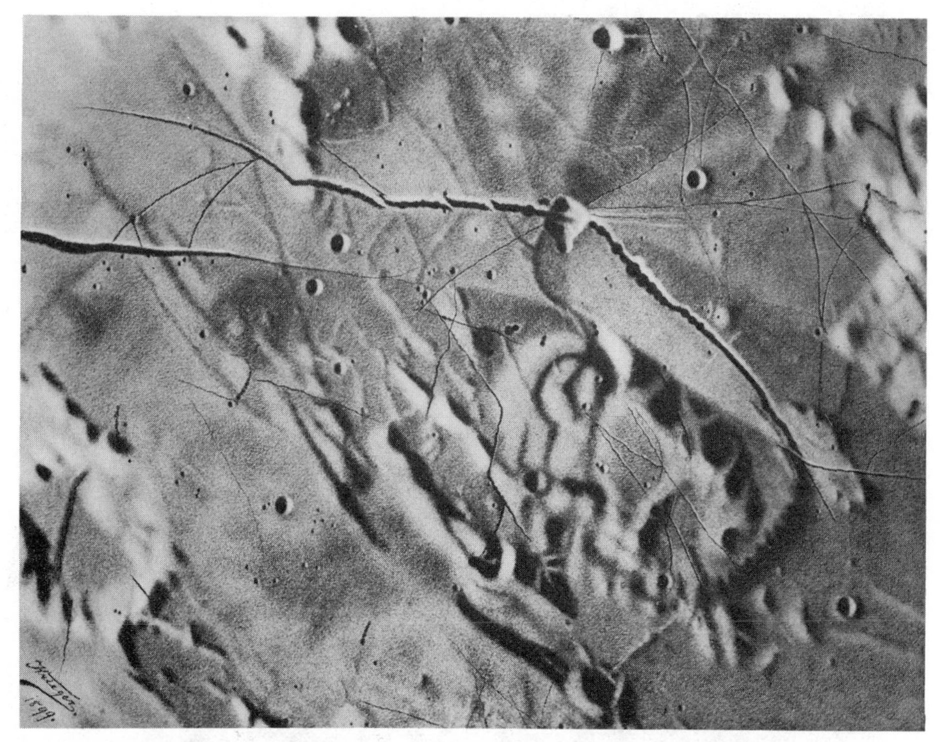

J. N. Krieger's chart of the Hyginus rille and its surroundings was based on 34 nights' work with a 10½-inch refractor. It is reproduced at a little less than two-thirds scale from his 1912 atlas. The 59 rilles in the picture have a combined length of two-thirds the Moon's diameter.

It is not quite certain how his attention first turned to astronomy, whether from books or through a fellow townsman nicknamed "the stargazer." But he early obtained a small telescope and began active observing, especially of the Moon. In 1886 Krieger visited Hermann Klein in Cologne for advice on his future. Klein was Germany's outstanding popularizer of astronomy at that time, the author of several good books for amateurs, and the editor of the journals *Copernicus* and *Gaea.* A skillful lunar observer himself, Klein had translated E. Neison's famous monograph, *The Moon,* from English into German. The Cologne professor warmly encouraged Krieger to go on with a career in selenography, and urged him to study the

The western part of Mare Tranquillitatis, from Plate 30 of Krieger's atlas, is remarkable for the complex pattern of ridges and rilles, 39 of the latter being charted here. The crater at the bottom is Arago, with a large low dome to its right. In the upper left, the small crater Moltke casts a long shadow.

mathematics that had been omitted in a spotty education. Klein also advised mastering the graphic arts and photography. Furthermore, he gave Krieger an introduction to Hugo Seeliger, the director of Munich Observatory, who became a lifelong friend.

By selling the brewery after his father's death, Krieger obtained the money to move to Munich, where he attended university classes in mathematics, physics, and astronomy. But it was not long before he knew that he had neither the background nor the taste for mathematics, and that his future in astronomy must be as an amateur. Krieger thereupon built an observatory, equipped with a good 10½-inch refractor, in the Munich suburb of Gern-Nymphenburg. Here he began his life's work, the task of accurately charting the Moon's surface in a systematic manner.

Such a project has been the ambition of many amateurs, but Krieger's long and careful preparations put him in a class by himself. Although the keen-eyed German showed much skill in drawing, this ability was not inborn but developed through persistent effort. From the start, he realized the value of charts that everywhere reach the same level of detailedness: "The selenologist who finds that some small area, randomly chosen, has been delineated with utmost minuteness will easily attach undue importance to it, perhaps basing an evolutionary theory on what was really faulty planning by the observer."

To get this uniformity, Krieger habitually worked with a power of 260, also reducing his 10½-inch aperture to 6½ inches. On first looking into the eyepiece with full aperture at this magnification, he could see fine details that became quite invisible a few minutes later owing to glare. In the best seeing, when the full objective was usable with very high powers, there was more fine structure than could be sketched.

Krieger's most important innovation was to use low-contrast prints of large-scale Moon photographs as the base for his drawings, thereby insuring accuracy in the placement and sizes of lunar features. This valuable technique has been surprisingly little used by other workers. At first, Krieger was provided by Klein with photographs taken with the Lick 36-inch refractor; later, he used mainly Paris pictures sent to him by M. Loewy and P. Puiseux.

The years 1890 to 1894 were spent in developing skills and techniques, and Krieger regarded them as a practice period, preliminary to his main effort. The 28 best of the 125 drawings from that time were published in 1898 at Trieste as the first volume of Krieger's atlas.

In 1895 he transferred the observatory from the northern side of the Alps to the southern, where he hoped to find clearer skies and steadier seeing at Trieste, near the head of the Adriatic Sea. The year 1896 was spent in readying the new Pia Observatory, which Krieger named after his wife. Finally, in 1897 he resumed his labor with great energy, securing 103 drawings in the first half year.

Krieger had now refined his method to assure utmost reliability. For each lunar area to be charted, a photographic print was chosen with a scale of about 12 feet to the Moon's diameter. At the telescope, additional features were entered on the same print during 10, 20, or more nights, with a different colored pencil each night, so that the color was a key to the date when a particular feature was noted. Next, all these markings were neatly transferred to a second print. On a night of good seeing, when the throw of shadow on the Moon matched that when the photograph had been taken, this print was compared with the moon and corrected. Finally, a finished copy was made on a third print, using lead pencil and charcoal, to serve for reproduction in the new atlas. In this laborious way, Krieger produced depictions of the Moon's surface that are startlingly superior to almost all other amateur efforts – in meticulous accuracy, aesthetic appeal, and legibility. Most later visual work fell short of his standards.

Krieger's hope to cover the entire visible face of the Moon in this fashion was never fulfilled, despite his single-minded energy. The year 1898 was one of ceaseless activity, yielding 458 observations on 92 nights. His days were spent preparing photographic bases, collating and finishing drawings, and doing the many chores of an active one-man observatory. This, too, was when he published the first installment of his atlas, and when he was involved in an acrimonious controversy forced upon him by Philipp Fauth, a famous lunar observer of that time (see Chapter 52). The same pace continued through 1899.

But this incessant toil, all-night vigils followed by full days in his office, finally broke down Krieger's health. The collapse came in 1900. His doctors told him to stop observing and to move to a gentler climate. He dismantled his observatory in January, 1901, giving the 10½-inch telescope to the Austrian naval observatory at Trieste. Desperately ill, he spent the next months at sanatoria in Italy seeking to finish the text for the second part of his atlas. From his sickbed he continued to dictate to a friend until he lost the power of speech. The end came at San Remo, in the Italian Riviera, on February 10, 1902, just after his 37th birthday.

The incomplete manuscript, together with hundreds of drawings ranging from finished products to rough sketches, eventually came into the hands of Krieger's old friend, Professor Seeliger. The latter wisely chose the very able Austrian selenographer Rudolf König to edit and reconstruct the atlas. The work appeared in 1912 as two magnificent volumes, published by the Vienna Academy of Sciences under the title *Joh. Nep. Kriegers Mond-Atlas – Neue Folge*. One volume is an album of 58 large pictures of lunar regions, some never completed. Each plate has a transparent overlay, bearing key numbers and letters for scores of formations of interest. The other volume is text, containing long and detailed descriptions of each region. Some of these were written by Krieger, but most of them were drafted by König with the aid of the original observing books and papers.

52. Philipp Fauth and his Moon atlas

NEARLY a quarter century after his death in 1941, Philipp Fauth's large *Mondatlas* was published in Bremen, West Germany. Its appearance caused a revival of interest in this German selenographer, who had been hitherto little known in English-speaking countries.

You can get surprisingly contradictory opinions about Fauth when you talk to people familiar with his story. Some call him a very great man, but there are many older European astronomers to whom his name is anathema. The popular articles about Fauth that

Philipp Fauth (1867–1941), photographed shortly before his death. Courtesy Ernst E. Both.

were published about the same time as his atlas hardly touch on the problem of his psychology. The question of the value of his atlas should be discussed separately.

Fauth was perhaps the most capable and versatile of all active visual observers of the Moon between about 1900 and 1940. He combined descriptive selenography with cartography, measurements of craters, and statistical studies. In 1936 he published *Unser Mond,* the best of all observing guidebooks to the Moon's surface. Nevertheless, this same man was also widely known for his crank cosmogonical theories and for vitriolic polemical writings. Fortunately, he managed to keep his telescopic work largely separate from his fantasies, and *Unser Mond* is practically free of them. He had the kind of complex personality that contemporaries found hard to appraise justly; no wonder that most articles about him are either tar and feathers or whitewash.

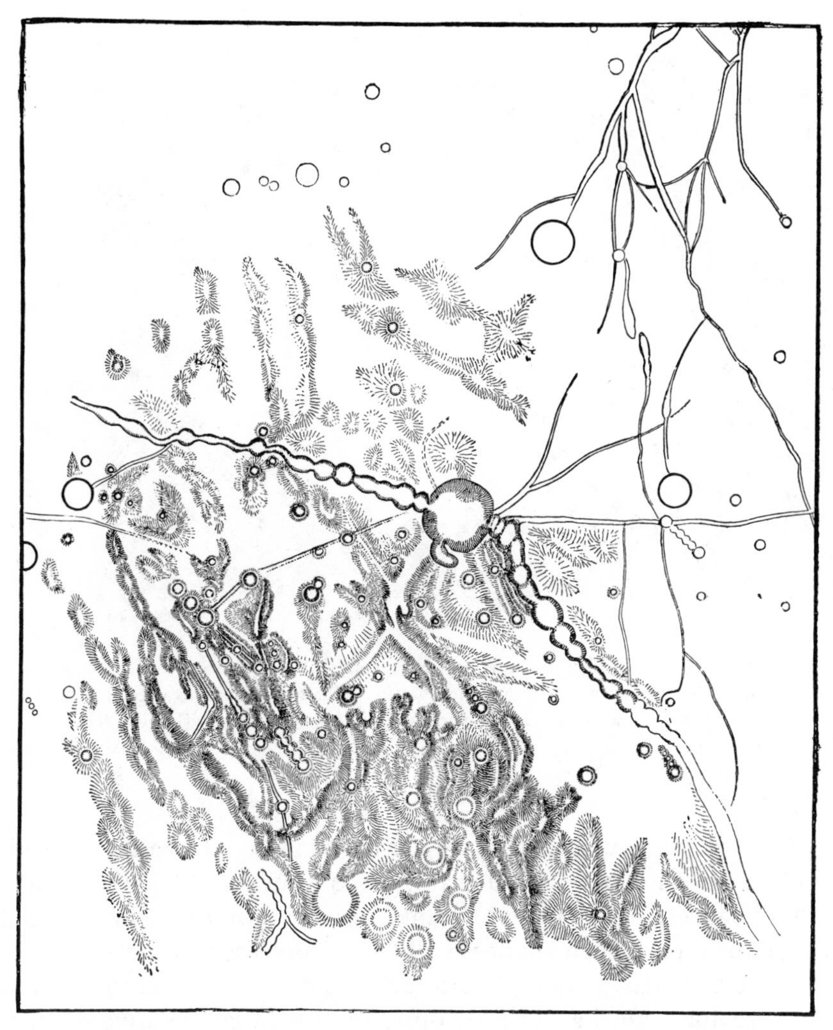

Philipp Fauth's map of the Hyginus rille, located near the center of the lunar disk. He eventually abandoned this hachure technique for indicating relief in favor of impressionistic contours. From Fauth's *The Moon in Modern Astronomy* (English translation, 1907).

A variety of circumstances have combined to give Fauth a diminished reputation that has reflected undeservedly on his legitimate lunar work. They include his conflict with organized astronomy and the failure of his writings to reach English-speaking astronomers.

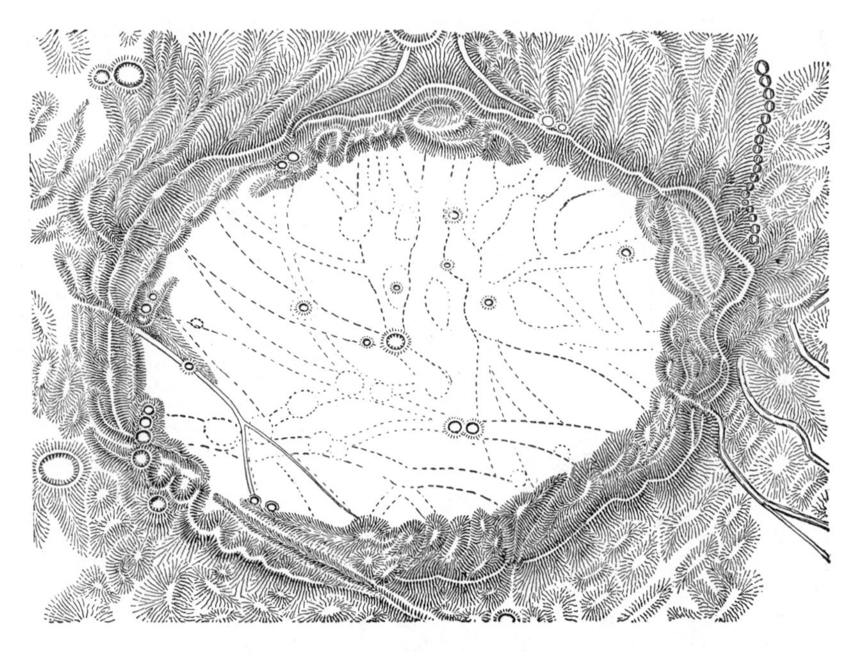

The lunar walled plain Plato as drawn by Fauth. Although some of the detail shown here is spurious, most of the features on Plato's dark floor can be seen in a large amateur telescope. From *The Moon in Modern Astronomy* (English translation, 1907).

The key to the whole story is the isolation in which Fauth worked, leading to an extreme independence of outlook. By vocation a schoolteacher, he remained an amateur astronomer all his life and never took part in the collective life at a professional observatory. Because Germany, unlike France or England, had no national amateur organization, he had little opportunity for cooperative observing and exchanging ideas with his fellows. Also, back in the early 1890's lunar studies were at a low ebb following the deaths of J. F. J. Schmidt and W. R. Birt, and young Fauth realized his superiority to most of his competitors.

It is easy to see how this lone wolf would get into irresolvable disagreements with professional astronomers. Particularly in the case of the Prague selenographer Ladislav Weinek (1848–1918). Fauth was so abusive in print that a rift widened between him and the academic world. Next came Fauth's advocacy of the *Welteislehre*

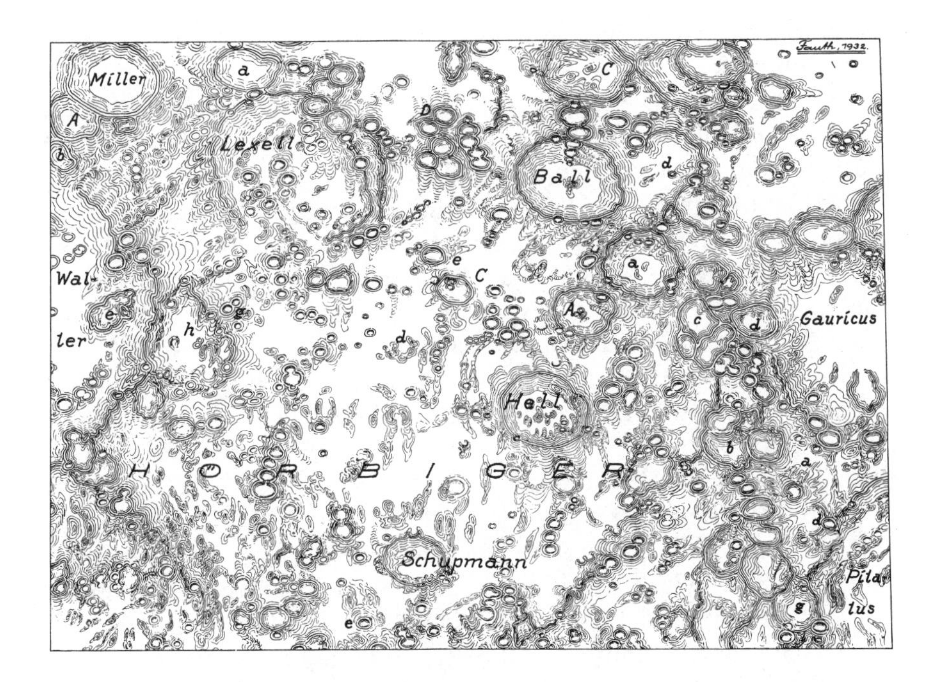

In 1932 Fauth published 16 large-scale charts of selected regions on the Moon. This one shows Hell Plain in the south-central portion of the lunar disk. The plain is actually the ruined and incomplete remains of an enormous crater that has never been officially named; Fauth dubbed it Hörbiger (see text). About 1¼ inches left of the crater Hell is Cassini's bright spot, one of the most brilliant lunar areas at full Moon.

(Cosmic Ice Doctrine), Hans Hörbiger's grotesque glacial cosmogony. According to Hörbiger, the Moon was covered with a layer of ice 140 miles thick. The Moon was spiraling in toward the Earth and upon crashing would cause a new ice age. Previous ice ages resulted from crashes of previous moons. The Milky Way was made up not of stars but of blocks of ice spiraling toward us. For many years, Fauth was involved in this strange movement, which carried on guerilla warfare against official astronomy in Germany and eventually became almost an item of faith among the Nazis, who linked it to their notions about racial purity. Disciples would disrupt scientific meetings with shouts of "out with astronomical orthodoxy! Give us Hörbiger!" Heinrich Himmler declared that the Führer himself "is a convinced supporter of this theory so much abominated by scientific hacks." These affairs all did lasting harm to Fauth's scientific reputation.

His serious selenographic work made surprisingly little impact on amateur astronomy in Great Britain and the United States. One reason seems to have been personal conflicts between him and some prominent members of the British Astronomical Association's lunar section. Characteristically, Fauth republished a set of British drawings as samples of "degenerate selenography." Another reason why his name remained relatively unfamiliar was the extreme rarity of Fauth's *Unser Mond* outside of Germany.

When the German amateur died on January 4, 1941, he left unfinished a major undertaking, his lunar atlas on 22 sheets, to a scale of 1:1,000,000. The basis of this project was the large collection of detailed drawings he had made since 1911 with an excellent 15½-inch apochromatic refractor, first at Landstuhl in the Rhineland and later at Grünwald in Bavaria. The charts were eventually completed by Fauth's son Hermann and published in 1964. The particular merit of this atlas is its richness in fine detail, for his 15½-inch telescope under favorable conditions had better resolving power than the best Earth-based lunar photographs today. Therefore, the student with some topographic problem – the pattern of Triesnecker rilles, or the ridge system near Linné – will find Fauth's atlas a valuable reference when used in combination with other sources.

Such applications raise the question of homogeneity, since some of the relief was drawn by Philipp Fauth, some by his son. There is a characteristic difference between their styles. The father drew with a surer stroke, whereas the son sometimes inserted ambiguous patterns.

But there is another reason for owning Fauth's *Mondatlas*. Ernst E. Both expresses it in his excellent *History of Lunar Studies:* "Fauth was the last of the great visual observers, and the very high standards he set for himself were never approached by his contemporaries. And, it seems unlikely that they will ever be approached by any individual."

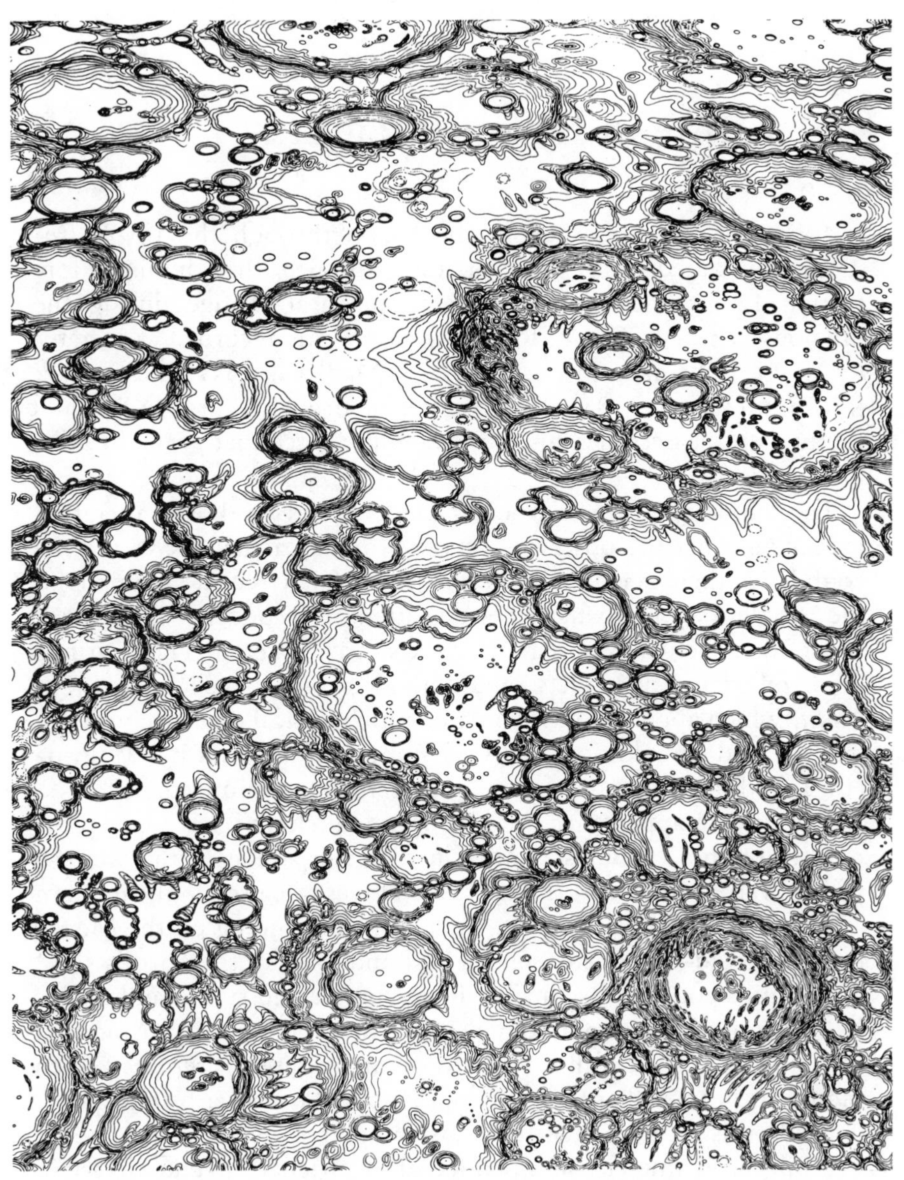

The southern part of the Moon as represented on one sheet of Fauth's great lunar map, left incomplete when he died in 1941. The original is on a scale of one to a million, corresponding to 11½ feet for the Moon's diameter. Tycho (bottom right) and Clavius (top right) are conspicuous. Fauth's contour lines are careful estimates rather than the result of detailed measurements. To appreciate the amount of detail in this map, and to judge the accuracy of its depiction, the reader should compare the view in his or her own telescope.

53. Linné in fact and legend

POPULAR WRITINGS about the Moon through the 1950's frequently stated, with more or less assurance, that the lunar formation Linné underwent a decided physical change during the previous century. We are told how Linné was recorded as a crater by W. G. Lohrmann in 1823, by J. H. Mädler in 1831, and by J. F. J. Schmidt in 1841–1843. But then it is reported that in 1866 Schmidt found its place occupied instead by a white patch that persists to this day.

The whole question of whether Linné suffered a drastic alteration hinges on what the observers of 1823 to 1843 actually saw and the confidence their evidence deserves. Many fairly recent appraisals of the problem seem to have been written at second or third hand, without any actual examination of the source material. These writers often relied for their facts on Edmund Neison's well-known *The Moon,* which appeared in 1876 when much critical evidence was still unpublished. In addition, the very important analyses by W. Prinz (1894) and P. Fauth (1901) have gone generally unnoticed.

The present appearance of Linné in small telescopes is a conspicuous white spot, about six miles in diameter, lying in the western part of Mare Serenitatis. (Lunar east and west are used in the modern sense, as they would appear to an astronaut standing on the Moon, not in the old-fashioned sense of east and west in the Earth's sky.) The white spot appears small at lunar sunrise and sunset and attains its maximum size about four days before and after full Moon. It shrinks slightly at full. This apparent variation, repeated each month, is not peculiar to Linné, but is shared by other bright spots.

Inside this white patch, large instruments show a small craterlet 1½ miles in diameter and 2,000 feet deep. Richard J. Pike describes it as "an extremely fresh but otherwise quite ordinary impact crater." The eastern wall of the craterlet is noticeably thicker and higher than the western, a fact noted by L. Brenner about 1900 and since confirmed by W. H. Steavenson and several other observers. This tiny crater is much too small to have been seen as such with the very modest instruments used for observing Linné in 1823–1843.

When viewed in a small telescope, Linné (arrowed) normally appears as a rather conspicuous bright spot on the western side of Mare Serenitatis. Courtesy Lick Observatory.

The craterlet is located on the summit of a low hill about three miles in diameter that casts a conspicuous shadow when grazingly illuminated by the rising or setting Sun. From my own observations of the shadow length, the hill was found to be about 300 feet high. This broad, gentle swelling deserves classification as a lunar dome, a suggestion made by Fauth and independently by H. P. Wilkins and P. Moore.

This description of the present aspect of Linné agrees in detail with Schmidt's records on about 200 nights in 1866–1874. Using the 6-inch refractor of Athens Observatory, he repeatedly glimpsed the craterlet as a minute black speck within the white spot, and on several dates he saw the shadow of the hill. We have strong evidence, therefore, that Linné in 1866–1874 looked the same as now.

Let us next consider the crucial observations of 1823–1843, by Lohrmann, Mädler, and Schmidt. In 1821 Lohrmann, the German amateur, began preparation of a detailed lunar map from observations with a 4.8-inch refractor (see Chapter 49). The first four of the 25 sections of this atlas were published with an explanatory text in 1824. On Section IV Linné is depicted as a crater, labeled A. Lohrmann on page 92 of his book tells us: A is the second crater on this plain [Mare Serenitatis]; it lies according to my observation in longitude +11° 27′ 22″, latitude +27° 42′ 06″, beside a ridge extending from Sulpicius Gallus. It has a diameter of rather over one German mile [4½ statute miles], is very deep, and can be seen in every illumination." We learn further from page xv of the appendix that Lohrmann measured the coordinates of Linné only once, on May 28, 1823. He had previously chosen the crater Conon as a reference point for charting this part of the Moon but then decided on Linné: "Conon cannot be seen distinctly near the time of full moon; A, on the other hand, is always visible as a bright spot in the gray Mare Serenitatis."

Except for Linné being termed a depression, these statements are quite consistent with the present appearance. The explanation of that discrepancy can be found on page 19 of Lohrmann's work: "I sketch and measure the larger mountains and craters under high illumination, so as to be able to chart in detail as large a portion as possible under the rare favorable conditions when the region is near the terminator." These words suggest that Lohrmann's only examination of Linné may have been made under a high sun, when it would have been easy for him to mistake a bright spot for a bright crater floor. Because of this likelihood that the term "crater" resulted from an incorrect inference, no safe argument for change in Linné can be based on Lohrmann.

Mädler's observations of Linné were made jointly with W. Beer at Berlin, using a 3¾-inch refractor. Charting Linné as a crater, he describes it on page 232 of their 1837 book: "The two main ridges unite near the deep crater Linné, which according to one measure by Lohrmann and seven by us lies in longitude +11° 32′ 28″, latitude +27° 47′ 13″. It is 1.4 German miles [six statute miles] in diameter . . . but indefinitely bounded at full moon." All seven of Mädler's positional measures of Linné were made in one night, December 12, 1831, and hence their number does not imply special attention to this formation on many dates.

There is reason to believe that Mädler made the same mistake we have ascribed to Lohrmann. Unlike modern observers, he paid particular attention to lunar features under high illumination, and he tells us explicitly that his brightness estimates were habitually so made. There are at least two other cases where Mädler charted and described as craters objects that are really bright spots: Lassell D (formerly called Alpetragius d) and Birt C. Furthermore, the verbal similarity between Mädler's and Lohrmann's accounts suggests that the earlier one influenced the later.

Finally, there remain Schmidt's records of 1841–1843. There is a remarkable contradiction between the two versions of them that he published. When Schmidt in 1867 announced to the Vienna Academy of Sciences the disappearance of the crater Linné, he said he had drawn this region 11 times between April 27, 1841, and August 17, 1843. Only five of these drawings showed Linné, but each time as a crater. Yet Schmidt's book of 1878 listed 19 drawings in those same years, on at least two of which Linné was represented as a bright spot! At the time of these observations, Schmidt was still a boy in his teens, having been born in 1825. He had not yet gained the skill and experience that later made him a leader in selenography, and he was then using only small telescopes. For these reasons, and because of his inconsistent testimony, little weight can be attached to his 1841–1843 observations.

This survey of the Linné history reveals how weak the evidence was for supposing that this object had ever been observed as a large

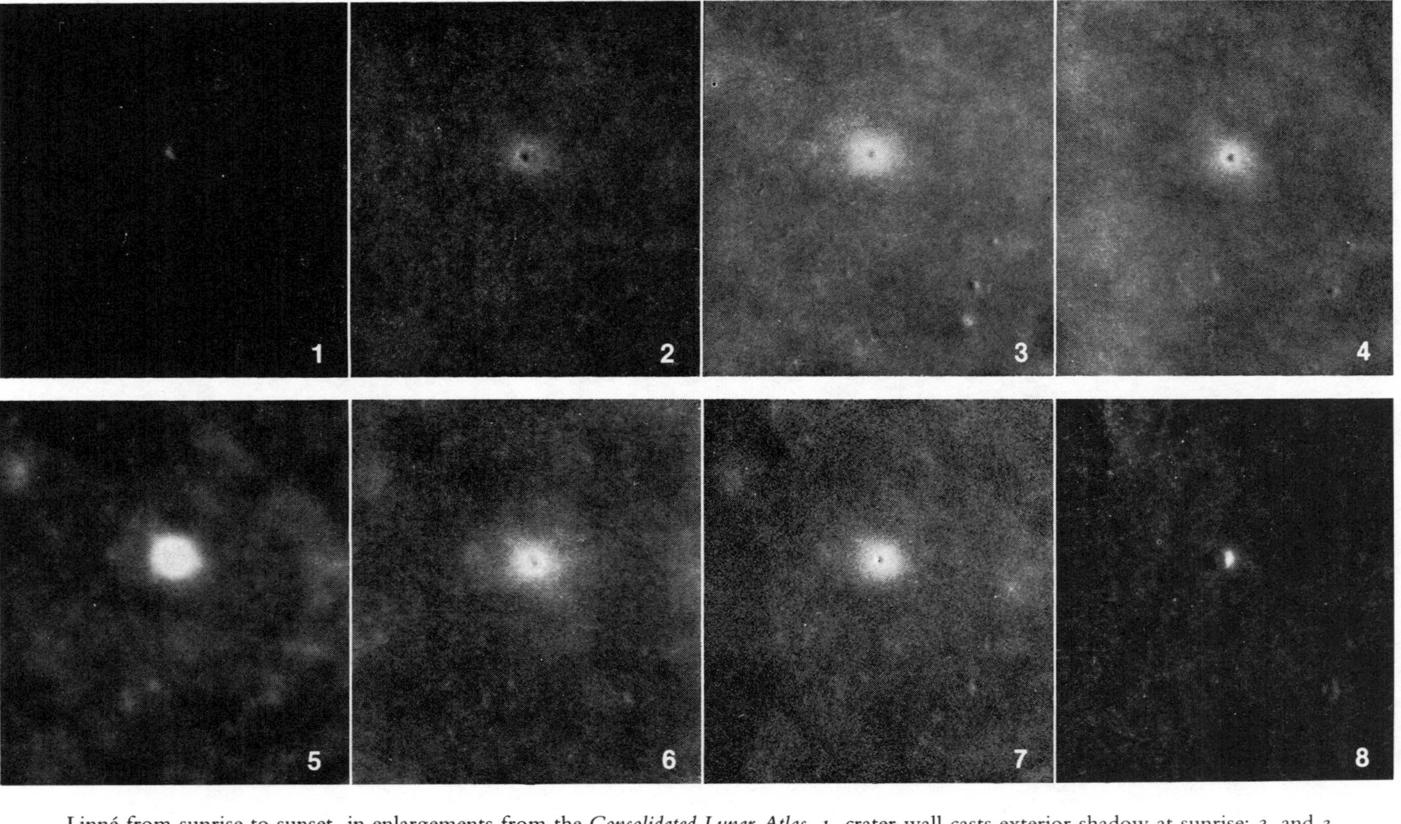

Linné from sunrise to sunset, in enlargements from the *Consolidated Lunar Atlas*. *1*, crater wall casts exterior shadow at sunrise; *2*, and *3*, white spot appears, shadow inside crater; *4*, interior shadow persists. *5*, near noon, no shadow, spot bright; *6*, interior afternoon shade; *7*, crescentic inner shadow; *8*, crater casts sunset shadow eastward. University of Arizona photographs, courtesy Ewen A. Whitaker.

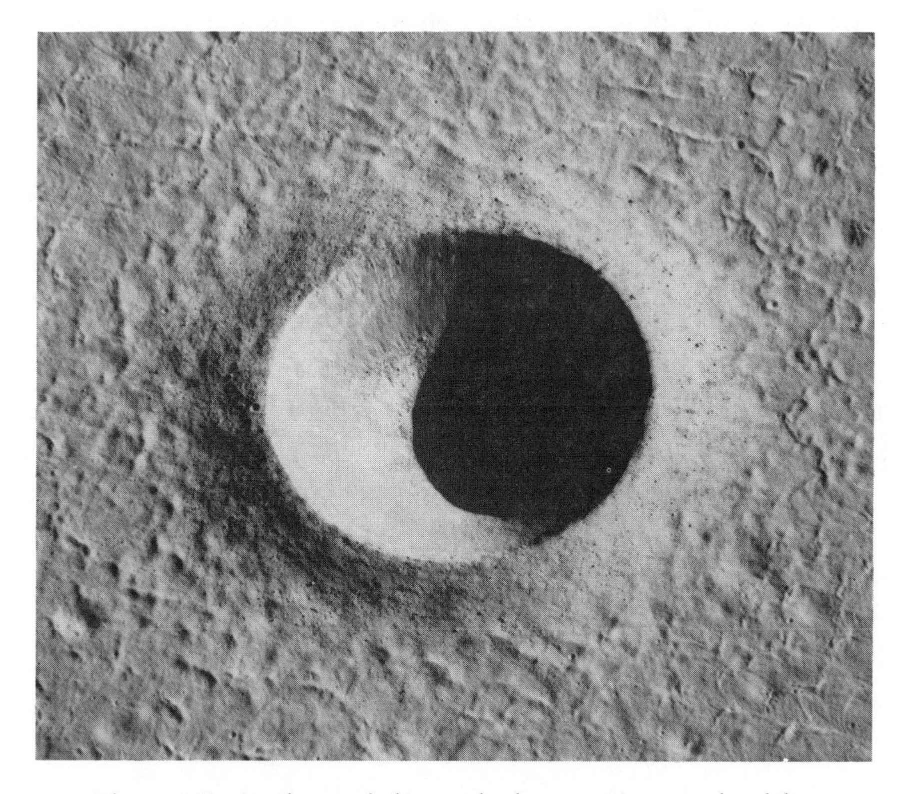

The crater Linné as photographed in 1971 by the panoramic camera aboard the Apollo 15 command-service module. From crest to crest, this young impact crater is 2,450 meters (8,040 feet) in diameter. Courtesy NASA.

crater. A deeper skepticism is warranted by several very early observations of Linné as a bright spot. It is so depicted in two drawings of Mare Serenitatis made in 1788 by the English artist John Russell, as was pointed out by A. A. Rambaut in 1904. Moreover, another drawing of 1788 by J. H. Schröter contains a light patch that is probably Linné. (This identification has sometimes been doubted, but it is substantiated by Schröter's description.)

There is another line of inquiry to be traced. Mädler was still alive in 1866 when Schmidt first announced that Linné had changed. His reexamination of the formation might be expected to decide the issue. The aged selenographer was living in Germany, after having retired from the directorship of Dorpat Observatory in Russia, and was recovering from a cataract operation in April, 1866. His first opportunity to look again at Linné was on May 10, 1867, with the heliometer of Bonn Observatory. "I found it of the same form and even with the same throw of shadow as I remember seeing it 37 years before," Mädler reported.

On that very same night Schmidt also observed Linné, with the 6-inch refractor of Athens Observatory: "Linné appeared as a conspicuous hill, casting a shadow, more striking than I have seen, at least since October, 1866. The hill may have been 3,000 feet in diameter and 500 feet high. The surrounding bright patch in the gray plain was very insignificant."

What Schmidt saw is valid today as a description of Linné just after the sun has risen there. If, then, this same appearance matched Mädler's memory of the 1830's, we have a further argument against a change having taken place.

Thus the famous Linné legend turns out to have no sure foundation. Instead of repeating it, future writers of observing guides might more realistically call attention to this formation as one of the best-studied examples of a numerous class of bright patches, each containing a small craterlet. Other prominent instances are Lassell D, Werner D, and Posidonius Gamma, all three being strikingly similar to Linné.

PLANETS AND OTHER SOLAR SYSTEM OBJECTS

54. An alleged satellite of Venus

HOW MUCH CONFIRMATION do you need to place an observational discovery beyond any reasonable doubt? This important and disturbing question is raised by the story of a satellite of Venus, which at one time was confidently accepted as a member of the solar system. No fewer than 33 observations of it by 15 different astronomers were recorded during the 17th and 18th centuries, beginning with F. Fontana at Naples on November 11, 1645. In the year 1761 alone, 18 observations of the object were made.

Some of these sightings were by well-known observers who even today have a reputation for reliability. Thus G. D. Cassini, director of Paris Observatory and discoverer of Cassini's division in Saturn's rings, wrote in his journal for August 28, 1686: "At 4:15 a.m., while examining Venus with a telescope of 34 feet focal length, I saw at 3/5 of its diameter to the east an ill-defined light, which seemed to imitate the phase of Venus, but its western edge was more flattened. Its diameter was very nearly 1/4 that of Venus. I observed it with attention for a quarter of an hour, when, on quitting the telescope for five minutes, I could not find it again, the dawn being too bright."

In 1773, the German astronomer Johann Lambert calculated the orbit of the satellite of Venus. He found that this moon revolved about its primary in 11 days and five hours. Its average distance from Venus was 66½ radii of the planet (250,100 miles – about the same distance as the Moon is from the Earth). Also, the orbit was rather oval, having an eccentricity of 0.195. Widespread interest was aroused when Lambert announced these results to the Berlin Academy of Sciences. The King of Prussia, Frederick the Great, proposed that the satellite be named after his friend, the French astronomer-mathematician Jean d'Alembert. The latter prudently declined the honor, explaining that his place on Earth was so insignificant he had no ambition for one in the skies.

Yet nothing can be more certain than that no such satellite exists. It was never seen after 1768. Veteran observers of Venus such as J. H. Schröter, William Herschel, and J. H. Mädler could not find it. More recently, E. E. Barnard, whose interest in satellites was

Francesco Fontana (1602?–1656), a lawyer-astronomer, was born and died in Naples, Italy. From *Rivista di Astronomia e Scienze Affini,* 1910.

marked, made many observations of Venus with telescopes up to the Yerkes 40-inch in size without finding any companion. Finally, it is almost certain that space probes, beginning with Mariner 2 which flew by Venus in 1962, would have detected any substantial satellite. Yet, even in the last century, the satellite of Venus found a few champions. In 1875, a curious book, *Der Venusmond,* was published in its defense by the German, F. Schorr. He argued that the many failures to see it were due to a brightness variation that normally caused the satellite to be so faint as to be invisible. Schorr's elaborate hodgepodge was more enthusiastic than critical, and it soon sank into the vast abyss of forgotten semiscientific literature without having cleared up the mystery.

What the old observers really saw was finally explained by Paul Stroobant, of Brussels Observatory, then at the start of his career as the most distinguished Belgian astronomer of his generation. In his memoir of 1887, Stroobant reprinted in full all the original statements of the observers of the satellite of Venus, described the earlier hypotheses to account for the observations, and subjected all this to a searching examination.

In the first place, Lambert's orbit was clearly impossible, as it

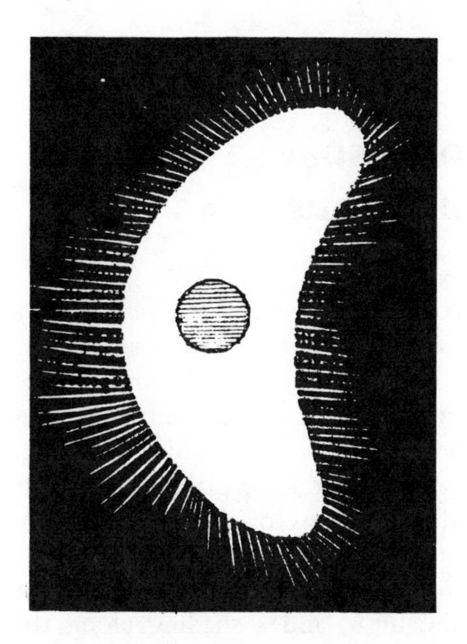

The legendary satellite of Venus appears as a shaded circle in these drawings by Fontana at Naples. Left: November 11, 1645; right: January 22, 1646. On another occasion he even saw two of these companions in his primitive telescope. From *l'Astronomie* for 1889, page 136.

required the mass of Venus to be 10 times greater than it actually is. Next, the descriptions of the satellite were highly contradictory. Several accounts spoke of a large disk showing the same phase as Venus. At least some of these observations must refer to "ghosts" caused by reflections within the telescopes, which would be particularly noticeable for so bright an object as Venus. Other observers had described the satellite as starlike, and Stroobant, by calculating the positions of Venus at the dates in question, could identify in seven cases what star the observer had mistaken for a satellite. It is possible, though not certain, that the "satellite" seen by Roedkiaer at Copenhagen on March 5, 1761, was the planet Uranus – 20 years before its recognition by Herschel.

There remained little else of the observations except some statements so vague as to inspire no confidence. So final was Stroobant's critique that the alleged satellite of Venus has now been almost forgotten.

55. The problem of visual observations of Venus

IN 1956 THE GERMAN ASTRONOMER Joseph Meurers wrote a small book called *Astronomical Experiments*. It is a systematic survey of scores of different laboratory experiments intended to simulate astronomical objects and phenomena under controlled conditions. It is also a disturbing book, as I found on reading the pages dealing with visual observations of detail on the planet Venus. He reports experiments making plain that many of the ordinarily observed disk features on this planet have no objective existence.

Amateur observers often record bright caps near the cusps of the crescent, a bright band along the limb, and a shading along the terminator. In addition, darkish collars bordering the cusp caps are often seen, as well as indistinct dusky and brighter patches on the disk. All these features represent only slight differences in contrast.

The preceding general description applies well to certain drawings made at Munich Observatory around 1896 by the Swiss-born astronomer Walter Villiger, who later was for many years chief of the astronomical division of the Carl Zeiss works in Jena. But, in addition to his Venus sketches, Villiger also drew illuminated balls of rubber or plaster of Paris. These were 55 millimeters (2.2 inches) in diameter and were viewed from a distance of 400 meters (about 1,310 feet) with a 5-inch refractor. Although each of these balls actually had a completely structureless surface, the drawings show cusp caps, bright limb bands, and the characteristic dusky patches. Note the close resemblance between these model drawings and Villiger's sketches of the actual planet made with a 5½-inch aperture.

Confirmation of these disquieting results is afforded by another series of careful experiments on Venus models carried out in 1952 by the well-known German amateur W. W. Spangenberg. He too used smooth, featureless spheres and viewed them with small telescopes at such distances that they appeared the same size as Venus

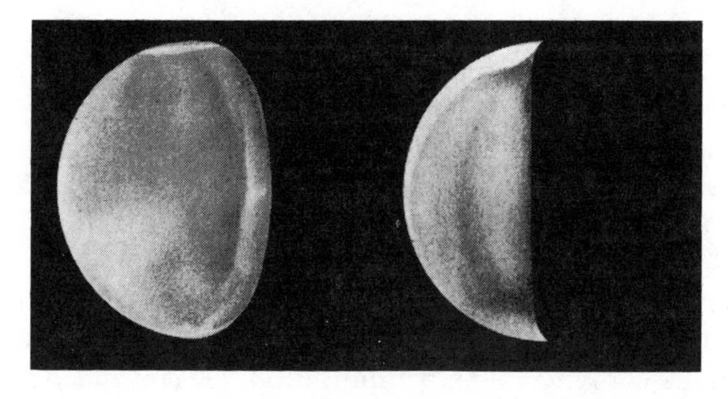

Two drawings of Venus made by Walter Villiger with the 10½-inch Munich Observatory refractor stopped down to 5½ inches. Left: November 16, 1895, at 18:15 Greenwich mean time; right: March 21, 1896, at 21:20. From the *Neue Annalen* of Munich Observatory, 1898.

Three sketches of models of Venus, as seen by Villiger through a 5-inch telescope. The one at the left was a rubber ball, the others a plaster-of-Paris globe. Since the models were featureless, the details in these drawings are entirely illusory.

observed with powers of 60 to 140. To simulate unsteady seeing, he placed an electric heater underneath the telescope objective.

Spangenberg's drawings of his models again suggest typical depictions of the planet itself. They add to the demonstration that purely illusory cusp caps and collars, limb brightenings, and dusky streaks can be recorded. Furthermore, the Villiger and Spangenberg experiments taken together demonstrate that independent observers may share a common pattern of optical illusions. Evidently, we are not always entitled to assume that a Venus feature is real simply because it has been similarly drawn by independent observers.

The visual observer of Venus, before he can secure wholly trustworthy results, ought to find out how to distinguish possible true features from subjective ones. A very promising clue is suggested by the experience of Audouin Dollfus of Meudon Observatory. Using a number of telescopes of different apertures and with powers of 20 to 1,200, he ascertained that the illusory markings become less conspicuous with increasing magnification. He concluded that when Venus is magnified enough to make its telescopic image six times the naked-eye diameter of the Moon (or about three degrees), spurious effects become inappreciable.

This finding throws interesting light on the often-stated fact that Venus markings are better seen with small telescopes than large. W. F. Denning, a very experienced British planetary observer, wrote in 1891: "Accounts are sometimes published of very dark and definite markings seen with only 2 or 3 inches aperture. Such assertions are usually unreliable. Could the authors of such statements survey the planet through a good 10- or 12-inch telescope, they would see at once they had been deceived. Some years ago I made a number of observations of Venus with 2-, 3-, and 4¼-inch refractors and 4- and 10-inch reflectors, and could readily detect with the small instruments what certainly appeared to be spots of a pronounced nature, but on appealing to the 10-inch reflector, in which the view became immensely improved, the spots quite disappeared, and there remained scarcely more than a suspicion of the faint condensations which usually constitute the only visible markings on the surface. . . ."

To this might be added the testimony of Franz Gruithuisen, who made many observations of Venus from 1813 to 1847. He was the first to report the cusp caps, and he examined them in many telescopes. However, he saw them best with his smallest instrument, one of only 1½-inch aperture!

Clearly, it is a wise precaution for amateur astronomers who make visual observations of Venus to carry out analogous observations of models. In this way the observer can find if his or her drawings contain subjective markings. By varying the experiments it should be possible to discover which observing techniques minimize the risk of self-deception. Similar experimental studies might well be made by observers of surface detail on the minute disks of Uranus, Neptune, or the four large satellites of Jupiter. Spangenberg pointed out that in many of his model tests he recorded the dark central cross characteristic of the Uranus drawings of Leo Brenner (see Chapter 21).

56. The disappearance of a marking on Mars

AROUND THE BEGINNING of the 19th century the world center of lunar and planetary studies was the little North German town of Lilienthal, near Bremen. Here were the observatory and optical shops of J. H. Schröter (1745–1816), chief magistrate of the neighborhood, who for nearly 30 years kept unflagging watch upon the Moon and planets. Here his assistant, K. L. Harding, discovered the third asteroid, Juno, and the famous F. W. Bessel began his career.

Fate was unkind both to Schröter and to his reputation. The decades of war that convulsed Europe finally engulfed him in April, 1813, as Napoleon sought to regain his slipping hold on Germany. Lilienthal was sacked and burned by the French army of General Vandamme, and Schröter lost at one stroke his observatory, records, magistrateship, and wealth. Unable to resume observing, he died three years later.

This enthusiastic amateur, the Percival Lowell of his age, has been underrated by later astronomers. His own work overlapped that of his more versatile, brilliant contemporary, William Herschel, whose fame has obscured all other observers of his time. Again, it has been Schröter's misfortune to be remembered more for his faulty rotation periods of Mercury and Venus than for being the first astronomer to devote a lifetime to telescopic study of the planets. And, finally, the row of fat volumes devoted to his findings on Mercury, Venus, Jupiter, Saturn, and the Moon, were written in a fantastically unreadable style that left their contents largely unnoticed.

J. H. Schröter's 9.3-inch reflector pictured at Lilienthal in 1793. Note the carriage at left, which provided rotation in azimuth, and the pulleys for altitude adjustment atop the tall mast. From A. v. Schweiger-Lerchenfeld's *Atlas der Himmelskünde,* 1898.

The Lilienthal astronomer published no book on Mars, although he had followed this planet intensively. It was long supposed that his manuscript of Martian observations had been destroyed in the ruin of his observatory. Finally recovered among the family papers of a nephew, the Mars writings were published in 1881 by the Dutch astronomer H. G. van de S. Bakhuyzen, together with 216 of Schröter's drawings.

A curious delusion pervades this book – that the Martian markings were mere wind-blown cloud formations in a thick atmosphere that veiled the true surface of the planet. Acting on this belief, Schröter from night to night carefully estimated the times when various dark markings crossed the central meridian of the planet, in an effort to measure Martian wind velocities. These markings can often be identified; hence Schröter's transit times could have been used in combi-

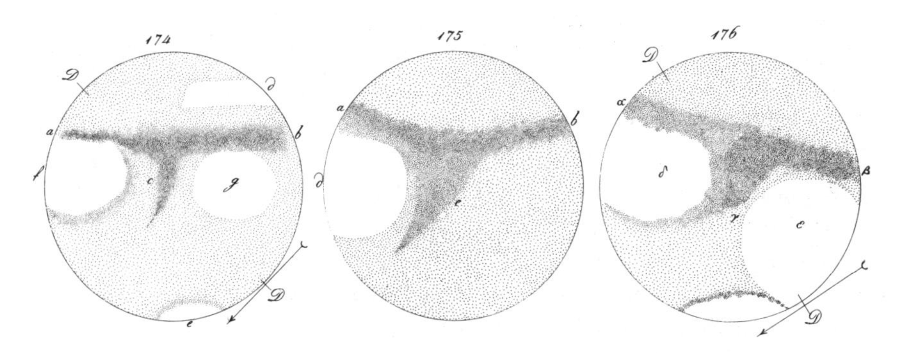

Three drawings of Mars made by Schröter with a 9.3-inch reflector during the favorable opposition of 1800. Left to right: On November 2nd the Arrowhead was near the center of the disk, extending downward toward the north polar cap; 3½ hours later that night, the Arrowhead had rotated leftward out of view and Syrtis Major was centrally located; on November 4th, the Arrowhead was again central. From Schröter's *Aerographische Beiträge,* Leiden, 1881.

nation with later observations to determine the rotation period of Mars with high precision.

No one ever did this, and in fact this valuable material seems to have been totally overlooked. Only Schröter's drawings were used by astronomers studying Mars' rotation period, and these were ill-suited for the purpose. The drawings nevertheless remain of value to persons looking for surface changes on Mars, for they were made with instruments respectable even by modern amateur standards. Schröter's first large telescope was a reflector of 7-foot focus, made by William Herschel; and in 1792 he acquired a 9.3-inch, f/17 Schrader reflector which Lalande called the finest telescope in existence.

While most of the features in the drawings reproduced here are identifiable on present-day maps, there is a very striking exception. This is a dark region nearly as large as Texas, marked on 16 of Schröter's sketches of 1798–1800, which is no longer to be seen on Mars. Two drawings made by Herschel in 1783 also show it. Bakhuyzen, who wrote in German, named it *Spitze B;* we might call it the *Arrowhead.*

Both Schröter and Herschel represent this now-vanished feature as long, appearing triangular much like the prominent present-day marking known as Syrtis Major. The base of the Arrowhead was near Martian longitude 225°, latitude −15°, at the edge of Mare Cimmerium; it stretched northward over the so-called canal Cyclops and Pambotis Lacus. These last two features may be the remnants of the Arrowhead. The disappearance of this vast tract – for two decades one of the most easily seen features on Mars – is the most striking change yet recorded on the surface of the red planet. Its explanation remains an intriguing puzzle.

57. Asaph Hall finds the moons of Mars

On a Saturday morning in 1856 or 1857, in the little Ohio town of Hudson, a stranger entered the library of Western Reserve College. The librarian looked up to see a tall young man of about 28 in rough clothes. He was so covered with dust that the custodian of the books took him for a tramp and was about to turn him out. But the visitor explained that he wished to consult Laplace's *Mécanique Céleste* to resolve some difficult problem in the motions of heavenly bodies. He identified himself as Asaph Hall and said that he had just walked the 15 miles from Shalersville, where he was a schoolteacher. The book was brought out, and Hall was allowed to make all the extracts he wanted. Afterwards he returned by foot to Shalersville. This anecdote, preserved by Hall himself, is an indication of the determination by which he rose from youthful poverty and obscurity to become one of the leading American astronomers of his generation.

Hall was born at Goshen, Connecticut, on October 15, 1829, son of an unsuccessful merchant who tried to retrieve his affairs by establishing a clock factory. The father used to load up a wagon with clocks and drive as far south as Georgia, selling them and finally the horse and wagon, and then return home for another trip. The boy

Asaph Hall photographed on July 5, 1871, six years before he discovered the two satellites of Mars, Phobos and Deimos. Both he and his son Asaph Hall, Jr., were leading staff members of the U.S. Naval Observatory in Washington, D.C. Courtesy U.S. Naval Observatory; supplied by Brenda Corbin.

was 13 when his father died on one of these clock-selling journeys, leaving the family in difficult circumstances. Young Hall became apprenticed to a carpenter.

By 1856 he had decided to become an astronomer. Having saved a little money, he married Chloe Angeline Stickney, and the pair proceeded to Ann Arbor, Michigan. There Hall studied astronomy at the University of Michigan under the able Franz Brünnow, from whom he gained a firm grounding in the practical side of the science. However, after one year Hall left, probably because his money ran out. It was at this point that the couple went to Shalersville to teach for a winter. But still intent on becoming an astronomer, he applied for and obtained a post at Harvard Observatory. Many years later Hall recalled: "My wife and I reached Cambridge in the last part of August, 1857. We had a kind reception from Professor W. C. Bond [the director]. Professor G. P. Bond [his son] was absent on a trip to New Hampshire. I was set to work making observations for time, and was shown how to use the transit circle, to read the chronograph sheets, to work out the instrumental constants, and to compare and rate the chronometers. Professor Bond was very kind and pleasant, so that under his guidance I made good progress. I worked hard, and spent most of my time at the observatory."

"After a month or six weeks Professor G. P. Bond returned. He seemed a little surprised to find an assistant in the observatory, and doing so much work. He had a free talk with me, and found out that I had a wife, $25 in cash, and a salary of $3 a week. He told me very frankly that he thought I had better quit astronomy, for he felt sure I would starve. I laughed at this, and told him my wife and I had made up our minds that we were used to sailing close to the wind, and felt sure we would pull through."

Hall stayed at Harvard Observatory for five years. It is good to know that his annual salary was raised to $400 after about a year. Meanwhile he was becoming a skillful calculator of comet and asteroid orbits and shared in varied researches of George Bond, now the director. But eventually a growing family and Civil War inflation forced Hall to take a better-paying job at the U. S. Naval Observatory.

Here Hall finally achieved financial security when in 1863 he was appointed professor of mathematics in the U. S. Navy, then the job title of senior staff members at the Naval Observatory. At that time the observatory had not yet moved to its present location on Massachusetts Avenue in the northwest part of Washington but was located on low land near the Potomac River.

The war came close to Hall several times. During the second battle of Bull Run, the rumble of guns could be heard at the observatory, and again in July, 1864, when the Confederate general Jubal Early was approaching Washington from the north. Hall and other observatory personnel were among the government employees sent to man emergency entrenchments until the Sixth Corps arrived. Another memory from those times was the night when Hall showed President Lincoln and Secretary of War Stanton the Moon and other objects with the 9.6-inch Merz refractor. A few nights later Lincoln returned unattended to ask questions and talk astronomy.

For most of his first decade at the Naval Observatory, Hall's work centered about this instrument. In 1875 he was put in charge of the recently erected 26-inch Clark refractor, and he continued in this responsible position until his retirement in 1891. With this large

The Washington 26-inch refractor was the largest in the world when the satellites of Mars were discovered with it in 1877, shortly after this picture was taken. At the eyepiece is Simon Newcomb; seated is Admiral Benjamin F. Sands, superintendent of the U. S. Naval Observatory. This historic telescope has been fully modernized and today continues to produce double-star measurements of high quality. U. S. Naval Observatory photograph, supplied by Brenda Corbin.

telescope he carried out an extensive series of double star measurements that are perhaps his most important astronomical legacy.

The main task of the 26-inch, however, was to provide positions of the fainter satellites in the solar system. Such observations were the best material for determining accurate masses of the planets, which were needed as a basis for better planetary tables and ephemerides. Since Mars had no known satellites, its mass could only be found rather roughly from its perturbations of the motions of the Earth and Jupiter. Naturally, Hall awaited with interest the close approach of Mars to the Earth in 1877 to see if the great Washington telescope would reveal a satellite. That year Mars' least distance from Earth was only 56.35 million kilometers (35 million miles). Hall later reported: "The search was begun early in August. . . . At first my attention was directed to faint objects at some distance from Mars; but all these proving to be fixed stars. I began to examine the region close to the planet, and within the glare of light surrounding it. This was done by keeping the planet just outside the field of view, and turning the eye-piece so as to pass completely around the planet. While making this examination on the night of August 11, I found a faint object on the following [east] side and a little north of the planet, but had barely time to secure an observation of its position when fog from the Potomac River stopped the work."

The "faint object" was in fact the outer satellite, Deimos. Poor weather prevented reobservation of it until the 16th when Hall again saw it on the following side of the planet and found that it was moving with Mars. On the 17th, while waiting and watching for the outer satellite, Hall discovered the inner moon, Phobos. He noted in the record book that both "Mars stars," though faint, were distinctly seen by the night assistant George Anderson and himself. On August 18th, he was joined in the dome by other astronomers – Simon Newcomb, D. P. Todd, and W. Harkness – and all made measurements of what were now recorded as the "Mars satellites." The official press announcement was made later that day, and Hall quickly became an international celebrity. He was always very modest about his accomplishment.

Some aspects of the double discovery remained generally unknown until about 1970, when Owen Gingerich examined Hall's unpublished correspondence. The morning after Hall discovered Phobos he disclosed his observations to Newcomb, the scientific

Mars and its satellites photographed on April 27, 1967, by D. Pascu with the U. S. Naval Observatory's 61-inch reflector at Flagstaff, Arizona. Phobos is at about 8 o'clock, fainter Deimos at 9 o'clock. A special filter was used to reduce the light from Mars' disk (seen inside the dark circle). South is up, east to the right. Official U. S. Navy photograph.

head of the observatory. In Gingerich's words: "Newcomb erroneously believed that Hall, in his modest conservatism, was reluctant to recognize the 'Mars stars' as satellites, and hence took for himself an undeserved credit for this recognition in the wide press coverage that followed. For many years Hall quietly harbored a grudge against Newcomb, who eventually offered his apologies."

The other unknown part of the story was Hall's fear that Newcomb's aggressive young protégé Edward S. Holden would take over the search. (Holden later became the first director of Lick Observatory.) In a letter of February 14, 1888, to E. C. Pickering of Harvard, Hall wrote: "In the case of the Mars satellites there was a

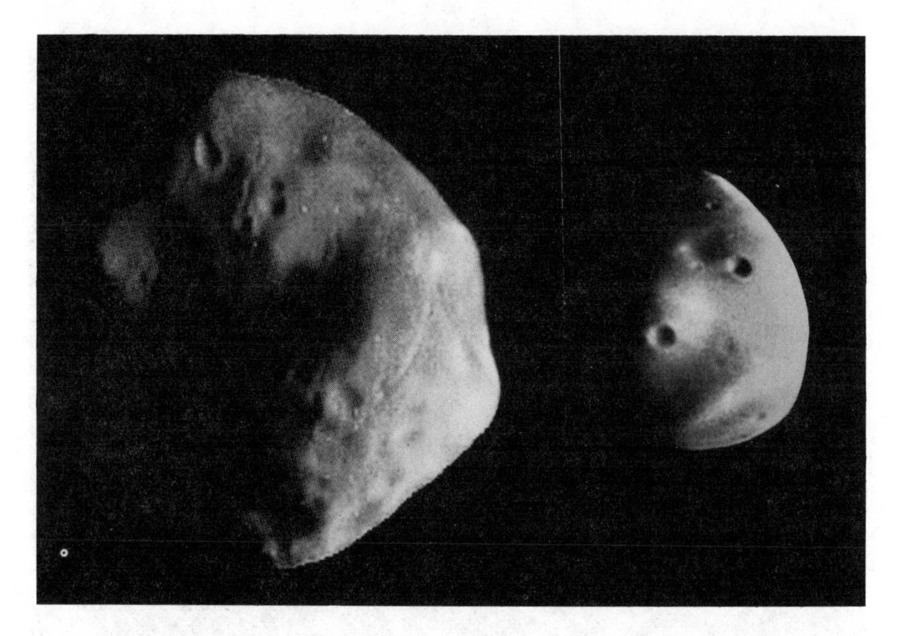

The two Martian moons Phobos (left) and Deimos are reproduced at about their correct relative sizes in these closeups taken by the Viking I orbiter at Mars in 1977. Rugged Phobos is shown from the south, with the crater Hall (about four miles across) near the terminator. Deimos, seen south up, exhibits dark craters up to ¾ mile in diameter. These bodies reflect only a few percent of the sunlight striking them and are much darker than many asteroids in the inner asteroid belt beyond Mars. NASA photographs.

practical difficulty of which I could not speak in an official Report. It was to get rid of my assistant [Holden]. It was natural that I should wish to be alone; and by the greatest good luck Dr. Henry Draper invited him to Dobb's Ferry at the very nick of time. He could not have gone much farther than Baltimore when I had the first satellite nearly in hand." As it was, Holden reported that he and Draper, using the latter's 28-inch reflector, had found a third and fourth satellite! Of course, neither was ever confirmed.

Hall reached the Naval Observatory's mandatory retirement age in 1891, and a few years later he returned to Connecticut where he died, full of honors, on November 22, 1907. He was unquestionably one of the greatest of the old-time visual observers, skilled in the manipulation of the filar micrometer, a solver of knotty problems in spherical astronomy, and an able calculator of satellite orbits. Hall was sufficiently modern to recognize the potential value of celestial photography for solar eclipses and transits of Venus, but he was conservative enough not to use it in his own astrometric work. How he would have stared in disbelief to hear that Phobos and Deimos, within a century of their discovery, would have new orbits determined from television photographs of them taken at close range by the Mariner 9 spacecraft!

58. Naming some early minor planets

OVERPRODUCTION CAN BE as troublesome in astronomy as in industry. A prime example is provided by the minor planets, or asteroids, which circle the Sun mostly between the orbits of Mars and Jupiter. Over 3,000 of these pocket-size planets have been given permanent identification numbers, and the addition of another usually makes little stir in the astronomical world.

At the time when only a few of these objects were known, however, the finder of a new one could count upon fame and honors. For example, the German amateur astronomer K. L. Hencke, who discovered 5 Astraea in 1845 and 6 Hebe two years later, was rewarded by the King of Prussia with a pension equivalent to $300 per annum, an adequate living then.

But as the count of known asteroids climbed into the hundreds, new ones became an embarrassment. Measuring their positions and, especially, the tedious orbital computations needed to keep these minute bodies from being lost became an ever heavier burden to astronomers. The asteroids became known as the vermin of the skies. The strain reached the breaking point in 1891, when Max Wolf introduced mass-production photographic search methods. Long-exposure photographs with fast, large-aperture telescopes caused such a rush of new discoveries that most of the bodies were

With this photographic equatorial telescope, Maximilian Wolf (1863–1932) revolutionized minor-planet search techniques. From *Atlas der Himmelskünde,* 1898, by A. v. Schweiger-Lerchenfeld.

promptly lost. Observers measuring precise asteroid positions and astronomers calculating preliminary orbits with pencil and logarithm tables could not keep pace. Only since the 1960's has the use of computers ended this bottleneck.

After a newly discovered asteroid has its orbit reliably calculated and has been observed again at another opposition, a permanent serial number is assigned by the International Astronomical Union. The discoverer then may exercise the traditional right to name the body. All of the principal planets known from antiquity carry the names of Roman gods and goddesses, and thus the first asteroids to be discovered were named from Greek and Roman mythology. The astronomer of a century ago usually had worked through a solid classical education and was primed with names of nymphs and other minor deities from Homer, Vergil, and Ovid.

Giuseppe Piazzi (1746–1826). Courtesy Mary Lea Shane Archives of Lick Observatory.

The first minor planet was discovered on January 1, 1801, by Giuseppi Piazzi, director of the observatory at Palermo, Sicily; it bears the name of Ceres, who was not only the Roman goddess of harvests but also the special protectress of the island of Sicily. Piazzi first proposed the name Ceres Ferdinandea, to honor his royal patron King Ferdinand III of Naples, a Bourbon whose avocation was to dress in rags and sell fish in the public market of his capital city. But other astronomers refused to accept the non-mythological addition to the name.

Many curious cases of this kind are collected in a 1955 mimeographed pamphlet, *The Names of the Minor Planets,* prepared by Paul Herget, who was director of the Cincinnati Observatory. It contains explanations of the names of the first 342 asteroids, mostly from the historical gleanings by R. C. Cameron, formerly of Cincinnati Observatory, and A. Paluzie-Borrell, Barcelona, Spain. This small publication contains interesting footnotes to astronomical history.

Many of the asteroid names reflect half-forgotten controversies. In 1850 a London astronomer, J. R. Hind, proposed the designation Victoria for the 12th minor planet, picked up by him that September 13th. This compliment to the Queen of England was strongly opposed by the American astronomers of the day, who were not averse to twisting the British lion's tail. The debate quieted when it was established that there actually had been a minor Roman divinity of the same name.

Asteroid 55 was discovered in 1858 at the Dudley Observatory, Albany, New York, during a heated dispute between its trustees and the director, B. A. Gould (see Chapter 13). This complex quarrel involved bitter newspaper charges of fraud, embezzlement, and incompetence; there were tortuous legal proceedings, "bands of hired ruffians," and even a robbery of the mails. No wonder astronomers agreed on the aptness of naming this asteroid Pandora, after the fabled owner of the box from which came all the evils that have afflicted the human race.

There is an unusual story told in connection with 139 Juewa. The first to see it was J. C. Watson, on October 10, 1874, at Peking, China. This American astronomer had gone to China to observe that year's transit of Venus, and there, as he continued his asteroid hunting, he found a slowly moving 10th-magnitude object in Pisces. "This being the first planet discovered in China," wrote Watson, "I requested Prince Kung, regent of the Empire, to give it a suitable name. In due time, a mandarin of high rank brought to me the document containing the name by which the planet should be known, coupled with a request – communicated verbally – that I would not publish the name in China until the astronomical board had communicated to the Emperor an account of the discovery and the name which had been given to the planet. This request was of course promptly acceded to; and I afterwards learned upon inquiry that if the knowledge had come to the Emperor other than through the astronomical board, organized specially for his guidance in celestial matters, some of these ministers would have been disgraced. The name determined upon by Prince Kung . . . Juewa . . . means literally the Star of China's Fortune."

The most successful discoverer of minor planets by the slow and difficult older method of visual search was the Austrian astronomer, Johann Palisa, who found 125 asteroids between the years 1874 and

Johann Palisa (1848–1925) photographed in Vienna in 1888. Courtesy Mary Lea Shane Archives of Lick Observatory.

1924. His first searches were at the small observatory attached to the Austrian naval academy at Pola, near the head of the Adriatic Sea and now in Yugoslavia. The fourth of his planets discovered there, 142 Polana, honors the site. But Palisa, who held the naval rank of commander, did not keep his professorship at this Austrian Annapolis for very long. The story is told that once, at an admiral's inspection, he appeared wearing a straw hat instead of the regulation headgear, and then strolled off peacefully to his observatory without paying his respects to the admiral. As punishment, Palisa was removed from Pola to the Vienna Observatory, where he worked happily for the rest of his long life and became vice-director (see Chapter 59).

There was to be a total eclipse of the sun visible from Africa on August 29, 1886, and Palisa needed money for his expedition. He therefore announced that he would sell for the equivalent of $250 the right of naming his latest minor planet discovery. The tab was picked up by Baron Albert von Rothschild, an Austrian member of the famous banking family, who named asteroid 250 Bettina after his wife.

Many other asteroid anecdotes like these can be found in Herget's collection. In 1968 he published a sequel extending to planet 1564. Today, explanations of asteroid names are routinely published by the International Astronomical Union through its Minor Planet Center, located at the Smithsonian Astrophysical Observatory in Cambridge, Massachusetts.

59. Tales of two lost planets

SINCE 1950 astronomers have followed the general rule that a newly discovered asteroid receives its permanent number designation only after it has been observed at two or more oppositions, and only after a satisfactory orbit has been computed. The result of this rule has been that asteroids with numbers higher than, say, 1400 run little or no risk of becoming lost, since good ephemerides can be calculated for decades ahead.

The situation was quite different a century ago, when less than 200 minor planets were known, and all observations of them were visual. In those days, a serial number was assigned immediately after discovery, even before any orbit could be computed, in the assurance that astronomers would follow the new object. Mishaps did occur, as in the case of 132 Aethra, which was not observed between its discovery year, 1873, and 1922. The longest stay in limbo for any minor planet ended in August, 1970, with the announcement that 155 Scylla had been recovered after having been missing for almost 95 years. The story involves an Austrian naval officer, a Massachusetts clergyman, a German schoolteacher, and a group of mathematical astronomers at the University of Cincinnati.

The Adriatic port of Pola, now in Yugoslavia, belonged in 1875 to Austria and was the main base of the Austro-Hungarian navy. On the summit of Monte Zaro, a low hill overlooking the harbor, stood the long one-story building that served as hydrographic institute, chart depot, and naval observatory; the latter was directed by Johann Palisa (see Chapter 58). For his asteroid searches Palisa used a 6-inch Steinheil refractor in the observatory's southwest dome. His method was a laborious comparison of the sky with Jean Chacornac's Paris star charts, on which he inserted any additional stars he saw. Sometimes Palisa made his own charts at the telescope. To plot all the stars that the 6-inch telescope showed in a zone five degrees long and 20 minutes of arc wide took about two hours, he tells us.

On the morning of November 9, 1875, Palisa was checking out a star field near Delta Arietis when at about 3 a.m. he came upon a 12th-magnitude star not on his chart. Two positions determined with a ring micrometer showed that in 40 minutes the intruder had moved eastward by about the amount to be expected of an asteroid in that part of the sky. Within hours Palisa sent a telegram to the Royal Observatory at Berlin. There Victor Knorre, using a 9-inch refractor, was able on the evening of the 9th to locate the planet and measure another position. Berlin Observatory forwarded the news to the editor of the *Astronomische Nachrichten* in Kiel, who announced Palisa's discovery of minor planet 155 in the November 16th issue of that journal. Full Moon in November, 1875, came on the 12th and hindered further looking for this relatively faint asteroid. Palisa managed to reobserve it on the morning and evening of November 23rd, but he failed to locate it again.

The next spring Palisa published the details of his observations in the *Astronomische Nachrichten*. He commented unhappily: "Planet 155 should be stricken from the list if no observations have been obtained at other observatories. According to Dr. Tietjen, the Pola and Berlin positions on November 9th and the Pola positions on the 23rd cannot be represented by an elliptical orbit. Thus it is possible that a different object was observed on the 23rd. This suspicion is reinforced by the fact that I could not find the planet during searches on December 5th to 9th." (In this quotation, astronomical dates have been converted to civil dates for uniformity with the rest of the chapter.)

However, Friedrich Tietjen was wrong, because the Berlin observation had an error in declination of approximately a minute of arc. The five observations did belong to the same asteroid, and L. Schulhof calculated an orbit for it. But the orbital elements were very uncertain because of the poor time distribution of the observations, bunched in two tight groups less than 15 days apart. Reliable predictions of its position could not be made, and 155 Scylla was lost.

The next man to observe Scylla did not recognize it. He was an American amateur astronomer named Joel Hastings Metcalf (1866–1925), who gained fame by making large refracting telescopes and by discovering 41 minor planets, a number of variable stars, and six comets (three of them within two days in 1921). At Taunton, Massachusetts, where he was the Unitarian minister, Metcalf regularly used a 12-inch refractor of his own construction for the photographic observation of minor planets. On October 10, 1907, he photographed a field in Cetus on which several planets were found, among them a 12.5-magnitude object provisionally designated 1907 AP. Metcalf photographed this planet again on October 31st and November 5th, but he did not get any further observations of it.

As it happened, two more positions of 1907 AP had been obtained at Heidelberg Observatory from photographs by A. Kopff taken on November 2nd and 8th. His find was reported to the Imperial Observatory at Vienna, where Palisa himself observed 1907 AP visually with the 27-inch refractor on the 5th and 7th of the same month.

But the opportunity was lost. Nobody computed an orbit. The possibility that these observations referred to Scylla does not seem to have occurred to anyone then. In fact, 1907 AP was for a time mistakenly supposed to be identical with another minor planet, 912 Maritima, that was found in 1919 by A. Schwassmann at Bergedorf Observatory.

By the 1960's it seemed highly probable that Scylla had been photographed from time to time without being recognized. After all, Palisa's visual magnitude estimate of 12 would correspond on the modern photographic scale to about 14, which is bright for present-day planet hunters. In 1966 C. M. Bardwell at Cincinnati Observatory's Minor Planet Center began an investigation of the cases of 155 and a number of other lost asteroids. As a first step, he reexamined the observations made in 1875, and his colleague Jay Carr discovered and corrected the error in the Berlin position. New elements resembling those found by Schulhoff were computed, but the orbital period and eccentricity (shape) were still quite uncertain. Nevertheless, the orientation of the orbit plane in space was fairly well determined.

Bardwell therefore proceeded to look through the orbital elements of numbered and unnumbered planets to see if any corresponded to an orbital orientation like that of Scylla. The unnumbered planet 1939 TK met this test. Also, a rough check of its period indicated that it would have had an opposition in the fall of 1875. Unfortunately, there were only three accurate positions of 1939 TK, made at Simeis Observatory in Russia. Its orbital elements thus were unreliable. Bardwell had to wait until some others of the thousands of unnumbered planets could be identified with 1939 TK.

For many years, one highly successful searcher for asteroid identities has been a German Catholic priest named Otto Kippes, who teaches in an elementary parochial school in Reckendorf, Bavaria. The process he uses is essentially as follows. When a minor planet is near opposition, it is retrograding (moving westward) among the stars on a path that is nearly a great circle. Hence, if two or more observations of an unidentified planet are made at that time, one can estimate what the planet's position would have been on the date of opposition. (This extrapolation can also be done from a single observation but with less certainty.) To test whether this planet is identical with a particular numbered one, the latter's opposition position is calculated from its elements to see if the two positions match. As a check, the direction of the path among the stars should be the same in both cases. Although some observatories use computers to hunt for asteroid identities, Kippes calculates with a pencil on slips of paper that he stores in cigar boxes.

The Cincinnati Observatory sent to Kippes ephemerides of 1939 TK and other unnumbered planets. From these Kippes ascertained that 1939 TK was identical with 1930 UN, 1941 HL, 1950 FL, and 1950 FN. He also found that 1939 TK was the same object as Metcalf's 1907 AP.

Now all of the pieces of the puzzle fell together. Using all the newly uncovered positions of 1939 TK, Bardwell obtained an accurate orbit for it. By calculating back to 1875 he found that the old Pola and Berlin observations were nicely represented. This proved that 1939 TK and Scylla were the same object. In the process, he noted an additional identity, with 1934 RU.

Thus, the long-errant Scylla is now on a tight leash as it travels

Vienna Observatory as it appeared around the turn of the 20th century; the large dome contained the 27-inch refractor. From A. v. Schweiger-Lerchenfeld's *Atlas der Himmelskünde,* 1898.

around the Sun once every 1,674 days in an elongated orbit having an eccentricity of 0.274 and an inclination to the ecliptic of 11°.

"A series of coincidences caused 155 Scylla to remain 'lost' for so long," Bardwell summarizes. "Each time the planet was reobserved the observations were practically ignored. No preliminary orbit was computed for 1907 AP or for 1939 TK until more than 25 years later. The photographs of 1941 HL were not measured accurately until recently, and the observations of 1930 UN were confused because of a mistaken identification of a third position. Also, before the advent of modern computers, no one would have undertaken the extensive calculations needed to recover Scylla."

Thirty-six years after his discovery of Scylla, Palisa found another asteroid that subsequently became lost. This story has a less happy ending.

The year was 1911, and for nearly three decades Palisa had been working at the Imperial Observatory in Vienna. This monument to the grandeur of the Austro-Hungarian monarchy was said to be the largest single observatory building in the world. Over 320 feet long, the brick structure was in the form of a cross, with instruments in the east-west wing and offices in the north-south one. To some viewers the instrument wing gave the impression of a fortress – from its few and small windows, the narrow slits for the meridian instruments which seemed intended for marksmen, and its terraces that looked like bastions. The chief instrument of the Imperial Observatory was the 68-cm (27-inch) refractor made by Howard Grubb, which for a few years after its erection about 1880 was the largest telescope of its type in the world. Its dome rose above the intersection of the two wings.

By 1911 Palisa had begun to work in cooperation with Max Wolf of Heidelberg, who was hunting asteroids photographically with a 16-inch refractor. The prints that Wolf supplied to Palisa made excellent star charts for use at the eyepiece of the 27-inch and greatly facilitated the recognition of faint minor planets. In October of that year, Palisa discovered no fewer than five new minor planets. By far the most interesting of these was a rapidly moving object provisionally called 1911 MT and later officially known as 719 Albert.

Palisa came upon it on the night of October 3rd, while he was using the 27-inch to compare a star field in northern Cetus with a Wolf photograph. His attention fell on a pair of 12½-magnitude stars, one or the other of which was evidently an asteroid, and he began to make micrometer measures of their relative location. Almost immediately clouds came over, and Palisa waited half an hour for the sky to clear. To his great surprise, the more southern of the pair had shifted southeastward in the star field. Thus the new object, though near opposition, was in direct motion, not retrograde as would be expected of an ordinary asteroid. Less than 30 minutes later the sky clouded up for the rest of the night.

The unusually rapid apparent motion of the new asteroid, 0°.75 per day, indicated that it was passing relatively near the Earth. News of the discovery was promptly telegraphed to the astronomical information center of Kiel, Germany, for distribution to other obser-

vatories. On the next evening in bright moonlight, Palisa managed to observe 1911 MT again. Three hours later, as the gibbous Moon was setting, C. H. Pechüle secured two measurements of the asteroid's position with the 36-cm (14-inch) refractor at Copenhagen Observatory.

That was the last time a human eye viewed 1911 MT directly. Only four positions had been measured, and standard methods of orbit calculation were ill-suited for dealing with such data. After several photographic searches had failed to find 1911 MT, the English astronomer A. C. D. Crommelin glumly mused: "It is to be feared that it will remain one of the unsolved enigmas of astronomy."

Our story now shifts to the Berkeley campus of the University of California, where Armin O. Leuschner (1868–1953) was becoming equally famous for his researches in celestial mechanics and for being a superb teacher. He had developed a powerful new method for the calculation of preliminary orbits of comets and asteroids that worked well even for the recalcitrant case of a body near the Earth. This method was successfully applied to 1911 MT by two of Leuschner's students, Eli S. Haynes and John H. Pitman. In April, 1912, they announced a preliminary orbit for the lost asteroid and provided a finding ephemeris for the previous summer and fall.

This work led to the discovery of images of 1911 MT on three plates taken October 11th at the Royal Greenwich Observatory. Also, other alleged images of the minor planet were found subsequently at Heidelberg, Union, and Lick observatories. Unfortunately, positions from the latter institutions could not be combined with the definite observations of 1911 MT and reconciled by a single orbit. It became clear to Haynes that some of the observations were mistakes, but he never did manage to separate the sheep from the goats.

When Albert came to opposition early in 1913, its predicted magnitude was as faint as 20. Nevertheless, Heber D. Curtis hunted for it with the 36-inch Crossley reflector of Lick Observatory. Three

---

Vienna Observatory's 27-inch refractor. With it Johann Palisa discovered asteroid 719 Albert, which was subsequently lost. From *Publications* of the Astronomical Society of the Pacific, *3*, 1891.

very small asteroids were photographed in the search area, but enough observations of each were obtained to show that it could not be the much-sought 1911 MT. Either the latter was too faint for the Crossley, or a usable search ephemeris was no longer possible.

Although lost, 1911 MT did receive the permanent number 719. Palisa, who had named it Albert to honor Baron Albert von Rothschild, a generous benefactor of the Vienna Observatory, must have felt a slight embarrassment at the miscarriage of his compliment.

60. Perrine discovers Jupiter's sixth and seventh satellites

IN ALL, THE PLANETS have more than 40 known natural satellites, of which the first 21 were detected visually – from our Moon in prehistoric times to Jupiter V in 1892. But, beginning in 1899, all subsequent satellite finds have been photographic, with the cameras being on either Earth-based telescopes or space probes.

The dividing point was the discovery of Saturn's ninth moon, Phoebe, by William H. Pickering at Harvard Observatory. He recognized this faint outer satellite in April, 1899, while examining plates that had been taken the previous August with the 24-inch Bruce refractor at Arequipa, Peru.

This accomplishment attracted particular attention at Lick Observatory, which in 1895 had acquired a 36-inch reflector from the English amateur Edward Crossley. Although this was the largest mirror telescope in the United States, it was virtually unworkable at first because of its poorly designed mounting, a "plumber's nightmare." After modification by James Keeler during his brief Lick directorship (1898–1900), the Crossley telescope took many superb photographs of nebulae and clusters, but it was still a difficult instrument to use. Hence a new mounting was installed for the Crossley in 1904. Its design was largely due to Charles Dillon Perrine, who was in charge of the instrument after Keeler's death. Part of the program planned for the remounted telescope was a photographic

Edward Crossley, industrialist, amateur astronomer, and Member of Parliament, is shown wearing the robes and chain of Mayor of Halifax County Borough, an office he held three times. Courtesy Calderdale Central Library, Halifax, England.

search for new satellites of the outer planets. Perrine was to play an important part in this project.

Perrine's name is less well known in America today than might be expected from his achievements. Born in 1867 at Steubenville, Ohio, he moved to California at an early age and became an enthusiastic amateur. In 1889 he took part in an expedition to observe the total solar eclipse of January 1st and also joined the newly founded Astronomical Society of the Pacific. Leaving behind a business career in San Francisco, Perrine in March, 1893, became secretary of Lick Observatory and moved to Mount Hamilton. Eager to use the instruments, he volunteered to take daily photographs of sunspots with the photoheliograph, a horizontal 5-inch refractor of 40-foot focal length.

But it was as a highly successful comet discoverer that he first made his mark. A list of his comets shows a rapid succession of finds: 1895 IV, 1896 I, 1896 VII (period 6.4 years), 1897 I, 1897 III,

C. D. Perrine was an enthusiastic observer who joined the Lick Observatory staff in 1893 and eventually became director of Cordoba Observatory in Argentina. Courtesy Mary Lea Shane Archives of Lick Observatory.

1898 I, 1898 VI, 1898 IX, and 1902 III. In addition, during the years 1897–1899 he recovered periodic comets d'Arrest, Pons-Winnecke, Holmes, and Tempel 2. Not only a discoverer, Perrine also made many positional observations and orbit calculations of comets.

His scientific stature grew with the award in 1897 of the Lalande prize of the Paris Academy of Sciences and with Lick staff promotions to assistant astronomer, acting astronomer, and astronomer. Perrine led the Lick eclipse expedition to Sumatra in 1901, and in the same year on Crossley photographs he discovered the apparent motion of the nebulosities surrounding Nova Persei. J. C. Kapteyn in the Netherlands explained this effect as due to the expanding shell of light from the nova illuminating successively more distant interstellar dust clouds.

In December, 1904, Perrine began his search with the Crossley telescope for outer satellites of Jupiter and scored immediately. He wrote: "The first photograph of the region about *Jupiter* was ob-

The 36-inch Crossley reflector, with which Perrine discovered Jupiter's satellites VI and VII. This success came just after the instrument had been put back into service following his renovation in 1902–1904 to improve its performance. Courtesy Mary Lea Shane Archives of Lick Observatory.

tained on December 3d and others on the 8th, 9th, and 10th. A comparison of these negatives showed an object of the fourteenth magnitude which was moving with an apparent velocity among the stars and not very different from that of *Jupiter*. *Jupiter* was retrograding slowly at that time. The suspected object was to the westward and moving a little faster than the planet." At first Perrine was uncertain whether the new body was in fact a sixth satellite of Jupiter or an asteroid, but further observations on January 2nd, 3rd, and 4th verified the former interpretation. Official announcement of the discovery was made by Lick Observatory on January 5th.

The first visual sighting of Jupiter VI was made at the U. S. Naval Observatory on the night of January 8th by John C. Hammond, who with the 26-inch refractor found a faint object in the expected position. It was proven to be the satellite by its motion with respect to a neighboring star. At Lick Observatory on January 28th, after Saturday night visitors finished viewing through the 36-inch refractor, Robert G. Aitken turned the instrument toward Jupiter. Although the planet was low in the west, he found VI and estimated its magnitude as 14.

Photographic observations of VI were made with the Crossley on 36 nights up to March 22, 1905, after which Jupiter became too close to the Sun in the sky. "A preliminary investigation of the orbit," wrote Perrine on March 30th, "shows the inclination to the ecliptic and to the planet's equator to be about 30°. It has a period of about two hundred and fifty days, its mean distance being about seven million miles."

An unexpected bonus was the discovery of another, much fainter, Jovian satellite while Perrine was examining his plates of January 2nd, 3rd, and 4th. Its nature was confirmed in February, and up to March 9th Jupiter VII had been photographed a total of 20 nights. As faint as magnitude 16, it was estimated as 35 miles in diameter by Perrine. Relative brightnesses of VI and VII are well shown in the accompanying photographs.

As if finding two new satellites wasn't enough, on July 4, 1905, Perrine married Belle Smith. They sailed from New York as members of the Lick Observatory eclipse expedition to Spain, for nearly four minutes of totality on August 30th.

Perrine did not discover any more satellites. In 1909 he left Lick

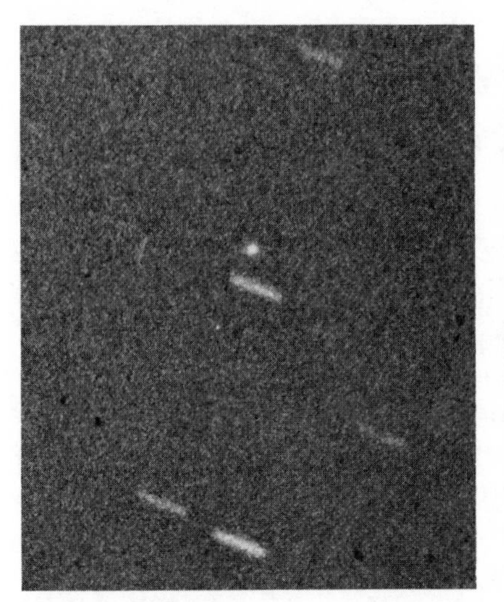
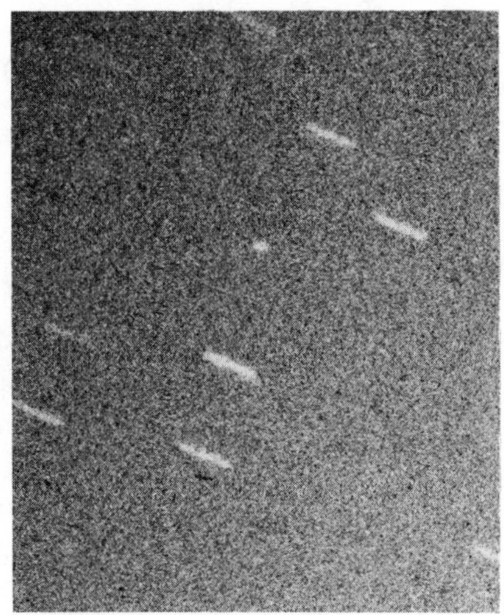

Jupiter's satellites VI and VII, which Perrine discovered in 1904–1905, appear as dots; background stars are streaks. These images are from a photograph obtained by Charles T. Kowal with the Palomar 48-inch Schmidt telescope on October 1, 1975. Courtesy Palomar Mountain Observatory.

to accept the directorship of Cordoba Observatory in Argentina. There he constructed a 30-inch reflector with which he made extensive observations of southern galaxies. But his plan to erect a 60-inch telescope lagged, for though the Argentine government appropriated finds for it in 1913, the instrument did not become operational until 1942, after Perrine's retirement. The delay in erecting the 60-inch telescope wasn't Perrine's only frustration at Cordoba. The article on this astronomer in the *Dictionary of Scientific Biography* refers tersely to growing difficulties: "Despite his scientific achievements, Perrine and his office became a target for nationalist politicians and he was attacked verbally by deputies in the Argentine Congress. In 1931 he was barely missed by a sniper's bullet and in 1933 the Argentine Congress passed legislation removing authority from the director of the observatory. He retired, under duress, in 1936." Perrine never returned to the United States but died on June 21, 1951, at his home at Villa General Mitre, a small town about 80 kilometers from Cordoba. He was 83.

61. 300 years of Jovian satellite eclipses

EVER SINCE GALILEO, it has been remarked that the planet Jupiter and its four bright satellites form a kind of miniature Sun and planets moving with a speeded-up time scale. Concerning this analogy, the great Dutch astronomer Willem de Sitter, who devoted much of his life to studying the motions of these satellites, wrote in 1931: "The interval of 321 years since the discovery of the satellites thus is, in the number of revolutions, equivalent to nearly 18,000 years of the four inner planets, and to more than 1,100,000 of the outer planets. During all this time the general aspect of the system has not changed, and especially the stability appears to be unimpaired. This enormous magnification of the time scale makes the system of the satellites of special interest for the study of secular and long periodic perturbations."

To trace these very slow changes in the motions of the Galilean satellites, we fortunately have more or less usable observations extending back to the 1660's. These data consist of the recorded times when a satellite disappeared on entering Jupiter's shadow or reappeared on exiting from it. These eclipses are spectacular phenomena to watch in a small telescope, especially ones of I, the innermost Galilean satellite, also called Io.

Imagine that you are using a small telescope to watch the beginning of an eclipse of Io. To the west of the elliptical banded disk of Jupiter, this satellite shines brightly, like a star of about magnitude 5½. Close scrutiny reveals that Io is gradually fading as it enters the planet's shadow. Dimming more and more quickly, it dwindles to a faint speck of light that vanishes from sight. So sudden is the final disappearance that it can be timed to within a few seconds. This sequence of events occurs rapidly because Io takes only 4½ minutes to travel its own diameter. Eclipses of satellite II, Europa, proceed nearly as briskly. Disappearances and reappearances of III, Ganymede, and especially IV, Callisto, are more leisurely, and timings of the first or last speck of light are correspondingly more uncertain.

During the 17th and 18th centuries astronomers observed satellite

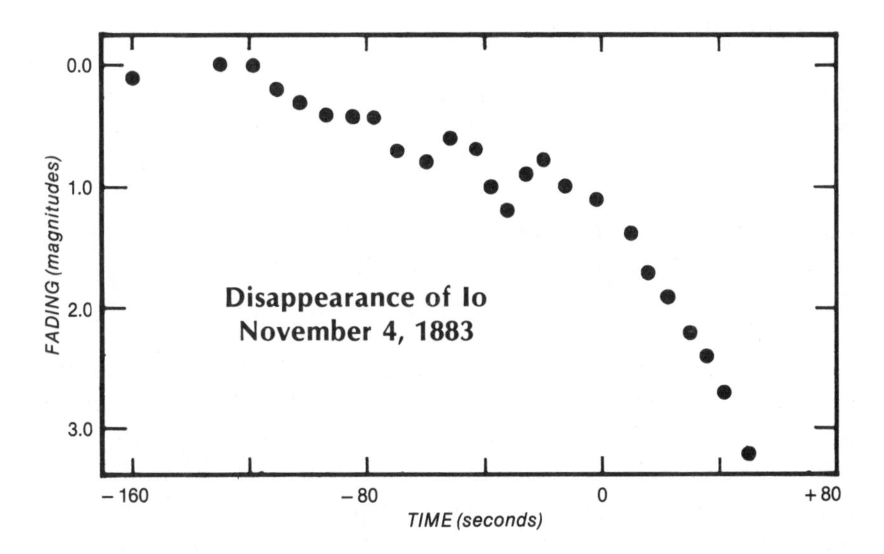

E. C. Pickering made these measurements of Io's fading as it entered into eclipse on November 4, 1883, to find when the satellite's brightness was halved. Times are counted from the predicted time of disappearance as given by the *American Ephemeris*. This is a plot of data from Harvard *Annals, 52,* 49, 1907.

eclipses by the thousand, because these phenomena provided a simple means of finding geographical longitudes. Suppose, for example, that the same reappearance of II was seen from both Paris Observatory and Mexico City. Each observer would record the local time of the event, and the difference between the two times would be equal to the longitude difference between Paris and Mexico City – provided that the two observers used telescopes of similar light-gathering power and both had equally clear skies. This method of longitude determination enjoyed great popularity for many years since it could readily be applied in the field by travelers or at temporary stations. Thus, in the 1780's Danish astronomers organized an extensive series of corresponding observations of Jovian satellite eclipses at Copenhagen Observatory and at Lambhus in Iceland, Godthaab in Greenland, and Tranquebar in India. At that time all three places were under the flag of Denmark.

But this popularity ended early in the 19th century as new and better methods of longitude determination came into general use.

These methods involved solar-eclipse contact times, occultations of stars by the Moon, transportation of chronometers, and especially the electric telegraph. By 1840 the rate of publication of satellite eclipse timings had dropped dramatically. Nevertheless, the Galilean satellites remained of great interest to theoretical astronomers because of the intricate and beautiful problems posed by their gravitational influences on each other (perturbations). These problems led to a revival of observers' interest in the 1870's and 1880's, when quite a few small institutions took up timing programs.

The director of Harvard Observatory, Edward C. Pickering, wrote in 1907 how he started a revolution: "The advantage of photometric methods over those previously employed in observing the eclipses of Jupiter's satellites suggested itself to me on June 9, 1878, while watching the light of the fourth satellite slowly diminishing, previous to its disappearance. The visual observation of this eclipse had been recommended to me by Professor David P. Todd, as it was the last eclipse of that satellite to be visible for some years in the longitude of Cambridge. The instrument employed was the East Equatorial [15-inch refractor], with its aperture reduced to 4 inches . . .

"When once suggested, the superiority of the photometric method to that of simple observation is obvious. Successive results can be rapidly obtained during the entire period occupied by the variation of light of the satellite which is undergoing eclipse, and if these results depend upon the comparison of its brightness with that of another satellite, or with that of Jupiter, all sources of uncertainty due to the telescope, the observer, the condition of the terrestrial atmosphere, and the altitude of the planet, are nearly, if not completely, removed. It was therefore decided to begin such photometric observations at once, and the first eclipse thus observed occurred on June 23, 1878."

In this way Pickering and his assistants began the great Harvard photometric series of 731 eclipses during the next 25 years, covering two complete orbital revolutions of Jupiter. These measurements were made with polarizing double-image photometers of several types, which enabled the observer to place an image of the satellite undergoing eclipse next to the image of another satellite and then make repeated determinations of their magnitude difference. An observer working alone could seldom make more than three observations a minute. But with a recorder to read the photometer settings and the chronometer, seven or eight observations a minute were possible, or even 13 when two recorders were used.

By fitting a template to a plot of the observations made at an eclipse, the time when the satellite was at half light could be ascertained. Pickering hoped for high accuracy by this method and entrusted the entire Harvard material to the English astronomer R. A. Sampson for working up. This Sampson did with great thoroughness in a series of memoirs that derived new orbital constants for the satellites (1909), provided elaborate tables for predicting satellite phenomena (1910), and yielded a comprehensive mathematical theory of the satellite motions (1921).

But the great increase in accuracy anticipated by Pickering was only partly won. Sampson found that in many cases the eclipses came too early or too late, as if the cloudy surface of Jupiter were in places 100 miles higher or lower than normal, in an unpredictable fashion. The result, in de Sitter's words, is that "observations of eclipses, however carefully made, cannot determine the time with a greater accuracy than ± 10 seconds, and this limit cannot be lowered by combining a great number of observations."

Other methods of observing the orbital motions of the Galilean satellites began to be tried. In South Africa, at the Royal Cape Observatory, David Gill in 1891 started a long and very accurate series of heliometer measurements of the relative positions of the four satellites on every possible night. At the same time, photographic observations of the satellites were undertaken at Pulkovo and Helsingfors observatories in Russia. But when R. T. A. Innes turned his attention to Jupiter's satellites, visual observations of their phenomena gained a fresh and lasting importance. Just what he did should be told in the context of his unusual astronomical career.

Innes was a self-educated Scotsman who emigrated to Sydney, Australia, in 1884, where he not only became a successful wine merchant but also published papers on the perturbations of the Earth's orbit by Mars and Venus. In 1894 he borrowed a 6¼-inch refractor from a Sydney amateur and in the course of 30 hours searching discovered two dozen new double stars. His growing interest in practical astronomy led to an invitation from Gill to come to South Africa and take a job as secretary of the Royal Cape Observatory, at a very modest salary. Innes made the most of this oppor-

The versatile Robert Thorburn Ayton Innes (1861–1933) became the leading South African astronomer of his time.

tunity. In addition to filling his duties as secretary, he found time to make numerous measurements of double stars with the Cape 7-inch refractor, to publish a reference catalogue of southern doubles, and to do several other pieces of useful work. For the rest of his life, double stars remained his primary interest (his new pairs totaled 1,628).

Such keenness was rewarded in 1903 by his being appointed director of the newly founded Transvaal Observatory in Johannesburg (later called the Union Observatory and still later the Republic Observatory). At that time this institution was largely concerned with meteorology, but in 1907 Innes acquired a 9-inch refractor for it. Two years later he persuaded the government to order a 26½-inch refractor, but it was not delivered until 1925. One of the first things he did with the 9-inch was to begin a special program of observations of Jupiter's satellites.

His program included not only eclipses of the satellites by Jupiter's shadow, but also their occultations by the planet's disk, their transits in front of the disk, and their shadow transits. It was hoped that this multiplication of observations would give a more accurate determination of their orbits than eclipses alone could. Furthermore, Innes introduced a new technique for making the observations. By using a power of 700 on the 9-inch he could see the tiny satellite disks distinctly. Thus, in observing, say, an eclipse disappearance of Io, he would time not only the moment of the last speck of light but also the moments when Io's disk was one-quarter, one-half, and

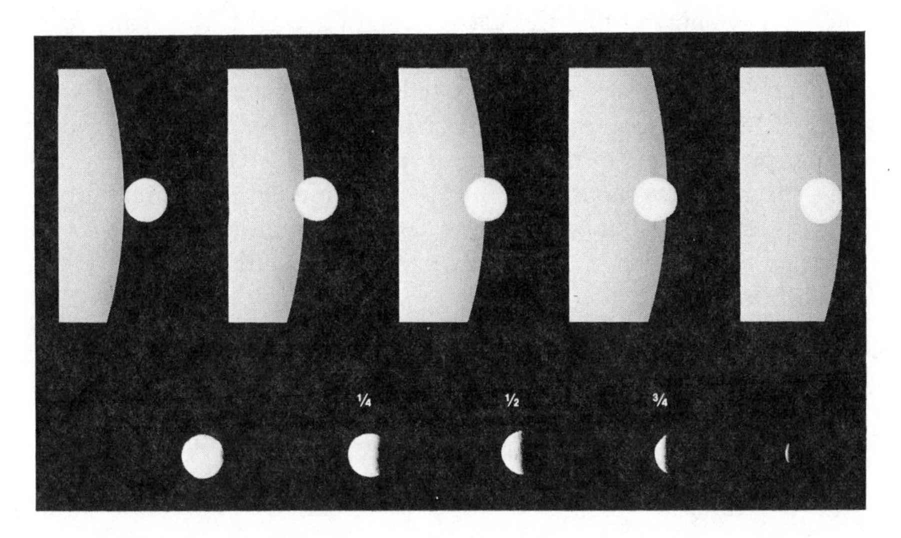

When observing phenomena of Jupiter's satellites, illustrated here by a transit ingress (above) and an eclipse (below), Innes sought to time all of the stages pictured.

three-quarters covered by shadow. Similarly, in observing a transit ingress, he would note the moments when the satellite disk was externally tangent to Jupiter's, one-quarter on, half on, three-quarters on, and internally tangent.

Between 1908 and 1926 Innes observed well over 1,000 different satellite events with the 9-inch. However, he appears to have lost some confidence in his fractional-disk timings, and in the later years confined himself largely to speck observations of eclipse disappearances and reappearances. Dirk Brouwer, who in 1928 published a penetrating analysis of Innes' observations, concluded that the fractional-disk method failed to yield the expected increase in accuracy. It may well be that despite Innes' remarkably acute sight and the excellent astronomical seeing in the high veld, a 9-inch telescope was too small. After all, the disks of the Galilean satellites are only about one second of arc in diameter. Nevertheless, Innes' observations were good enough for him to demonstrate variations in the Earth's rate of rotation. In an important paper published in 1925 he pointed out that both his satellite timings and the observed times of transits of Mercury showed that the Earth's rotation was slower from 1910 to 1923 than the average for the 19th century.

Today, astronomers who make refined studies of the orbital motions of Jupiter's Galilean satellites have several techniques of much greater power. One of these is measurement of satellite positions on long-focus photographs. Another is the photoelectric observations of eclipses or occultations of one satellite by another. In the future, systematic radar-range measurements of the satellites may give the best orbital data of all.

Yet the old method of visually clocking eclipse disappearances and reappearances will retain much attraction for amateur astronomers, both for the interesting phenomena to watch and for the possibility of making what amounts to a precise measurement with the simplest equipment.

62. Comets true and false

THE FIRST TIME I EVER HEARD of Jean August d'Angos was in Admiral Smyth's *Celestial Cycle* (see Chapter 12). There you may read, in the account of the star cluster M92 in Puppis: "The unlucky Chevalier d'Angos, of the Grand-Master's observatory at the summit of the palace at Malta, mistook this cluster for a comet: from which, and some still more suspicious assertions, my excellent friend, Baron von Zach, was induced to term any egregious astronomical blunders – Angosiades." Smyth was kinder than some other astronomers when he merely called the chevalier a blunderer. J. F. Encke described another of his feats more outspokenly: "D'Angos had the audacity to forge observations that he never made, of a comet that he had never seen, based on an orbit he had gratuitously invented, all to give himself the glory of having discovered a comet."

D'Angos was a Frenchman of noble lineage born at Tarbes in the Pyrenees on May 13, 1744. His career was varied, for besides being known as an astronomer and physician, he also served as a captain in the Navarre regiment of the pre-Revolutionary French army and became a Knight of Malta. (This was the original knightly order, and not its present-day namesake.) By 1780 he was well enough known to be named a correspondent of the French Academy of Sciences, and was elected an associate in 1796. The grand master of the Knights of Malta invited d'Angos to the island and built an observatory with fine instruments for him in 1783. Here the chevalier engaged in chemistry as well as astronomy, and a mishap with some phosphorus burned the observatory and its records in 1789. Soon after, d'Angos returned to France, and he finally died at Tarbes on September 23, 1833.

The comet of which Encke speaks is Comet 1784 II. On May 14th, Charles Messier at Paris received a letter from the chevalier, dated April 15th, announcing the discovery four days earlier of a faint, tailless comet in Vulpecula. D'Angos said that he measured its position on the 11th, but could convince himself that it was a comet only on the 13th, when the sky clouded up before he could determine its coordinates. The letter gave a second position observed on April 15th. Some time later a second letter reported the comet's orbital elements. Messier was never able to find the comet, despite careful search, and no other astronomer reported observations of it.

Suspicion became attached to this comet in 1806, when the Paris astronomer J. K. Burckhardt wanted to recalculate its orbit. For this he needed three observations of its position, but Messier had received only two. A letter to d'Angos brought the reply that his records had all been lost except for a meteorological journal. Burckhardt, a tireless computer, thereupon took the two observations and, making various assumptions as to the comet's distance from the Earth, calculated several sets of elements. None bore the slightest resemblance to the orbital elements that d'Angos had sent to Messier.

About a dozen years later, matters came to a head with the discovery that an obscure German periodical of 1786 contained a note by d'Angos, giving 14 observations of his comet extending to May 1st. This contradicted what d'Angos had told Burckhardt, and moreover the newly found observations could not be represented by the chevalier's orbit. The suggestion of fraud was now stronger. Then in 1820 Encke found that with d'Angos' elements he could reproduce the 14 positions, provided that in the calculations he used *exactly 10 times too large* a value for the comet's distance from the Sun. The

The positions reported by the Chevalier d'Angos for his comet on April 11 and 15, 1784, are shown as disks here in C. F. Goldbach's star atlas of 1799. Goldbach used double underlinings to indicate double stars; a few star designations have been added. The conspicuous asterism at lower right is the constellation Delphinus.

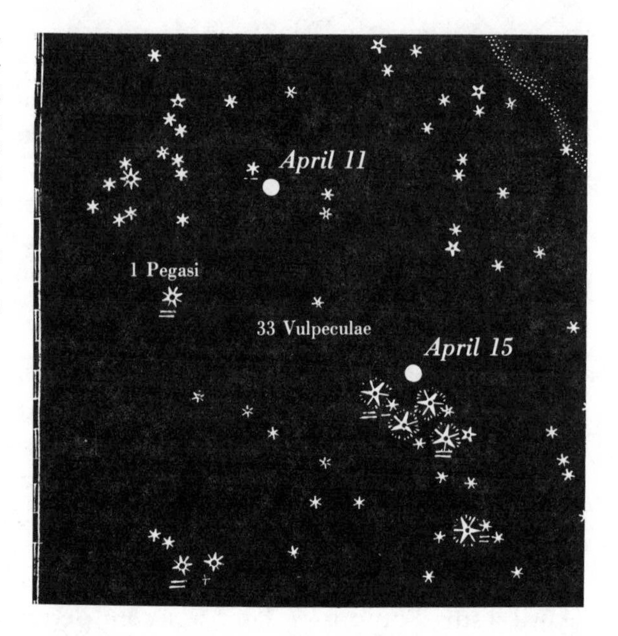

conclusion was unmistakable, said Encke, that the chevalier had fabricated his observations of the comet, making this numerical error in the process.

But this is not an end to the story of d'Angos. In 1798 he wrote to the astronomers of Paris, saying that he had seen a comet cross the Sun's disk on January 18th of that year. On this date the Sun was at the descending node of the comet of 1672, according to its value in J.-J. Lalande's *Astronomie.* (The descending node is where a comet or other body crosses southward through the plane of the Earth's orbit.) Since d'Angos claimed that he had also seen a comet transit the Sun in 1784, he asserted that both objects were returns of the comet of 1672, as the three dates were exactly fitted by a 14-year period. An apparently fatal defect to the picture was promptly pointed out by the Dresden astronomer, J. G. Köhler: The value of the node of the comet of 1672 given in Lalande's *Astronomie* was a misprint, being 60 degrees in error. This comet of 1672 could not possibly have passed in front of the Sun in January, 1798, as claimed. J.-J. Lalande, on learning of this disclosure, denounced d'Angos in words that not even the strong-minded von Zach could bring himself to repeat.

Nevertheless, the very enthusiasm with which Encke and von Zach attacked the chevalier and all his works makes their bias evident. The possibility arises that d'Angos may have been too sweepingly condemned.

This was the opinion of C. F. Gauss, the famous German mathematician and director of Göttingen Observatory. In a letter of November 13, 1846, to H. C. Schumacher, editor of the *Astronomische Nachrichten,* Gauss pointed out that Encke had not really proven his case against the reality of the comet of 1784. What Encke had demonstrated was that the orbital elements published by d'Angos were vitiated by a numerical error, not that all the observations were forged. The possibility could not be excluded that d'Angos had really discovered a comet on April 11, 1784.

And what about the comet transits of the Sun in 1784 and 1798? All that can really be said is that they were not returns of the comet of 1672. Despite the unlikeliness, the chevalier could have observed transits of other comets. The cases brought against d'Angos are all inconclusive, though the circumstantial evidence seems compelling. It is not impossible that he was a victim of character assassination.

Only a relatively few comet discoveries are as suspect as d'Angos', but many have been lucky hits. In the early 1880's Dudley Observatory at Albany, New York, had just been rebuilt and reequipped by public subscription. A delegation of citizens was visiting the director, Lewis Boss, when someone remarked that comets were being discovered at other institutions and that Albany should not be left behind. Jokingly, Boss turned to his assistant and said, "You see, Mr. Wells, you must discover a comet." He did just that within a week! Wells' find was 1882 I, a fine naked-eye object that remained in view for five months.

A number of curious cases are on record of what might be called collusion between comets. Take, for example, the events at the Vienna Observatory on the night of November 16–17, 1890. The astronomer on duty was a 31-year-old assistant named Rudolf Spitaler, who later became a professor at Prague. At 2:30 a.m. he received a telegram from T. Zona, director of the observatory at Palermo, Sicily, announcing the discovery on the previous night of a fairly bright comet in the constellation Auriga. Spitaler pointed the 27-inch Vienna refractor at the approximate sky location given in the telegram, and on his first glance into the eyepiece he saw a cometary object. He then set to work making repeated filar-micrometer measurements of its position relative to a neighboring star. To his surprise, the motion was much slower than it should have been for Zona's object, and the comet was fainter than the telegram had led him to expect. It finally occurred to Spitaler that he had been observing a new comet, so he promptly explored the neighborhood and found Zona's comet little more than a degree away.

This experience was not unlike that of George Van Biesbroeck, at Yerkes Observatory on November 17, 1925. He had for some time been engaged in a series of observations of Comet Orkisz 1925c, detected that April and having since faded to magnitude 13. When Van Biesbroeck set the 40-inch refractor on the predicted position, he saw a fine 8th-magnitude comet in the field of the 4-inch finder! Officially designated 1925j, the new object was widely observed and turned out to have a hyperbolic orbit, one that would never bring the comet back to the Sun's vicinity.

A variation on this theme is provided by the story of Honoré Flaugergues' comet, 1826 III. Flaugergues was an amateur astronomer of Viviers, in southern France, who became well known for his numerous observations of eclipses, comets, and Jupiter's satellites. But he is chiefly remembered as the first person to see the great comet of 1811. To his contemporaries, he was enough of an astronomer to be elected a corresponding member of the Paris Academy of Sciences, and to be offered the directorship (which he declined) of Marseilles Observatory in 1810.

Flaugergues received a letter from Jean Gambart of Marseilles, telling of the finding of a new comet on March 9, 1826. (It turned out to be a return of Biela's Comet, a famous visitor that broke apart about 1846 and eventually "evaporated.") Using a very small telescope, he located it with some difficulty on March 29th, "under the left arm of Orion." On several other evenings to April 5th, he made rough sketches of its location with respect to field stars. Realizing that the comet was being more effectively studied by other astronomers with better optical means, he did not examine his notes carefully until months later.

But then, on reading a detailed report in von Zach's *Correspondance Astronomique,* Flaugergues realized that his own crude positions gave quite a different path across the sky, and he belatedly announced his

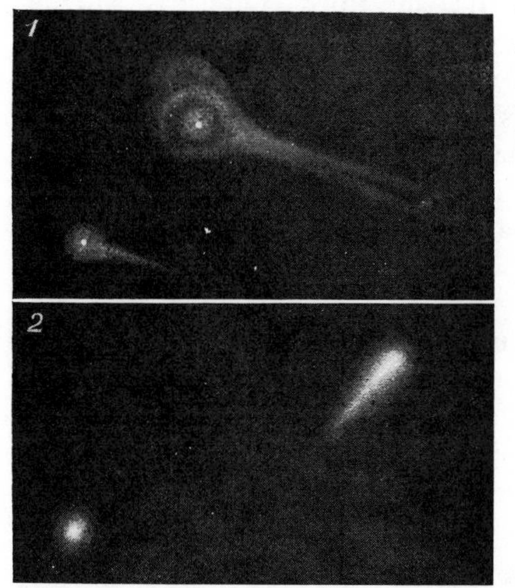

Two views of Biela's comet after it split apart. The top drawing is by F. G. W. Struve in 1846, the bottom one by A. Secchi in 1852 during the comet's next trip past the Sun. From A. Berget's *Le Ciel,* 1923.

discovery of a new comet. No one else ever reported seeing this object, now called 1826 III. But early in 1914 a young German astronomer, W. Hassenstein at Strassburg, looked into this story. Borrowing Flaugergues' record books from the archives of Paris Observatory, he studied the original sketch maps of 1826 III and its surrounding stars. His surprising conclusion was that on every night the French astronomer had misidentified the star fields, which actually matched configurations along the path of Biela's comet! Thus 1826 III had no existence of its own, and Flaugergues' observations really did refer to Biela's comet!

There are other examples of mistaken comet discoveries. For example, the "comet" of A.D. 1006 is now known to have been a supernova, a star that blew up inside our galaxy. There is also an extensive twilight zone of spurious and borderline comets. In fact, as Brian G. Marsden of the Smithsonian Astrophysical Observatory has pointed out to me, considering the number of spurious cometary reports he receives, many widely publicized 18th- and 19th-century comets seen only by their discoverers may have been nonexistent. Some of the less familiar cases of questionable comets are worth reviewing.

One example is known only from a short note in the *Astronomische Nachrichten* by Julius Franz, the director of Breslau University Observatory. It is dated July 25, 1911, and titled "On a Fast-Moving Cometary Object." Here is an excerpt: "While searching in a perfectly clear sky for Comet Kiess on July 22, 1911 . . . I chanced upon a rapidly moving nebulosity . . . near the ecliptic. It was evidently a comet very near the earth. In six minutes its right ascension increased by 3^{m}, while the declination remained practically unchanged. The object was of magnitude 6 and about 6′ in diameter. In appearance it resembled Comet Kiess, which I saw soon afterward.

"From the orbital velocity that a comet would have at the earth's distance from the sun, if moving along a parabola, and from the known orbital velocity of the earth, I found that the comet's distance from Earth was 2.7 sin α times the moon's distance, where α is the unknown angle that its direction of motion made with the line of sight. No conclusions can be drawn about the orbit. On the next two nights the sky was cloudy." A footnote by the editor, H. Kobold, stated that unsuccessful searches had been conducted at Kiel Observatory on July 23rd and at Bergedorf Observatory on the 24th. Although only Franz saw this object, his account appears trustworthy, since he was a highly competent professional astronomer with much practical experience in comet observing. His failure to track the object for more than a few minutes is perhaps explained by his preoccupation with Comet Kiess. This observation very likely does refer to an otherwise undetected tiny comet passing very close to Earth.

Less satisfactory is the odd business of Comets 1905d and 1905e, reported from Lowell Observatory. On December 15, 1905, the Eastern Hemisphere center for astronomical news, then at Kiel Observatory, received this cablegram from the Western Hemisphere center at Harvard Observatory: "Lowell telephones: A comet was discovered by Slipher on photograph November 29.704 Greenwich mean time. Right ascension 341° 01′, north polar distance 101° 28′. Pickering." Not only was nothing said about brightness or physical appearance, but the message gave no clue to the direction and rate of motion during the intervening weeks. A flurry of cables elicited the information that the daily motion was nearly two degrees, but whether this was west-northwest or east-southeast was ambiguous from the single photograph. No Comet 1905d could be found on

December 17th in a visual search at Bamberg Observatory and in a photographic search by Max Wolf at Heidelberg.

Then, on December 24th, another cablegram from Harvard to Kiel relayed word that a second comet had been discovered *on the same plate,* at right ascension 338° 30′, north polar distance 98° 42′. This new object was said to have two tails. No other observation of it is on record.

Presumably the Lowell plate was one taken in the search for a trans-Neptunian planet that was begun that year with a 5-inch lens of 35 inches focus. Today it is standard practice for observers of comets to take their plates in pairs, as a precaution against photographic defects, which can sometimes be nearly indistinguishable from small comets. There seems little doubt that 1905d and 1905e were simply plate blemishes viewed by an overenthusiastic eye.

Our final case, the dubious Comet Reissig of 1803, is more puzzling. Its discoverer and sole observer was a German living at Kassel, where his father was a well-known optician and court instrument-maker to the Elector of Hesse-Kassel. In a letter to J. E. Bode, director of Berlin Observatory, Reissig told the following story.

On the morning of February 2, 1803, he saw in a small comet-seeker a star of magnitude 5 or 6 near 36 Ophiuchi that he had not noticed on January 28th. "The star or comet appeared without appreciable nebulosity and slightly enlarged when viewed at 400× [presumably with another telescope]." Two mornings later the object was seen farther west; on the 7th it was observed with difficulty due to the full Moon. On the 9th it occulted 25 Scorpii at 3:02 a.m., and had separated from this star by 4:09 a.m. This was the last date on which Reissig reported seeing his object. He sent to Bode a small chart of the track, and also four crude positions.

In printing these observations, Bode commented that the four positions did not lie on a simple curve, presumably because of their low accuracy. He surmised that the object was evidently a very distant comet in retrograde motion around the Sun. An anonymous contributor to *Nature* magazine (who was probably J. R. Hind) called attention in 1876 to this case and calculated a parabolic orbit from Reissig's positions. Although no very satisfactory solution was possible, it appeared that the comet had passed very close to the Earth on January 29th, in a direct orbit inclined only 0° 55′ to the ecliptic. In discussing the *Nature* article with me, Marsden suggested that the Reissig object, if it really existed, was an asteroid passing very near the Earth, not a comet.

Unfortunately, like d'Angos, Reissig's veracity is open to question. In 1801, Lalande offered a prize of 600 francs for the first comet to be discovered in the 19th century. This object was detected on July 12, 1801, at the head of Ursa Major by Jean-Louis Pons at Marseilles and practically simultaneously by three other French astronomers. Later, Reissig claimed to have been first. In a letter to Bode he said he had seen a small comet during a brief break in the clouds on June 30th, between the head of the Great Bear and the Giraffe. However, using A. W. Doberck's elements of the Comet of 1801, I find that on June 30th it was in Andromeda. Contemporary astronomers did not accept Reissig's assertion, and the 600 francs went to Pons. If the 1801 discovery was a fabrication, then the 1803 one is indeed suspect.

63. An early American meteorite

EVEN THOUGH PREHISTORIC MAN made weapons from meteoritic iron, the study of meteorites did not become a recognized branch of astronomy until around the year 1800. The idea of stones or lumps of metal coming down from the sky was unattractive to most scientists of the enlightened 18th century, who dismissed it as another popular superstition. A case in point is the shower of stony meteorites that fell on July 24, 1790, at Barbotan, France, and was attested to by the mayor and city councillors. Concerning this episode, the French scientist Pierre Bertholon wrote: "How sad it is to see a whole municipality attempt to lend credibility, through a formal deposition, to folk tales that arouse the pity of not only physicists but of all sensible people!"

But the tide was already beginning to turn. In 1794 the German scientist E. F. F. Chladni published a penetrating survey demonstrating that the meteorites were in fact cosmic bodies that had descended as fireballs through the atmosphere. Chladni's early training as a lawyer was revealed by the cogency of his argument. Finally, the controversy ended dramatically when the widely observed fall of over 3,000 stones at L'Aigle, Normandy, on April 26, 1803, was investigated and fully confirmed by a representative of the Paris Academy of Sciences.

It was an exciting breakthrough to realize that certain strange crusted lumps in museum collections were actually specimens of extraterrestrial material. Early chemical analyses of these bodies showed only the same elements already familiar on Earth. Thus, decades before any spectroscopic analysis of the Sun or stars, there was evidence of a uniformity of chemical composition in the cosmos.

But in the first decade of the 1800's, scientific ideas often took much time to cross the Atlantic from Europe to the United States. President Thomas Jefferson, who had a keen interest in things astronomical, was slow to accept the existence of meteorites. He greeted with frank disbelief a report that in December, 1807, stones had showered from a Connecticut cloud. According to his friend, Senator S. L. Mitchill of New York, the president said that five words were enough to sum the case: "It is all a lie!" Either Mitchill's anecdote is apocryphal or Jefferson quickly reconsidered, for on February 15, 1808, he wrote to another correspondent about the Connecticut happening:

"Sir, – I have duly received your letter of the 8th instant, on the subject of the stone in your possession, supposed meteoric. Its descent from the atmosphere presents so much difficulty as to require careful examination. But I do not know that the most effectual examination could be made by the members of the National Legislature, to whom you have thought of exhibiting it. . . . I should think that an inquiry by some one of our scientific societies, as the Philosophical Society of Philadelphia for example, would be most likely to be directed with such caution and knowledge of the subject, as would inspire a general confidence. . . ."

The scientific inquiry recommended by Jefferson was in fact carried out by Nathaniel Bowditch of Salem, Massachusetts, the country's leading astronomer of that day and author of the *American Practical Navigator*. His report was published in the *Memoirs* of the American Academy of Arts and Sciences, Boston, Massachusetts, as "An Estimate of the Height, Direction, Velocity and Magnitude of

The memorial to Nathaniel Bowditch (1773–1838) in Mount Auburn Cemetery, Cambridge, Massachusetts. The book and globe mark Bowditch's contributions to celestial mechanics and navigation. Photograph by Leif J. Robinson.

the Meteor that Exploded over Weston in Connecticut, December 14, 1807. With Methods of Calculating Observations made on such Bodies."

This paper, which seems to be the first scientific study of an American meteorite fall, refers to a spectacular daylight bolide that appeared over Vermont, traveled southward along a nearly horizontal trajectory, and exploded over Weston, a village in Fairfield County of southwestern Connecticut. To triangulate its path through the atmosphere, Bowditch used selected observations made at Rutland in central Vermont, at Wenham in northeastern Massachusetts, and at Weston.

The Wenham observer was a Mrs. Gardner, who happened to be facing a southern window of her home at 7 a.m. when she saw the fireball through thin clouds. It looked so much like the Moon, but in evident motion, that her first startled thought was, "Where is the moon going to?" She recovered her composure sufficiently to make some accurate estimates of the direction of the phenomenon with respect to terrestrial landmarks. Remaining in sight for half a minute, the meteor seemed even brighter to her than the Moon, although it was about 140 miles distant.

The most vivid description was by Judge Wheeler at Weston, where the sky was dappled with broken clouds that morning. He used the formal third person to report:

"The attention of Judge Wheeler was first drawn by a sudden flash of light, which illuminated every object. Looking up he discovered in the north a globe of fire, just then passing behind the cloud, which obscured though it did not entirely hide the meteor. In this situation its appearance was distinct, and well defined, like that of the sun seen through a mist. It rose from the north, and proceeded in a direction nearly perpendicular to the horizon, but inclining, by a very small angle, to the west, and deviating a little from the plane of a great circle. . . . Its apparent diameter was about one half or two thirds the apparent diameter of the full moon. Its progress was not so rapid as that of common meteors and shooting stars.

"When it passed behind the thinnest clouds, it appeared brighter than before; and when it passed the spots of clear sky, flashed with a vivid light, yet not so intense as the lightning of a thunder storm. Where it was not too obscured by thick clouds, a waving conical train of paler light was seen to attend it, in length about 10 or 12 diameters of the body. In the clear sky a brisk scintillation was observed about the body of the meteor, like that of a burning fire brand carried against the wind. It disappeared about 15 degrees short of the zenith, and about the same number of degrees west of the meridian. It did not vanish instantaneously, but grew, pretty rapidly, fainter and fainter, as a red hot cannon ball would do, if cooling in the dark, only with much more rapidity.

"The whole period between its first appearance and total extinction was estimated about 30 seconds. About 30 or 40 seconds after this, three loud and distinct reports, like those of a four pounder near at hand, were heard. – Then followed a rapid succession of reports less loud, so as to produce a continued rumbling. This noise continued about as long as the body was in rising, and died away, apparently in the direction from which the meteor came."

From these and other reports, Bowditch calculated that the disappearance of the bolide occurred at a height of 19.5 miles over the Earth's surface at north latitude 41° 19′.3, west longitude 73° 28′.5. He further deduced that the luminous head of the meteor must have been at least 491 feet in diameter. And since the meteoritic stones picked up around Weston weighed 225 pounds per cubic foot, he concluded that the entire bolide must have weighed not less than 6,000,000 tons! Bowditch's views on the nature of this object give a curious vignette of the rival theories of meteorite origins at that time: "The greatness of the mass of the Weston meteor does not accord either with the supposition of its having been formed in our atmosphere, or projected from a volcano of the earth or moon; and the striking uniformity of all the masses that have fallen at different places and times (which indicates a common origin) does not, if we reason from the analogy of the planetary system, altogether agree with the supposition that such bodies are satellites of the earth."

According to Frederick C. Leonard's catalogue of meteorite falls, about 150 kilograms of material were recovered from the Weston event of 1807. He classifies the several fragments as brecciated, spherulitic chondrites: stony meteorites containing small, nearly spherical silicate particles.

STARS AND STELLAR SYSTEMS

64. William Herschel and the Sun

Few major astronomical discoveries have remained barren for so many years before springing into life as did the telescopic detection of sunspots in 1610 by Galileo and Fabricius. For two centuries afterward these temporary dark markings attracted only casual attention, though from many astronomers. The spots seemed of astronomical importance only in finding the rotation period of the Sun and the direction of its axis. To most astronomers of the 18th century the Sun was merely a convenient skymark for the determination of time and geographical latitude.

The beginnings of solar physics can be traced to Sir William Herschel. In his famous experiment of 1800 the great English astronomer used a prism to spread sunlight into a spectrum; then he measured the rise in temperature of a sensitive thermometer exposed to different portions of the colored band. To his surprise a marked heating effect occurred beyond the red end of the visible spectrum. Often this is nowadays referred to as the discovery of infrared radiation; Herschel, however, concluded that the visible light and the heat of the Sun must be essentially different in nature. This misinterpretation of an ingenious experiment misled physicists for a generation, thanks to the prestige of Herschel's name. In 1796, at a time when he was thinking much about variable stars, Herschel suggested that the Sun itself might vary in brightness and that such changes might explain alterations in the Earth's climate. Looking for some index number to characterize solar radiation, he proposed using the price of wheat in different years. A modern social scientist might call Herschel's notions of economics oversimplified, but clearly here was a pioneer attempt to study solar-terrestrial relations in a modern spirit. Both these papers of Herschel's were anticipations of things to come, inspirations that awaited physical knowledge in one case, and statistical data in the other before they bore fruit.

The same cannot be said of Herschel's memoir of 1795 on the constitution of the Sun, which is a curious survival of older fancies. From time to time, Herschel noted that the dark umbrae of sunspots appeared to lie at a lower level than the solar photosphere (the Sun's surface as we see it). For example, "In the year 1783, I observed a fine large spot, and followed it up to the edge of the sun's limb. Here I took notice that the spot was plainly depressed below the surface of the sun; and that it had very broad shelving sides." From a number of such observations, he came to the belief that the Sun is encased by a "shining atmosphere" in which occasional gaps occur, revealing the cool, dark "real body of the sun itself."

Analogies were collected to support this idea. To an observer on the Moon, the Earth's surface would be seen only between cloud areas, and this cover would be in part luminous, because of aurorae. Herschel continued: "Nay, we have pretty good reason to believe, that probably all the planets emit light in some degree; for the illumination that remains on the moon in a total eclipse cannot be entirely ascribed to the light which may reach it by the refraction of the earth's atmosphere. . . . The unenlightened part of the planet Venus has also been seen by different persons . . . this faint illumination must denote some phosphoric quality of the atmosphere of Venus."

Herschel did not hesitate to carry his speculation further: "The sun, viewed in this light, appears to be nothing else than a very eminent, large, and lucid planet . . . all the others being truly secondary to it. Its similarity to the other globes of the solar system with regard to its solidity, its atmosphere, and its diversified surface; the rotation upon its axis, and the fall of heavy bodies, leads us on to suppose that it is most probably also inhabited, like the rest of the planets, by beings whose organs are adapted to the peculiar circumstances of that vast globe. . . . [I] am persuaded that the foregoing observations, with the conclusions I have drawn from them, are fully sufficient to answer every objection that may be made against it."

Solar inhabitants! This fantasy was actually published in the *Philosophical Transactions* of the Royal Society. The suggestion that the Sun was a cool, planetary body with a glowing atmosphere was not new; John Flamsteed in 1681 and J. E. Bode in 1772 had proposed similar ideas.

But in an English courtroom in 1787, only eight years before Herschel wrote, the holding of this theory was alleged as a proof of insanity. This was the trial of Dr. Elliot for a murderous assault on Miss Boydell. A friend of the accused man offered as evidence of mental derangement a letter Dr. Elliot had written to him the year

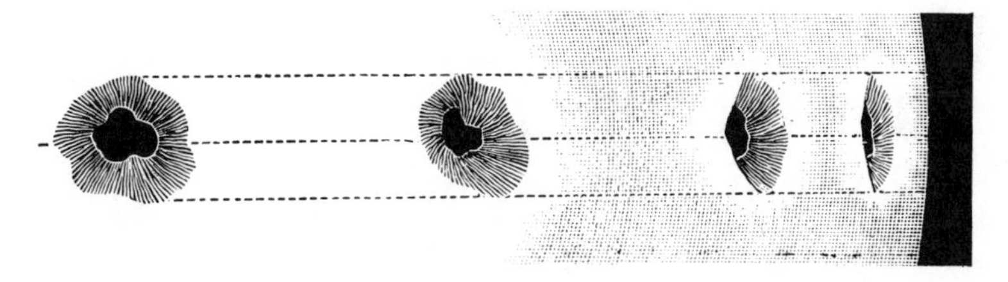

As rotation carries a sunspot toward the Sun's limb, the dark umbra seems to become displaced (relative to the lighter penumbra) toward the center of the solar disk. This phenomenon was discovered in 1769 by A. Wilson, and it causes the center of the sunspot to appear depressed, as described by William Herschel. The central depression is actually an illusion. The umbral displacement is due to different absorptions of light at different depths within the Sun's atmosphere, an effect that becomes manifest near the Sun's limb. From the classic *A Text-book of General Astronomy* by Charles A. Young, 1900.

before. This stated "that the sun is not a body of fire, as hath been hitherto supposed, but that its light proceeds from a dense and universal aurora, which may afford ample light to the inhabitants of the surface beneath, and yet be at such a distance aloft as not to annoy them. . . . Vegetation may obtain there as well as with us. There may be water and dry land, hills and dales, rain and fair weather; and as the light, so the season must be eternal, consequently it may easily be conceived to be by far the most blissful habitation of the whole system!"

These strange notions still recur from time to time among persons who invent private cosmologies without benefit of modern astronomy. Shortly after World War II, a German named Gottfried Bueren became an enthusiastic champion of the cold-Sun theory and publicly wagered 25,000 marks that he could not be proved wrong. A group of leading West German astronomers took up this challenge and submitted the proof. When Bueren refused to accept it the case was taken to court, and in 1953 the astronomers won the 25,000 marks with costs.

65. Richard Carrington and a "singular appearance" on the Sun

One major difference between today's astronomy and that of a century ago is the greatly increased cost and complexity of the tools for research. Today, a major telescope can be built and effectively worked only through a great cooperative effort and with the financial backing of a wealthy foundation or governmental agency. But in the 19th century an individual like Lord Rosse or William Lassell could erect and use the largest telescope in the world at his own expense (see Chapters 27 and 76).

In the England of the 1850's, amateurs and professionals were on much the same footing instrumentally, and many lines of research that would be regarded as professional activities today were being pursued at a score of private observatories. Thus Lassell was cataloguing galaxies, J. R. Hind was discovering minor planets and computing comet orbits, Admiral Smyth and W. R. Dawes were measuring double stars, and J. Baxendall was detecting new variables.

Among the ablest in this group of English private astronomers was Richard C. Carrington (1826–1875). Son of a well-to-do brewer, he began his astronomical career as an assistant at Durham University Observatory. There he mastered the skills of an astronomer, but after three years at the poorly equipped institution he decided to establish his own observatory. This he erected at Redhill, south of London, where observing began in 1854. His first program was the formation of a catalogue of 3,735 stars north of declination +81°, whose positions he observed with a transit circle of 5-inch aperture. The completed work was published in 1857 at the expense of the Admiralty, and it won Carrington the gold medal of the Royal Astronomical Society.

Even this enormous undertaking did not occupy all of Carrington's energies. Simultaneously, he began an intensive series of sunspot observations, to clear up the vexing problem of the Sun's rotation rate. In the focus of his 4½-inch refractor he placed a cross of gold wires set at 45 degrees to the east-to-west motion of the sky, and projected on a white screen a solar image 11 inches in diameter. With the telescope held stationary, he allowed the Sun to drift across the field and recorded the times at which the solar limbs and sunspots contacted the crosswires. In this way, Carrington could determine accurately the position of each spot with respect to the center of the solar disk.

Observations on this plan were made on every possible day from November, 1853, to March, 1861. Carrington's masterful analysis of the results, published in a large quarto volume, was a milestone in solar investigation. He determined first the orientation in space of the Sun's axis of rotation, with such precision that his values are still employed in computations for the annual *Astronomical Almanac.* Next, he derived the solar rotation period for different heliographic latitudes, finding it to decrease markedly from the Sun's equator toward its poles. This important discovery explained many discrepancies among the less thorough rotation studies by earlier astronomers.

It was during the course of these observations that Carrington witnessed a remarkable event on the Sun. On the morning of September 1, 1859, he had completed his routine drawing of the disk and at 11:18 was timing the drifts of sunspots. Suddenly two intensely bright white patches appeared in a large spot group on the northern part of the Sun. He wrote:

"My first impression was that by some chance a ray of light had penetrated a hole in the screen attached to the object-glass . . . [but by] causing the image to move by turning the R. A. [right ascension] handle, I saw I was an unprepared witness of a very different affair. I thereupon noted down the time by the chronometer, and seeing the outburst to be very rapidly on the increase, and being somewhat flurried by the surprise, I hastily ran to call some one to witness the exhibition with me, and on returning within 60 seconds, was mortified to find that it was already much changed and enfeebled."

The bright patches faded out of sight five minutes after they were first seen, vanishing as two rapidly fading dots. During the interval they had moved from positions A, B to C, D on Carrington's drawing, reproduced on the next page. The scale of his picture is about 11 inches to the Sun's diameter, so the large spot group was roughly as large as Jupiter.

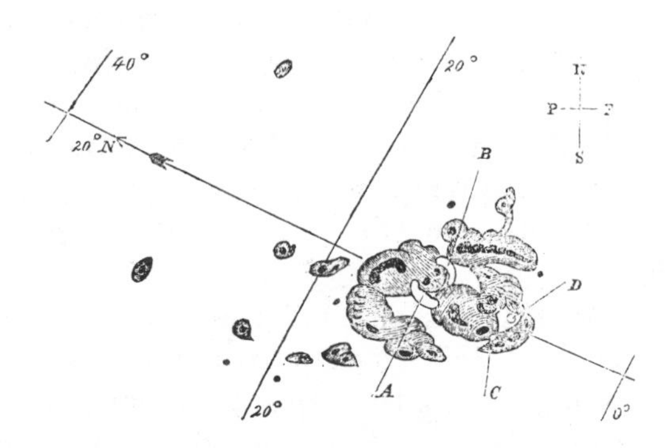

On Richard Carrington's drawing of the Sun, made September 1, 1859, he marked by A and B the places where a brilliant eruption first appeared. Five minutes later the flare faded from sight at C and D. From *Monthly Notices* of the Royal Astronomical Society.

Fortunately, a second English amateur, R. Hodgson at Highgate, had at that time been watching the same part of the Sun with a 6-inch refractor and solar diagonal. His independent testimony tells of "a very brilliant star of light, much brighter than the sun's surface, most dazzling to the protected eye, illuminating the upper edges of the adjacent spots and streaks." Hodgson also gave the duration as five minutes and noted that at about the same time a disturbance in the Earth's magnetic field was recorded on the instruments at Kew Observatory. On the following nights, an intense auroral storm was witnessed in many parts of the world. As we now know, had radio been in use in the year 1859, a worldwide blackout of shortwave communications would have taken place.

To present-day solar astronomers, Carrington's "singular appearance" is famed as the first solar flare ever reported. Furthermore, it was one of the very rare "white-light flares" – those intense enough to be visible without special equipment. It was a decade before another solar eruption was recorded. Not until the turn of the century, when the Sun's disk began to be regularly observed in the red light of hydrogen, did the flare phenomenon become familiar. In modern flare catalogues the event of September 1, 1859, is among only a handful ever to be seen in white light. D. H. Menzel once interpreted Carrington's flare as an exceptionally brilliant loop prominence seen projected on the Sun's disk. The Harvard astronomer pointed out

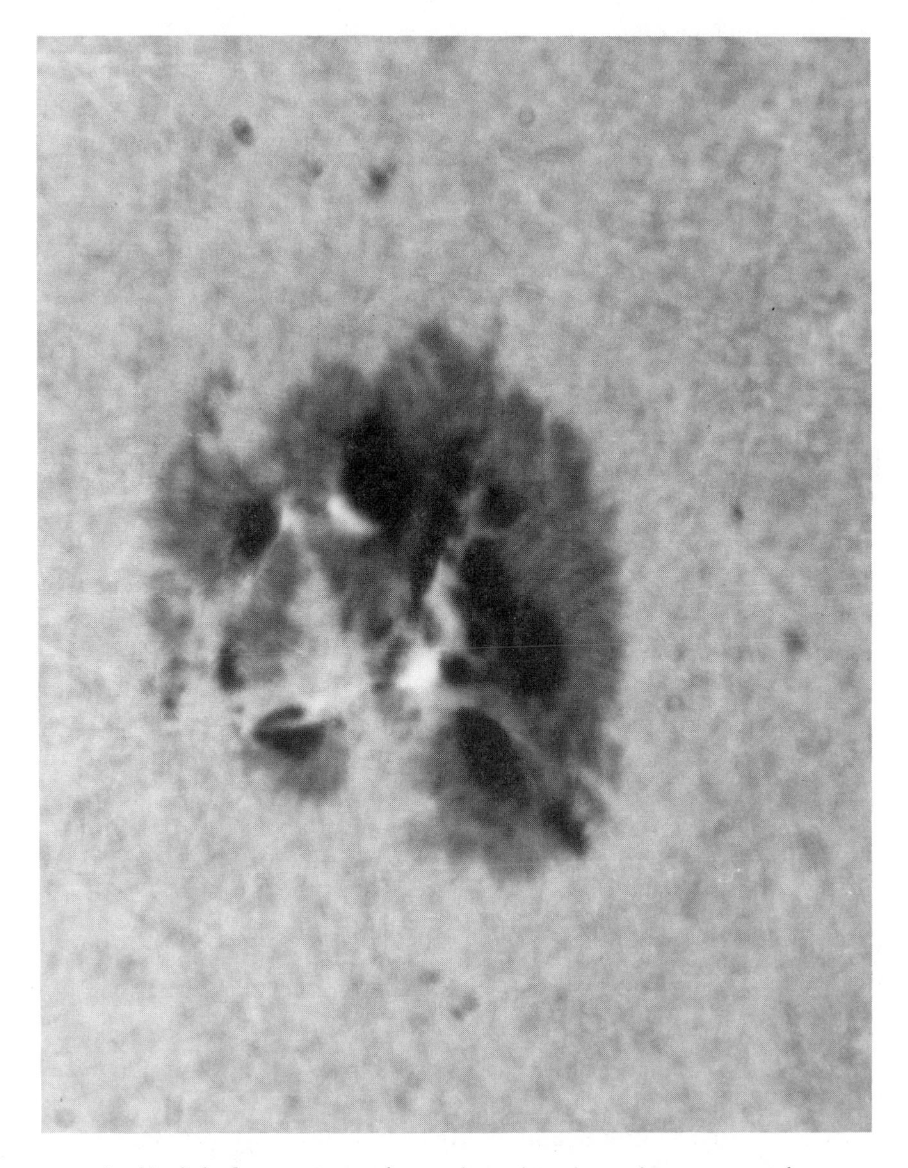

A white-light flare was captured at maximum intensity on August 7, 1972, by H. Mauter and F. Hegwer with the Vacuum Tower telescope at Sacramento Peak, New Mexico. The flare's white patches were about four seconds of arc across. Courtesy Sacramento Peak Observatory.

that if sufficient material in the solar atmosphere (corona) is trapped and compressed in the loop, it will shine as an ordinary flare, observable in hydrogen light. In extreme cases, a white flare such as Carrington saw could result.

66. The gradual recognition of helium

IN THE MID-19TH CENTURY the exciting new science of astrophysics was centered on the Sun. Heinrich Schwabe's announcement in 1844 of an 11-year rise and fall in the numbers of sunspots was followed in 1852 by Edward Sabine's surprising discovery that the Earth's magnetic-field disturbances followed the same cycle. The new art of photography began to be used for daily charting of the solar surface, and the spectroscope revealed such familiar elements as hydrogen, sodium, iron, and magnesium as vapors in the solar atmosphere. Total eclipses of the Sun were attracting special attention. Visual observations during the eclipses of 1842, 1851, and 1855 had proved that the brilliant crimson "flames" seen at the Moon's edge during totality were in fact appendages of the Sun, and these prominences were photographed in 1860.

The eclipse of August 18, 1868, provided the first opportunity to study prominences with spectroscopes. With a totality lasting longer than five minutes, the central line ran from the Red Sea across India, Malaya, and New Guinea. Two British expeditions were sent to India, Maj. J. T. Tennant observing at Guntur and Lt. John Herschel (Sir John's third son) at Jamkhandi. From Paris Observatory Jules Janssen went to Guntur and G. Rayet to Malaya. Norman Pogson of Madras Observatory set up his instruments at Masulipatam on the Indian east coast.

All these parties made spectroscopic observations. Lt. Herschel watched a cloud blot out the Sun only a quarter minute before totality began. But soon a break revealed an immense horn-shaped prominence at the northern limb, and his first glance at it through

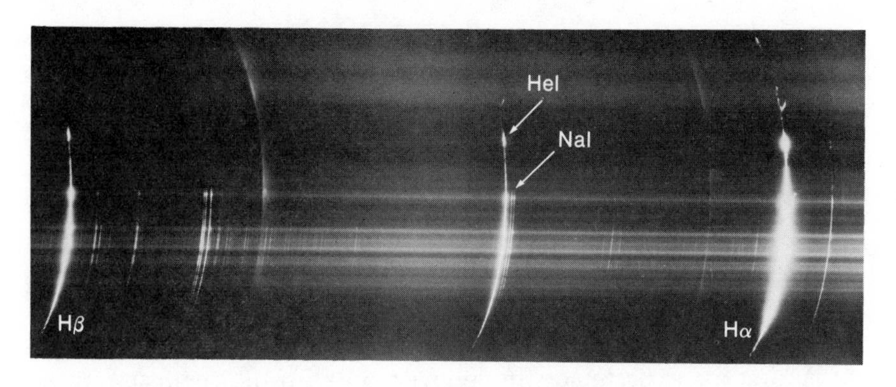

At the total eclipse of March 7, 1970, the brightest lines in the solar flash spectrum were recorded much as Lt. John Herschel described them. The D_3 line, now known to be from helium, is near the center, with the much weaker pair of sodium lines to its right. Courtesy Kwasan Observatory.

his spectroscope caught three bright lines: red, orange-yellow, and blue! During the remaining few minutes of totality he satisfied himself that these were probably hydrogen-alpha, the unresolved pair of sodium D lines, and hydrogen-beta. The "solar mountains" were manifestly gas clouds shining by their own light.

Other observers also viewed the emission spectrum of the "great horn." Tennant recorded five bright lines, Janssen six, and Rayet nine. Each of these men saw the orange-yellow emission and attributed it to sodium.

The great brilliance of the prominence lines suggested to Janssen that the darkness of an eclipse was not needed to make them visible. Clouds after totality postponed experimenting until the next morning. Then, placing the spectroscope slit just outside the Sun's edge revealed the emission lines in full daylight. In his report to the French Academy of Sciences, Janssen told of monitoring the spectra of the changing prominences day after day for the next three weeks.

Months before the 1868 eclipse, Norman Lockyer in England had designed and ordered a special spectroscope for observing prominence spectra in daylight. He received it by October, and on the 20th he too saw the bright lines. It is a famous coincidence that Lockyer's announcement was read to the French Academy on October 26th only minutes before Janssen's report from India. A medal bearing the pictures of both men was issued by the academy to commemorate their discovery, in happy contrast to the unequal treatment meted out a few years earlier to U. J. J. Le Verrier and J. C. Adams for their predictions of Neptune.

Now that an eclipse was no longer needed to study the prominence lines, many astronomers observed them more or less thoroughly, so that careful wavelength measurements became available. The orange-yellow line was put at 5874.9 angstroms by Lockyer (1869) and at 5874.3 by H. C. Vogel (1872), some distance from the sodium D_1 and D_2 lines at 5896 and 5890, respectively. This D_3 line, as it came to be called, was clearly not due to sodium. After a series of laboratory experiments in cooperation with the chemist Edward Frankland, who had suspected that the line might arise from hydrogen under extreme conditions, Lockyer concluded that D_3 was caused by a previously unknown chemical element, which he named *helium*.

For the next quarter century helium remained a mysterious hypothetical gas known only from the D_3 line in the spectrum of the solar chromosphere and prominences. It narrowly escaped recognition in some laboratory experiments carried out in 1891 by W. F. Hillebrand of the U. S. Geological Survey. He found that uraninite (a uranium mineral), when boiled with dilute sulfuric acid, yielded a gas with the chemical properties of nitrogen. Yet the spectrum of the gas contained some lines that did not belong to nitrogen, and a colleague suggested that a new element might be present. For some reason Hillebrand disbelieved this idea and did not pursue the possibility.

The identification of helium was left for Sir William Ramsay in England. Together with Lord Rayleigh, in 1894 he had discovered argon in the Earth's atmosphere, a gas hitherto not distinguished from nitrogen. While looking for other sources of argon, he learned of Hillebrand's work, and in 1895 secured from him a sample of cleveite (a Norwegian variety of uraninite). Ramsay repeated the American's procedure for extracting the gas, which he then purified and sealed in a discharge tube. Looking through a small hand spectroscope, he was startled to see not the many-lined argon spectrum, but an intense yellow ray with a few fainter ones in the red and green. A quick check showed that the yellow ray was not coincident with the D line of sodium. The wavelength was measured and found to be that of D_3. Helium had been captured!

"A beautiful confirmation of the identity was soon afterwards afforded," wrote Agnes M. Clerke in her classic *Problems in Astrophysics* (1903). "The golden line seen in the laboratory was perceived by Runge to have a faint close companion, and he declared that, unless the solar D_3 were also double, cleveite-gas should be regarded as different from helium. The challenge was taken up on both sides of the Atlantic. Professor Hale on 20th June [1895], and Sir William Huggins independently on 10th July, succeeded in resolving the prominence-ray into a delicate, unequal pair, and our possession of helium as a truly indigenous element was rendered incontrovertible."

At a dinner in his honor, Ramsay was toasted for "having run helium to earth." No other discovery of a chemical element from astronomical spectra has been confirmed. The hypothetical nebulium that was invoked to account for certain bright lines in gaseous nebulae were shown by Ira Bowen in 1927 to be ionized oxygen and nitrogen at an extremely low density. As Henry Norris Russell put it, "Nebulium has vanished into thin air."

67. Algol and scientific conservatism

In the years 1879–1880, a now-obscure British physicist, David Hughes, demonstrated a novel experiment before leading scientists of the Royal Society, including its president W. Spottiswoode and Sir George Stokes. Hughes used a spark transmitter to produce signals that were detected 500 yards away by a coherer connected to a telephone. The distinguished spectators were unaware that they were witnessing a pioneer demonstration of radio, seven years before Heinrich Hertz and two decades before Guglielmo Marconi. Instead, the committee decided that the effects were due to electromagnetic induction.

This episode is difficult to explain, for physics had developed to the point where the time was ripe for radio to emerge. In many

John Goodricke, the discoverer of the periodicity of Algol's light variations, is seen here in a pastel portrait done in 1785 or 1786 and now in the possession of the Royal Astronomical Society, with whose permission it is reproduced here. Photograph from Ann Ronan Picture Library.

other cases, the climate of scientific opinion was clearly not yet ready for the reception of some new idea. It is understandable that a novel hypothesis may through its very novelty be beyond the powers of observational test, or too isolated from other knowledge to attract study by theoreticians. Hence the idea may have an obscure currency for many years before a timelier announcement gains general scientific acceptance.

Astronomical history offers many examples of premature discovery. Such modern concepts as galactic rotation and interstellar absorption were already being discussed a century ago, and a method of measuring the radial velocities of stars (how fast they are moving toward or away from us) had appeared in print before 1800.

The story of the famous variable star Algol (Beta Persei) brings in the names of several astronomers as prophets before their time. The first man known to have noted that it is occasionally fainter than normal was Geminiano Montanari of Bologna, in 1667. He was one of the earliest astronomers to be interested in variable stars, at a time when the only known representatives were Mira (whose light changes in a period of some 330 days) and a few novae. Perhaps the foreignness of the notion that stars could vary in only a few days or hours was one reason why no further attention was paid to the star.

In fact, the variability of Algol had to be rediscovered in 1782 by the English amateur John Goodricke, an 18-year-old deaf-mute. His systematic observations soon showed the periodic nature of the light changes, a temporary decrease in brightness occurring every 69 hours. In reporting these results to the Royal Society, Goodricke advanced two explanations: "If it were not perhaps too early to hazard even a conjecture on the cause of this variation, I should imagine it could hardly be accounted for otherwise than either by the interposition of a large body revolving around Algol, or some kind of motion of its own, whereby part of its body, covered with spots or such like matter, is periodically turned toward the Earth."

Goodricke's binary interpretation was advanced when the existence of binary stars was only conjectural. Christian Mayer of Mannheim Observatory had, a few years earlier, been ridiculed when he announced that certain bright stars were attended by visible satellites. It is well known that William Herschel was very slow to reach a decision that visual binaries existed, a conclusion that he did not publicly acknowledge until 1802. Herschel inspected Algol telescopically and reported it as single even with highest magnification. The tide of astronomical opinion was clearly against the binary hypothesis, and it remained so for many years.

But in 1787, Goodricke's paper was studied by a young Swiss, Daniel Huber, who later became professor of mathematics at Basel. He first demonstrated that star spots could not satisfactorily explain Algol's variation. Proceeding to an analysis of the light curve, he deduced that its shape could result from eclipses of Algol by a dark secondary star with a diameter 0.71 that of the primary and separated from it by 5.635 of the latter's radii. Huber even calculated the average density of Algol as 0.292 that of the Sun. This remarkable investigation was the first crude solution of the light curve of an eclipsing binary. Disregarded by contemporaries, the paper was forgotten until it was rediscovered in this century by the German historian of astronomy, Ernst Zinner.

Independently, a somewhat similar discussion of Algol was carried through by a little-known English clergyman, William Sewell, in 1791. He sent his results to the Royal Society for publication, but their significance went unrecognized and the manuscript was discovered among the archives at Greenwich Observatory only in 1957, by Olin J. Eggen.

It is disconcerting that the correct insight of Goodricke, Huber,

examined it for the other stars nor indeed for all the
stars in your paper, having had as yet no time.
The last observation I made on Algol's greatest
obscuration was Jany 19 - 7^h - ' Mean time & upon
the supposition of the period's being 2^d - 20^h - 49'
I find it ought to happen
Aug. 15 - 12½^h +, which agreed with my observation
on that day to a quarter of an hour.
Dr. Shepherd informs me that you have
observed 200,000 stars pass the field of your
20 feet Telescope in an hour. This indeed gives
us a magnificent idea of the immense number of fixed
stars, that are scattered over the vast abyss
of space.
Be assured of my perfect esteem & remain
Sir
Your most obed^t hu^l Servt
J. Goodricke

Part of Goodricke's draft of a letter to William Herschel on September 2, 1784, tells about a correct prediction of an Algol minimum. The original letter is in the York University Library.

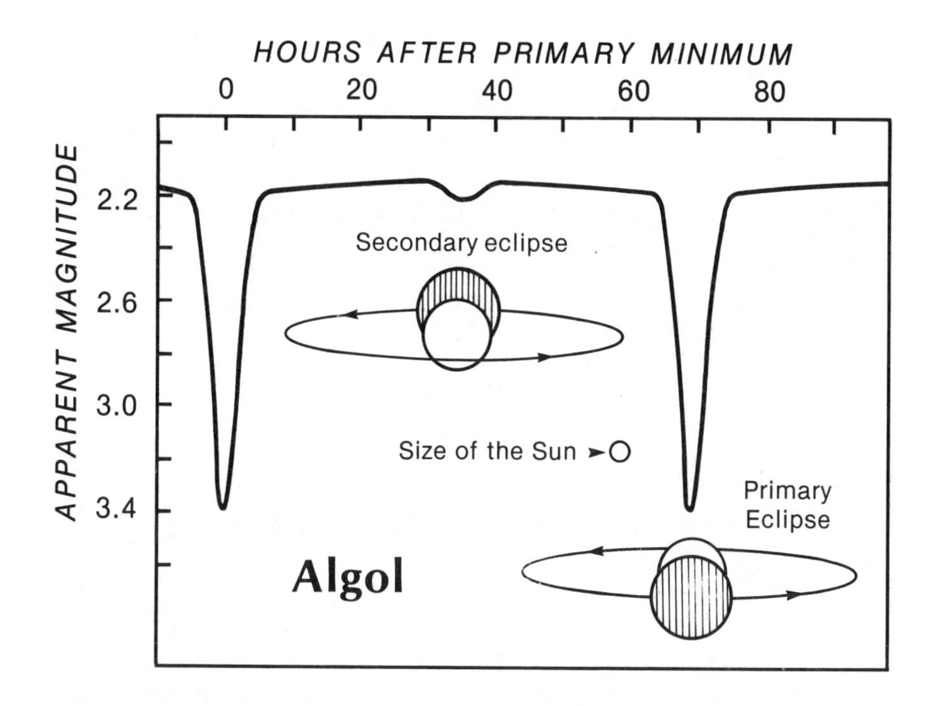

A modern light curve of Algol shows the character of its light variations. For nearly 90 percent of the time the brightness changes are imperceptible to naked-eye observers. The deep primary minimum corresponds to the partial eclipse of the bright component star by its dim companion. Half a period later the dim star is partially occulted by the bright one, causing the secondary eclipse.

and Sewell went for nothing, and that nearly a century was to elapse before their ideas took permanent root. From the viewpoint of 1791, a conservatively minded astronomer could marshal convincing doubts. To him the binary nature of Algol would seem an assumption not capable of being verified observationally. Also, at that time only four short-period variables were known – Algol, Beta Lyrae, Delta Cephei, and Eta Aquilae – and the light curves of the last two, which are pulsating Cepheid variables, did not suggest eclipses.

The growth of a favorable climate of opinion came much later. Additional discoveries of Algol-type stars during the 19th century finally suggested that they form a distinct class of variable, and meanwhile visual binaries became familiar objects. Goodricke's hypothesis began to be mentioned more often, by T. S. Aldis in 1870 and by Angelo Secchi in 1877. Four years later E. C. Pickering of Harvard Observatory reexplored it theoretically. The final justification came with H. C. Vogel's spectroscopic observations at Potsdam Observatory in 1889. They demonstrated that Algol's radial velocity varies in the same period as its brightness, with the bright star in the farthest part of its orbit at minimum light, exactly in accord with the eclipse hypothesis.

68. The story of Groombridge 1830

IN AN ARTICLE written in 1816, the German astronomer Bernhard von Lindenau gave a list of the astronomical observatories then existing in England. In first place, of course, was the Royal Observatory at Greenwich, then directed by John Pond. Next came Sir William Herschel's reflectors at Slough, near Windsor, and the well-equipped private observatory of King George III at Richmond. It was followed by the handsome Radcliffe Observatory founded in 1772 at Oxford. The list was concluded by 20 amateur observatories, mostly in and around London, of which easily the most important was that of Stephen Groombridge at Blackheath. A direct trail leads from its very able, self-trained proprietor to important 20th-century astrophysical research.

Groombridge was a successful retail merchant who became an active astronomer in middle age. Born in 1755 at Goudhurst, about 40 miles southeast of London, he moved to Blackheath in 1802. This suburb, famed for the oldest golf course in England, is just south of Greenwich. Here he built a small observatory attached to his house on Eliot Place, which was only two-thirds of a mile from the Royal Observatory.

Because Groombridge's interests centered on positional astronomy, in 1806 he obtained from the celebrated instrument maker Edward Troughton a reversible transit circle – the first effective

The English amateur Stephen Groombridge (1755–1832) was a founder of the Royal Astronomical Society in 1821. He was as devoted to music as he was to astronomy.

device of this kind in England. It consisted of a 3½-inch telescope mounted in the plane of the meridian, turning on a horizontal east-west axis that was supported at its ends by massive stone piers. For determining star declinations two graduated circles four feet in diameter were attached; four micrometer microscopes permitted readings to seconds of arc. Right ascensions were obtained by timing transits of stars behind five vertical threads in the field of view. In the expert opinion of Sir George B. Airy, this transit circle "at the time of its erection, and for several years afterwards, was the finest in the world." It still exists, on display at the Science Museum in London.

After several months spent mastering this instrument, Groombridge at the age of 51 embarked enthusiastically on an ambitious project of cataloguing all the stars brighter than magnitude 8½ between the north celestial pole and declination +38°. So energetically

The transit circle used by Groombridge in his observatory at Blackheath. On each stone pier is seen one of the micrometer microscopes for reading the two vertical graduated circles. Between the piers is a tripod to support a plumb line, used to determine the nadir point of the circles. From J. A. Repsold's *Zur Geschichte der Astronomischen Messwerkzeuge*, 1908.

did he work that by 1817 he had single-handedly made 24,000 observations of right ascension and 26,000 of declination. This output is all the more remarkable in that Groombridge continued his active business career until 1815.

Several contemporary descriptions of the Blackheath amateur at work have survived. The first is by his friend Thomas Firminger, who had been an assistant at Greenwich from 1799 to 1807. He told Airy:

"From the time that Mr. Groombridge erected his four-feet circle, up to the time above-mentioned of my leaving the Royal Observatory, he always made all his registered observations himself. . . . His uniform practice was, to write down his observations, after reading off the microscopes, upon a slate; and he usually kept two or three slates by him for that purpose, carefully examining his observations and registering them at his leisure. His observatory being close to his parlour, he frequently left his dinner, stepped into it, made his observation, noted it down on his slate, and then returned to his family and friends.

"He had a most accurate eye, both in observing and in reading off his observation, and was one of the most accurate and expeditious men I ever saw in the manipulation of his instrument. I do not recollect him ever to have made a mistake, or to have entertained a doubt on the accuracy of his observation, so far as reading off or time was concerned. . . . Mr. Groombridge was a man of no ordinary talents; and, although he had not in his early days acquired any knowledge of the mathematics, he nevertheless had a very ready and clear conception of all that is necessary in a good practical astronomer, and was most indefatigable in the pursuit of his favourite science."

Another glimpse comes from the geodesist Major-General Thomas Colby: "I never was in his [Mr. Groombridge's] observatory except once, and that once only for a few minutes. He was then observing, and I remember perfectly his making some multiplications of numbers, consisting of three or four places of figures, in a peculiar manner; setting down only the product. . . . From these calculations, he obtained the mean result to set down for the reading of the microscopes corrected for some instrumental errors. I also remember remonstrating with him on the advantage of setting down the readings of the microscopes in their simple form, as a security against error. And I also then recalculated some of the means he had taken, using the ordinary mode for multiplying, and in these cases I found him perfectly correct."

Groombridge had made considerable progress in the huge labor of reducing his observations when in 1827 he suffered a stroke from which he never fully recovered. This worthy man died at Blackheath on March 30, 1832. Finally, in 1838 Groombridge's catalogue of 4,243 circumpolar stars was published at government expense, under the editorship of Airy, who was now Astronomer Royal.

This catalogue has been of great value to later astronomers, providing as it does accurate early positions for all the naked-eye stars north of declination +38° and for about 1,000 stars fainter than magnitude 8.0. One indication of the lasting usefulness of Groombridge's observations was the publication in 1905 of a revised edition of his catalogue by F. W. Dyson and W. G. Thackeray of Greenwich Observatory.

Only four years after the first edition of this catalogue had appeared an inconspicuous star that Groombridge had observed five times suddenly became of special interest to astronomers. In a letter to the editor of the German journal *Astronomische Nachrichten* dated June 14, 1842, the director of Bonn Observatory, F. W. Argelander, noted: "Last year I accidentally found a remarkable star, whose proper motion significantly exceeds all others known, amounting to 7″ of a great circle per year. This 7th-magnitude star on the border between Canes Venatici and Ursa Major is No. 1830 in Groombridge's catalogue of circumpolar stars. I have frequently observed it last year and this in right ascension, and am waiting for the declinations that I have requested Mr. Nicolai of Mannheim Observatory to make. . . ."

Argelander wrote again to the *Astronomische Nachrichten* on December 8th of the same year. He noted that the star had been observed by Lalande (1794), Groombridge (1810), Bessel (1828), and Nicolai and himself (1842), leading to a proper motion of 7.059 seconds per annum. And in a third letter, dated March 30, 1843, Argelander made a penetrating comment about Grb 1830: "The difference in direction between its proper motion and solar motion is very great, 102° 10′, a circumstance which points toward a large parallax."

This prediction of the star being nearby was fully justified by later work. The most recent measurement of the star's parallax, obtained by W. D. Heintz of Swarthmore College from 1,130 photographs taken between 1938 and 1983, yields a value of 0.122 arc second with

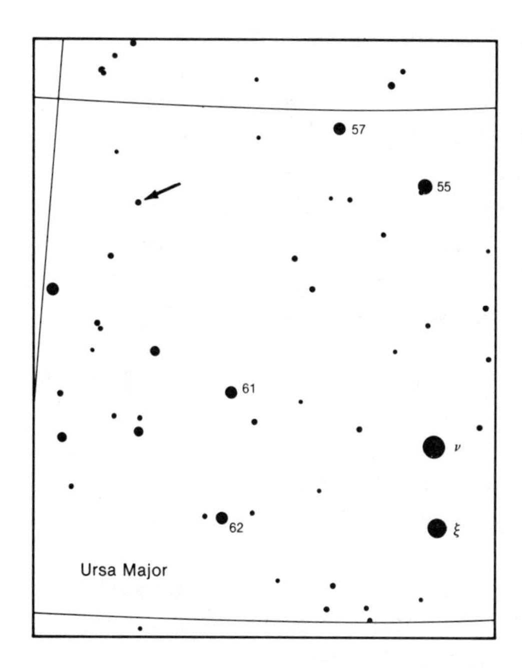

At visual magnitude 6½, the star Groombridge 1830 (arrowed) is easily visible in binoculars. This finder chart shows a 10-by-12-degree area of Ursa Major with north up and stars to magnitude 7.8 plotted. The labeled naked-eye stars Nu (ν) and Xi (ξ) Ursae Majoris lie about halfway between the Bowl of the Big Dipper and the Triangle of Leo.

only three percent error. This corresponds to a distance of just 27 light-years. Since the star's apparent magnitude is 6.45, the distance gives an absolute magnitude for Groombridge 1830 of 6.7; it is intrinsically six times fainter than the Sun.

This nearby high-velocity star has attracted much attention from astrophysicists. It is a subdwarf of spectral class G8, somewhat cooler and yellower than the Sun. The weakness of lines in its spectrum indicate that elements heavier than helium are 30 times less abundant than in our Sun. O. J. Eggen has pointed out that Grb 1830 is one of a group of four old halo-population stars, widely scattered in the sky, moving through space together with equal and parallel velocities.

The observed radial velocity of Grb 1830 is −98.2 kilometers per second, the negative sign indicating that it is approaching the Earth. The star's velocity perpendicular to the line of sight can be calculated as 275 kilometers per second from its proper motion and parallax. Combining these two velocity components, the total space motion of this star relative to the Sun can be determined as 290 kilometers per second. Grb 1830 will continue to approach the Sun until about the year 11,200, when it will have a parallax of 0.130 second of arc and appear 0.1 magnitude brighter than it does now. At that time its proper motion will have increased to 7.98 seconds of arc a year from the present 7.07. Thereafter, Grb 1830 will recede. Roger W. Sinnott has calculated the behavior of this star in the remote future, on the assumption that it is in uniform motion along a straight line through space. For our descendants 20,000 years hence, Grb 1830 will be a 6.5 magnitude star in Virgo, still with a sizable parallax of 0.120 second of arc. But 100,000 years from now it will have faded to magnitude 9.1 in Lupus, and its parallax will have diminished to 0.035 second of arc. One million years into the future, Groombridge's star will have receded into insignificance, being a 14th-magnitude object lost in the rich Milky Way of the southern constellation Norma.

Grb 1830 still holds mysteries. Up until 1968 no one suspected that it was double. Peter van de Kamp, who was then director of Sproul Observatory, tells the story: "A close companion appears on eight exposures of 2½ minutes each, taken by Michael D. Worth on four consecutive plates with the Sproul 24-inch refractor on the morning of February 27, 1968. . . . The companion appears about two magnitudes fainter than Groombridge 1830, i.e., of apparent magnitude 8.5." Measurements of the photographs placed the new object 1.7 seconds of arc from its primary in position angle 166°. (When the plates were later remeasured, the position was revised to 2.2 seconds and 165°.) Previous plates showed no sign of the object. Van de Kamp wrote, "It is difficult to avoid the obvious explanation that the observations reveal a flare-up of a companion of Groombridge 1830, which normally is several magnitudes or more fainter than the primary."

Six days after the event, further photographs showed no trace of the companion. But in early April, it was seen by two experienced visual observers – Charles Worley at the U. S. Naval Observatory and Heintz with the Sproul 24-inch telescope – as a difficult 12th-magnitude object. Other observers that month were unable to detect it, as if it had faded further. There have been no later sightings. On the basis of the 1968 observations, the companion was given the variable-star designation CF Ursae Majoris.

Groombridge 1830 (the left pair of stars in this double exposure) was announced to be a close double star based on this photograph taken on February 27, 1968, which was thought to show a companion in the process of flaring. However, in 1984 Wulff Heintz reexamined the plate and concluded that the "companion" is likely to be an instrumental effect. Courtesy Sproul Observatory.

In 1984 Heintz reevaluated the supposed flare object. He called attention to the fact that on the Sproul plates it does not appear stellar but "resembles a brush or moustache," as if it were caused by diffraction or some other effect inside the telescope. Other investigators have found tentative evidence for orbital motion by Grb 1830, but the presumed orbit disagrees with a companion being at the position of the flare object in 1968. "The suspected companion continues to be an utterly confusing issue," wrote Heintz. Some 175 years after Stephen Groombridge measured it, the books on this star are still far from closed.

69. The eagle eye of William Rutter Dawes

MANY ASTRONOMICAL OBSERVATIONS are useful for only a few years before they are completely superseded by vastly better ones. This is especially true in fields that rely on instruments developed with rapidly changing technology. In marked contrast, the best visual measurements of moderately close double stars made 100 to 150 years ago are still good by modern standards. The measures by Wilhelm Struve, E. Dembowski, and W. R. Dawes are valued highly by today's calculators of binary-star orbits. All three of these famous pioneers used relatively small telescopes. The accuracy of their micrometer observations of doubles is explained not by large or very perfect equipment but rather by an acute and highly trained eye, scrupulous care, and sound judgment. These qualities were especially marked in the English amateur astronomer William Rutter Dawes (1799–1868). "Probably the best astronomical observer in England in his day" was the considered opinion of Dawes by his fellow countryman W. F. Denning, himself an outstanding amateur of the next generation.

Dawes acquired his taste for astronomy from his father, who had gone out to Australia in 1787 as the astronomer on Governor Philip's first expedition to Botany Bay. When Dawes was born in London on March 19, 1799, his father was a mathematics teacher in a boys' school. He hoped his son would become a clergyman in the Church of England, but young Dawes' searching mind did not accept dogma easily, so he trained as a physician at St. Bartholomew's Hospital in London.

It was after he moved to Liverpool in 1826 that his interest in double stars began. In a letter written many years later to Sir John Herschel, Dawes recalled: "Having obtained the loan of a volume of Rees's *Encyclopaedia,* I had copied out the list it contained of Sir Wm. Herschel's Catalogues of Double-Stars, arranged in classes and constellations; and [using] a capital little refractor of only 1.6 inches

William Rutter Dawes was probably the best observer in England in his day. He acquired ever-larger refractors: a 1.6-inch, a 3.8-inch Dollond, a 6⅓-inch Merz (1846), a 7½-inch Clark (1854), an 8-inch Clark (1855), and an 8¼-inch Clark (1859). From *The Observatory, 36,* 1913.

aperture, and a copy of the French edition of Flamsteed's Atlas, which was presented by Dr. Maskelyne to my father . . . , I worked away on almost every fine night, when uncertain health would permit, and found and distinctly made out . . . *Castor, Rigel,* ϵ^1 *and* ϵ^2 *Lyrae,* σ *Orionis,* ζ *Aquarii,* and many others, of which I made correct diagrams in a book now lying before me. . . . The difficulty was often to get to bed in summer before the Sun extinguished the sight of the game."

In Liverpool, Dawes turned from medicine to religion as a career when he came under the influence of the celebrated Rev. Thomas Raffles, who for many years preached at the Independent Chapel on Great George Street. In this way Dawes came to take charge of a small congregation of the same denomination in the town of Ormskirk, about 15 miles north of Liverpool. There he took up astronomy in earnest about 1829, acquiring an equatorially mounted 3¾-inch Dolland refractor and a filar micrometer. During the next few years, with this small but good telescope, he secured over 600 measures of doubles. He introduced several refinements that made his work more accurate than that of his mentor John Herschel. One was the insertion of an achromatic Barlow lens to double the magnification without increasing the apparent thickness of the micrometer threads. Another was the use of a diagonal prism, to make the apparent direction of the line joining a pair of stars either vertical or horizontal, for controlling systematic errors in position angle (the direction in the sky of the secondary star from its primary). Both of these devices are used in refined form by present-day double star observers.

But while Dawes was becoming widely known and respected in the astronomical world, his personal life came apart. His wife, who was considerably older than he, died, and his always fragile health broke down, causing him to relinquish his Ormskirk congregation. Needing some way to earn a living, in 1839 he obtained the post of assistant at George Bishop's private observatory in London.

Bishop had amassed a large fortune in the wine business, and at about the age of 50 became a patron of science. In 1836 he erected an observatory with a 7-inch refractor near his residence at South Villa in Regent's Park. No astronomer himself, Bishop wisely hired the best available observers to use his excellent equipment. At various times during the next quarter century, the post of assistant was held by such able men as J. R. Hind, Norman Pogson, Eduard Vogel and C. G. Talmage.

Here Dawes continued his double star work from 1839 to early in 1844 with great success. However, there seems to have been some friction between the wealthy patron and his independent-minded astronomer. One indication is in a letter from Dawes to his friend George Knott many years later: "The observation of α Piscium which you quote as 'Bishop 1842.9' is *mine*. Mr. B. never did and never *could* observe at all, not even a transit; but after I left his observatory *he put his own name to all my observations!!* 1844.039 was the date of the last observation made there by me."

However, a major upturn in Dawes' fortunes began with his second marriage, in 1842, to the widow of an Ormskirk lawyer, John Welsby. He was able in 1844 to leave Mr. Bishop's service and to settle in a country house at Cranbrook in Kent, about 40 miles southeast of London. His friend Sir John Herschel lived nearby at Hawkhurst. At Cranbrook Dawes erected his own observatory with

Dawes made this rough sketch to show the inner dusky ring of Saturn that he independently discovered in 1850, on November 25th. Shortly thereafter his sighting was confirmed by William Lassell, who dubbed the newly recognized feature the crepe ring, the name by which it is known today. From *Monthly Notices* of the Royal Astronomical Society, *11*, 1851.

an excellent 6⅓-inch Merz refractor and returned enthusiastically to double star observing. About this time he began an active interest in planetary work. His health was still uncertain, and for a considerable period constant and intense headaches caused him to consider abandoning astronomy.

In 1850 Dawes was an independent discoverer of the crepe ring of Saturn. This faint inner ring was seen on November 11 and 15, 1850, by W. C. Bond with the Harvard Observatory 15-inch refractor, but before word of this crossed the Atlantic, Dawes had found it for himself. In a letter to Knott, Dawes first pointed out that William Lassell with a 24-inch reflector had very closely scrutinized Saturn on November 21st, a very fine night, without any suspicion of the dusky ring. "On the 25th of November," wrote Dawes, "I detected for the first time a light within the ansa of the ring at both ends while examining the planet with my Munich refractor of 6⅓ inches aperture. While I was endeavoring to make out what it could possibly be I was interrupted by some visitors. . . . The next fine night, the 29th November, I attacked it vigorously, and made it all out, though scarcely able to believe my eyes or my telescope . . . On December 2 Mr. Lassell came to see me . . . and the next night, the 3rd, being fine, I prepared to show him this novelty, which I had told him of and explained by my picture; but, naturally enough, he was quite indisposed to believe it could be anything he had not seen in his far more powerful telescope. However, being thus prepared to look for it, and the observatory being darkened to give every advantage on such an object, *he was able to make it all out in a few minutes*. . . . So true are the words of Sir W. Herschel himself, 'When our particular attention is once drawn to an object, we see things at first sight that would otherwise have escaped our notice.' "

A major service of Dawes to astronomy was his bringing the skill of Alvan Clark as a telescope maker to the attention of astronomers. In 1851 the American optician sent to Dawes a list of close and difficult double stars he had detected with instruments of his own construction. As Clark's first important customer, the English amateur purchased a 7½-inch refractor in 1854, an 8-inch in 1855, and an 8¼-inch in 1859. With these instruments he brought his lifetime total of double star measurements to nearly 2,800. The last installment of these results was published in Vol. 35 of the *Memoirs* of the Royal Astronomical Society, where the English amateur gave this description of the now-famous Dawes limit:

"It is a point of considerable interest to determine the *separating* power of any given telescopic aperture. Having ascertained about five and thirty years ago, by comparisons of the performance of several telescopes of very different apertures, that the diameters of star-disks varied inversely as the diameter of the aperture, I examined with a great variety of apertures a vast number of double stars, whose distances seemed to be well determined, and not liable to rapid change, in order to ascertain the separating powers of those apertures, as expressed in inches of aperture and seconds of distance. I thus determined as a constant, that a one-inch aperture would just separate a double star composed of two stars of the sixth magnitude, if their central distance was $4''.56$; – the atmospheric circumstances being moderately favourable. Hence, the separating power of any given aperture, *a,* will be expressed by the fraction $4''.56/a$."

While this formula is very familiar to amateurs today, its narrow range of validity is often overlooked. Determined with small refractors and Dawes' eye, the law does not tell what another observer can see with a large reflector or a catadioptric telescope.

In 1857 Dawes moved to Haddenham in Cambridgeshire where he spent the rest of his life. There he gained great regard for the free medical service he gave to the poor of the town. Eagle-eyed though he was at the telescope, he was so near-sighted that he could pass his wife on the street without recognizing her. Perhaps it was at Haddenham that the episode of Dawes' dog occurred.

A fine retriever regularly accompanied Dawes on his evening walk to the observatory, carrying the key in its mouth. Upon returning home with Dawes, the dog would surrender the key to the housekeeper. But one night Dawes was interrupted at the observatory by a call for urgent medical aid. Not wanting to take the observatory key with him he gave it to the dog to carry home. The next day Dawes could not find the key in its usual place, and he was told the dog had arrived without it. Then the astronomer held up a bunch of keys to the dog and said "Key, Dash, key." The retriever ran off to the garden and dug up the key that he had buried under a cabbage.

Dawes' wife died in December, 1860, and soon afterward his health worsened greatly. Despite heart disease and his continuing headaches and asthma, he managed to do a little observing until 1865. One gratification was his election that year as a fellow of the Royal Society, a scientific honor rarely given to an amateur. Dawes died at Haddenham on February 15, 1868, in his 69th year.

70. Mizar, Alcor, and a planet that wasn't

ONE OF THE RICHLY STORIED STARS in the sky is 2nd-magnitude Mizar, the next to last bright star in the handle of the Big Dipper. To the naked-eye observer it attracts attention because of its close 4th-magnitude neighbor, Alcor, distant only 12 minutes of arc or a third of the full Moon's width. Alcor is so easily seen without optical aid that it is surprising to hear the often repeated tale of the ancient Arabs regarding it as a test of acute vision. This statement sounds suspiciously like some blunder uncritically copied from book to book and kept alive by repetition.

Mizar holds a unique place in double star astronomy. It appears to have been the first double discovered with a telescope, around the year 1650, by the Italian astronomer Giovanni Battista Riccioli. Even a very small telescope and modest magnification will split Mizar into two stars of magnitude 2.3 and 4.0, separated by 14 seconds of arc. This pair was also the first double star photographed, when George P. Bond in 1857 secured a successful impression of it on a collodion plate with the 15-inch Harvard refractor.

The brighter component of the pair – Mizar A – was also the first spectroscopic double star detected. In 1889 Antonia C. Maury noted on Harvard Observatory photographs of its spectrum that the lines appeared sometimes double and sometimes single. This phenomenon indicated that Mizar A was a binary, its components being nearly equal in brightness and revolving around each other in a period later shown to be 20.5 days. The new companion is much too close to A to be made out visually with even the largest telescopes. In fact, when in 1925 F. G. Pease measured the spectroscopic pair at Mount Wilson Observatory with a 20-foot interferometer attached to the 100-inch reflector, the distance between the two stars was only about 0.012 second of arc – 10 times smaller than the visual telescopic limit.

Mizar is thus at least a triple system. The visual pair AB has remained almost unchanged in the three centuries it has been known. The very slow increase in position angle indicates that the orbital period of this pair is many thousands of years. At the distance of the Mizar system, about 70 light-years, their separation on the sky corresponds to about 28 billion miles, or 10 times as far as Neptune is from the Sun. The interferometer measures indicate that the spectroscopic pair has a separation half that of the Sun and Mercury.

Moreover, in a sense Alcor is connected with the Mizar system, for it is one of the few dozen other scattered stars that form the Ursa Major moving cluster – stars that are traveling together across the sky much like a flight of wild geese.

Is there yet another component in the Mizar system? This question was raised in 1939 by the French astronomer G. Oriano. On comparing the very numerous micrometer measurements of the AB pair made since 1822, he believed he had found a 57-year oscillation in the relative location of these two stars. This effect he attributed to an invisible companion moving around one of them, probably B, in an orbit presented almost edgewise to us. So nearly on edge was it thought to be that A. V. Nielsen in Denmark suggested that Mizar B might even be an eclipsing variable star. He cited a curious half-forgotten observation in support of this conjecture.

The Berlin amateur J. H. Mädler, well known for his lunar and planetary observations, in 1840 was appointed director of the university observatory at Dorpat, Russia (see Chapter 6). He was successor to the famous Wilhelm Struve, who had gone to Pulkovo Observatory. Mädler became a very active double star observer with the Dorpat 9.6-inch refractor. On April 18, 1841, he turned this instrument on Mizar just before sunset, and to his great surprise found no trace of the B component. It should have been seen, for check observations on other doubles showed even fainter stars readily. Later the same night Mädler looked at Mizar again, and B was once more shining with full brilliance. He commented that this behavior could be explained if Mizar B were an eclipsing star of the Algol type but with a much longer period. Moreover, this Mädler observation was made at a time when Mizar B and its hypothetical companion were practically in the same line of sight, according to Oriano's orbit.

But the neat picture contains a fundamental defect. If we intercompare the very accurate photographic measurements of the double star Mizar AB made over the last four decades, there is no indication of Oriano's 57-year oscillation. He had depended on the much rougher visual measurements, which are affected by systematic errors for which he made no allowances. Probably they account for his result. Even though we put Oriano's discussion to one side, however, Mädler's strange observation remains an unsolved puzzle, a small skeleton in the astronomical closet.

Nearly midway between Mizar and Alcor but somewhat south of the line connecting them is an 8th-magnitude star called Sidus Ludoviciana, although it is most unusual for an ordinary star this faint to be known by other than some catalogue number. The name was given two centuries ago by J. G. Liebknecht, who was professor of theology and mathematics at the German university of Giessen.

Alcor and Mizar as they appear in a small telescope; the circled field has a diameter of ½°. SL denotes Sidus Ludoviciana, once thought to be a planet. Courtesy Dennis di Cicco.

(One of his books, dated 1721, was *Pharus, or a Dissertation on Fiery Wonders in the Sky, Collected from Every Age,* which sounds rather like an early treatise on flying saucers.)

On December 2, 1722, Liebknecht was exploring the heavens with a nonachromatic refractor six feet long – not much of a telescope even in those days – when he chanced upon this star. He thought it a new object, and from some inaccurate measurements of its position decided that it was in motion. Believing himself the discoverer of a new planet, he named it Sidus Ludoviciana after his sovereign, the Landgrave Ludwig of Hessen-Darmstadt. Liebknecht also wrote a pamphlet announcing his find and sent copies to all the leading astronomers of Europe.

Their reaction was unfavorable. L. Zumbach von Koesfeld, then a prominent astronomer, physician, and musician of Kassel, announced that Liebknecht's measures were too crude to warrant any trust. In a further pamphlet, J. F. Weidler of Wittenberg told how he had looked at Mizar with a superior telescope 22 feet long on four nights in February, 1723, and saw the "planet" as a telescopic fixed star, with no indication of motion. The sharpest criticism came from L. P. Thümmig, professor of philosophy at Halle. In his pamphlet, he said that he had observed the controversial object many times and that it was obviously an ordinary star. It was scarcely necessary, he added, for Professor Liebknecht to announce every telescopic star as new and to give it some special name. In reply, the angry Liebknecht wrote a 16-page rejoinder that heaped vulgar abuse on Thümmig, and the latter published a rebuttal that brought the squabble to a close.

There is a strange fact about this controversy to which Wilhelm Olbers called attention a century later. None of the disputants, not even Weidler with his 22-foot refractor, mentions that Mizar is a visual double star. If they really did not see Mizar as double, it is hardly a tribute to the quality of the telescopes of that period.

71. Visual star colors and a list of red stars

WHAT IS THE COLOR OF A STAR? To answer this question, a contemporary astronomer would cite measurements made with a photoelectric photometer that has seen the star through different color filters. Through the 1950's most astronomers would have evaluated the color by photographing the star on blue- and yellow-sensitive photographic emulsions, then comparing the magnitudes measured on the two plates. It has been a long time since the colors of stars were determined simply by looking!

In the latter years of the 19th century, many astronomers, professional and amateur, studied star colors visually. In particular, searching for very red stars was a common occupation. The Irish amateur John Birmingham compiled a list of 658 of them in 1876, which was revised and extended in 1888 to 766 objects by the English clergyman-astronomer T. E. Espin. When the British Astronomical Association was founded in 1890, one of its observing sections was devoted to stellar colors and was very active for a few years.

Astronomical fashions can change abruptly, and amateur interest in colored stars is now largely history. There are some visible vestiges of the older period in the venerable Norton's *Star Atlas,* which was first published in 1910. Even its latest edition labels certain red stars by the letter "R," and occasionally gives their numbers in the Espin-Birmingham catalogue.

The stars themselves have not changed, and their hues can still fascinate the amateur who explores the heavens for his own pleasure. Even to the unaided eye, the white of Vega, the orange of Aldebaran, and the red of Antares are conspicuous, and with practice many intermediate hues become recognizable among the 1st- and 2nd-magnitude stars. Fainter objects, however, appear a nondescript gray or a subjective blue, as a rule. But a 3-inch or 4-inch telescope will extend a color survey to about magnitude 6, while a 6- or 8-inch reflector is highly effective for picking out red stars in Milky Way fields. The larger the aperture used to view a star, the more distinct and vivid its hue. Reflecting telescopes are better than refractors for this purpose, because they are fully achromatic (free of false color).

It is still an interesting experiment to spend an evening or two gauging the tints of a number of the brighter stars by inspection, preferably with the aid of binoculars. Certain precautions are helpful. Avoid stray light, which can falsify colors by contrast or fatigue effects. Do not attempt to view stars too near the horizon, where the atmosphere can cause a spurious reddening. A modern list of unusually red stars is given on page 372. Most are variable in brightness, a common property among the redder stars. To locate the fainter objects, a detailed star chart will be needed.

Verbal descriptions of color have their limitations; they are highly personal and often indefinite, and terms like *fiery red* invite exaggeration. It is much better to use a numerical scale for estimating star colors, a proposal first made in 1872, by the Athens astronomer J. F. J. Schmidt. He explains his idea as follows: "I leave aside the greens, blues, and purples seen only in double stars, as well as the greenish tinge of many single ones, and confine myself to the series which begins at pure white, runs through all grades to yellow, and passes eventually to red. Actually my observations do not show any perfectly white or perfectly red stars. . . . In my scale I set pure white = 0, pure yellow = 4, deep golden yellow = 6, and all of my red stars are graded between 6.5 and 9."

Many slight modifications of the original Schmidt scale were later suggested. The German amateur H. Osthoff of Cologne spent 25 years studying star colors visually. For practical use, the definitions of the color steps he proposed in 1912 are helpful:

0 White
1 White with a slight admixture of yellow
2 Yellow and white in equal parts
3 Light yellow
4 Pure yellow
5 Deep yellow
6 Yellow with a slight admixture of red
7 Red and yellow in equal parts (orange)
8 Orange red
9 Red

A good idea of how an old-time color observer of the best class went about his work is given by Osthoff. He describes the systematic estimates of star colors with a 4-inch refractor in preparation of his catalogue published in 1900: "The observing room was always entirely darkened. I covered my head and the eye end of the telescope with a dark cloth. The observations were written down in the dark, and the color was always expressed in one figure, without decimals. Only under the most pressing circumstances was the lantern opened during the time of observation, and then only to look at the star chart. I always estimated an unknown star with respect to a known star. At the conclusion of the observations, still during the night, or at the latest the next morning, I glanced over my notes and

Some well-known red stars

Star	R.A. (2000)	Dec.	Color (B-V)	Mag.	Notes and comments
R Lep	4h 59m 36s	−25° 46′ 09″	1.8	5.5–10.5	Hind's "Crimson Star," described by Schönfeld as "intense blood red"
α Ori	5h 55m 10s	+ 7° 24′ 26″	1.8	0.4–1.3	Betelgeuse
Y CVn	12h 45m 08s	+45° 26′ 25″	2.5	5.5–6.0	Named "La Superba" by Secchi
α Sco	16h 29m 24s	−26° 25′ 55″	1.8	0.9–1.8	Antares
μ Cep	21h 43m 30s	+58° 46′ 48″	2.4	3.6–5.1	William Herschel's "Garnet Star"
TX Psc	23h 46m 23s	+ 3° 29′ 13″	2.6	5.3–5.8	19 Piscium
Some very red stars					
R Scl	1h 26m 58s	−32° 32′ 36″	3.9	6.5–8.1	
U Cam	3h 41m 48s	+62° 38′ 56″	4.3	8.1–8.6	Argelander: "extraordinarily red"
W Ori	5h 05m 24s	+ 1° 10′ 39″	3.4	6.2–7.0	
SAO 172106	6h 39m 31s	−30° 02′ 19″	3.4	7.8	
X Cnc	8h 55m 23s	+17° 13′ 53″	3.4	6.5–7.0	
RY Dra	12h 56m 26s	+65° 59′ 37″	3.3	6.8–7.3	
HD 113842	13h 07m 28s	−60° 16′ 09″	3.6	7.2	
V Pav	17h 43m 19s	−57° 40′ 04″	3.7	5.6–7.5	
T Lyr	18h 32m 20s	+36° 59′ 56″	3.7	8.3–8.9	Secchi: "intense"
V Aql	19h 04m 24s	− 5° 41′ 06″	4.2	7.4–8.0	Schmidt and Vogel: "intense fire-red"
HD 189256	19h 57m 13s	+44° 15′ 39″	3.4	7.8	
RS Cyg	20h 13m 23s	+38° 43′ 44″	3.3	6.5–9.3	

From left to right, the data include: the star's name; celestial coordinates; photoelectrically measured color index in magnitudes (the larger this number the redder the star); visual magnitude or, if the star is variable, its magnitude when brightest and faintest.

identified the stars. Before the beginning of each evening's work I looked over the program, but did not take any heed of the observations already made. When there was bright moonlight, unsteady air, or a too cloudy sky, I made no estimation of color. I looked long and fixedly at each star until the impression of its color no longer fluctuated. The average time required for a color estimate was 2.21 minutes."

72. The visual Milky Way

VIEWED FROM A MOUNTAINTOP on a very clear, dark night, the Milky Way is a glorious sight quite unlike the pallid, featureless band that city dwellers know. Under such favorable circumstances, it becomes a richly textured pattern of bright star clouds, complex mottlings, softly luminous arms and dark lanes, with broad dim outliers. A sharp-eyed skywatcher who sets out systematically to learn this pattern will find, as experience grows, that the features of the Milky Way are remarkably complex and delicate. During the last years of the 19th century, study of the naked-eye Milky Way flourished briefly as an important branch of astronomy. But today this work is another example of a facet of our science that became nearly forgotten because technical progress led elsewhere.

Strictly speaking, the visible Milky Way is a subjective phenomenon. Since Galileo's time we have known its light comes from multitudes of faint stars. Those too dim to be individually recognized furnish, through the imperfect imagery of the rod receptors in our eyes, small overlapping blurs on the retina. The resulting picture is the end product of a very complicated process. Thus the Milky Way of the visual observer is by no means the same as the photographic Milky Way or that determined from star counts.

The real father of visual study of the Milky Way was F. W. A. Argelander (1799–1875) of Bonn Observatory (see Chapter 80). Although he did not contribute directly, his famous *Aufforderung* (1844) stimulated others to begin work. This influential tract was an appeal to amateur astronomers to undertake naked-eye observations of vari-

The southern winter Milky Way, as drawn in 1926 by Antonie Pannekoek. This vista extends from Canis Minor and Orion at right to Vela at left (where it adjoins the plate on page 376). In all these pictures the equator of our galaxy runs horizontally across the center. Reproduced from *Annalen* of Bosscha Observatory, 1928.

able stars, meteors, the zodiacal light, and the Milky Way. It gave valuable suggestions on techniques, including the earliest explanation of Argelander's step method for estimating the brightness of variables.

One of the first to take up the invitation was Eduard Heis of Aachen, Germany. His famous star atlas, fruit of a quarter-century's labor, shows the detailed course of the Milky Way north of declination $-35°$. Heis introduced a much used 1-to-5 scale of brightness and in his atlas plotted the contours for each brightness level. One device of Heis is worth borrowing by Milky Way observers today. To shield his eyes from extraneous light, he looked through a stubby cardboard cylinder, blackened within, that was 12 inches in length and diameter.

Another Argelander disciple was the famous selenographer J. F. J. Schmidt (see Chapter 50). For many years he was engaged in plotting the intricacies of the naked-eye Milky Way, but he published nothing. Long after his death, two of his exquisite pencil drawings were discovered in the archives of Potsdam Observatory and were published in 1923 as Vol. 14, Part 2, of the *Annalen* of Leiden Observatory. Unfortunately, Schmidt left no record of his procedure: Only a marginal note states that his Milky Way charts were based on observations of 1864 to 1879. His drawings are extraordinarily rich in detail, some of which, though absent from all other visual charts, can nevertheless be confirmed on modern small-scale photographs.

Also a skillful mapper of the Milky Way was C. Easton of Rotterdam, the Netherlands, who published his representation in 1893. Almost simultaneously, a similar study was completed by Otto Boeddicker, a German astronomer who was Lord Rosse's assistant at Parsonstown, Ireland. He spent six years in mapping the Milky Way to declination −10°, producing a striking rendition notable for its many faint, narrow branchings. This chart was published by lithography in *The Milky Way,* London, 1892.

This early work was largely in vain, due to the great difficulties 19th-century engravers and printers encountered in reproducing the delicate nuances and shadings of the original drawings. Even Boeddicker's handsome lithographs do less than justice to their originals, says a competent critic. The intervention of a workman's hand between the artist's original and the printed version made a scientifically faithful reproduction impossible.

Greatest of all naked-eye observers of the galaxy was, I think, the versatile Dutch astronomer Antonie Pannekoek of Amsterdam. As a young man he began his first experimental observations in 1889, stimulated by F. Kaiser's translation of Argelander's *Aufforderung* into Dutch. His elaborate study of the northern Milky Way was published in 1920 as Vol. 11, Part 3, of the Leiden *Annalen.* In 1925–1926 he was able to extend this survey to the southern heavens on the occasion of an eclipse expedition to Sumatra. The resulting second monograph appeared as a Bosscha Observatory publication. Pannekoek's charts cover the entire Milky Way. The three shown

The Milky Way from Vela (right) to Norma. The Southern Cross lies to the right of the famous dark Coal Sack near the center. To its left, the bright stars are Alpha and Beta Centauri. From Bosscha *Annalen*, 1928.

here illustrate the somewhat different styles he used in portraying its northern and southern portions.

A special part of his program was a careful comparison of the surface brightnesses of different places in the Milky Way by means of step estimates, much as stars would be intercompared. He chose 18 standard points, for which he decided on step values ranging from 6.4 for a spot in the bright Scutum star cloud to 0.4 for a dark region in Cepheus. Then other areas were estimated in terms of these 18 points. In this manner the Amsterdam astronomer compiled a catalogue of the brightnesses at 128 places in the northern Milky Way, each one observed several times. The mean error of a single estimate was only ± 0.3 step. His second monograph contains a further catalogue of 189 places, some of them coincident with earlier positions.

The last of Pannekoek's 1926 sketches extends from Norma (right) past Scorpius to Aquila. This is the southern summer Milky Way for Northern Hemisphere observers. The open cluster Messier 7 is conspicuous between the stinger of the Scorpion and the Teapot of Sagittarius. From Bosscha *Annalen,* 1928.

To identify particular points, Pannekoek used notations such as γ Aquilae–¼–ζ Aquilae, which means the position one quarter of the way from Gamma Aquilae to Zeta. Similarly, β Lyrae –γ Lyrae –δ Lyrae indicates the centroid of the triangle formed by these three stars. During the actual observations Pannekoek wrote down his records in total darkness. Thus it was necessary for him to memorize beforehand the stars that would be needed as bench marks. During his southern work in 1925–1926 he dictated his observations to his wife. This procedure was followed both for the brightness estimates and for the descriptive notes. The next morning he would make a drawing based on these notes and his still vivid recollection. Then

The star clouds and dust lanes of Scorpius, Ophiuchus, and Sagittarius dominate this picture centered in the direction of our galactic nucleus. One can almost imagine seeing the central bulge of our galaxy. The photograph was taken by Allan E. Morton from Arizona's Mount Graham, a candidate site for the proposed 15-meter National New Technology Telescope.

the drawing would be filed and not consulted again until the final map was prepared, thus avoiding bias.

The most recent chart of the naked-eye Milky Way was made by Harvard astronomer Sergei Gaposchkin during a 1956–1957 visit to Mount Stromlo Observatory in Australia. There his main program was spectroscopic observations with one of the large telescopes. The long exposures gave him abundant opportunity to inspect the Milky Way through the opening in the totally darkened dome. "In drawing or perceiving the fine filaments of the Milky Way," writes Gaposchkin, "it is essential that the eye should be exposed to the darkness for at least 15 minutes, and that the angle of vision should be not less than 60°." Since the latitude of Mount Stromlo is −35°, all of the heavens south of declination +55° can be seen above the horizon there at one time or another. Thus Gaposchkin was able to prepare a drawing showing the entire circuit of the Milky Way, apparently making use of observations at some other site to show features north of +55° declination. This delineation was published with a description in *Vistas in Astronomy*, Vol. 3. It is less detailed than some earlier maps but has the special advantage of showing the whole length of the galactic equator on a single sheet.

The present-day amateur who is blessed with superior observing conditions but is short on equipment might find naked-eye observations of the Milky Way an interesting project. He or she could study a limited region – say Cygnus – in great detail by Pannekoek's methods. Although personal preferences differ, probably a shaded pencil drawing is the best medium for portraying the Milky Way. A whole new approach is offered by short-focus, wide-field cameras or fisheye lenses. Such equipment can narrow the gap between the visual and the photographic Milky Ways.

73. The visual Orion Nebula

THE INTRODUCTION of a powerful new observing technique often ends a chapter in astronomical history. This happened when Henry Draper took the first successful photograph of the great Orion nebula on September 30, 1880 – a 57-minute exposure with the 11-inch Clark refractor of his private observatory at Hastings-on-Hudson, New York. He soon followed this picture with others of 104 and 137 minutes exposure, which showed clearly the vast wreaths and swirls of luminous gas so familiar to us on modern photographs of this splendid object.

These pictures and similar ones taken by A. A. Common in England brought to a close a long period during which many of the world's best-known astronomers spent enormous labor in mapping visually the details of Messier 42. Once it was abandoned, this work soon became nearly forgotten. Yet it has interest for present-day amateurs, whose views of the nebula match the best of the old drawings more closely than recent photographs, where the bright inner parts are usually heavily overexposed.

Although the Orion nebula can be discerned with the naked eye, the earliest extant reports of it date from just after telescopes came into astronomical use. The nebula was seen in 1611 by Nicolas Peiresc in France and soon after by J. B. Cysat in Switzerland. While observing the comet of 1618, the latter compared its appearance to "the tightly grouped small stars, embedded in a bright, white cloud" that his telescope showed in the middle of Orion's sword. These sightings were forgotten for centuries, so the discovery of the Orion nebula was long attributed to Christiaan Huygens in 1656. Using a 2⅓-inch nonachromatic refractor, the famous Dutch astronomer that year made the earliest known drawing of the object. This crude depiction was published in his 1659 book about Saturn.

While such well-known observers as Charles Messier and J. H. Schröter devoted much attention to the Orion nebula toward the end of the 18th century, the scientific study of it really began only with John Herschel in the 1820's. One reason for the delay lies in the history of printing, for until the invention of lithography around 1800 there was no really satisfactory way to reproduce drawings of the nebula. The old woodcuts and etchings could not represent the soft texture and range in surface brightness, and so they were inaccurate and misleading. One consequence was a widespread 18th-century belief that the nebula undergoes large and rapid changes in its overall appearance.

In 1826 Herschel (he did not become Sir John until 1831) reported to the Royal Astronomical Society a careful study of the Orion nebula. He used the 18¼-inch reflector of 20 feet focal length that he had built in 1820 under the supervision of his aged father, William (see Chapter 25). The usual magnification was about 150, obtained with a single-lens ocular. Nebulous matter, John Herschel conjectured with surprising foresight, was probably "a self-luminous or phosphorescent material substance in a highly dilated or gaseous

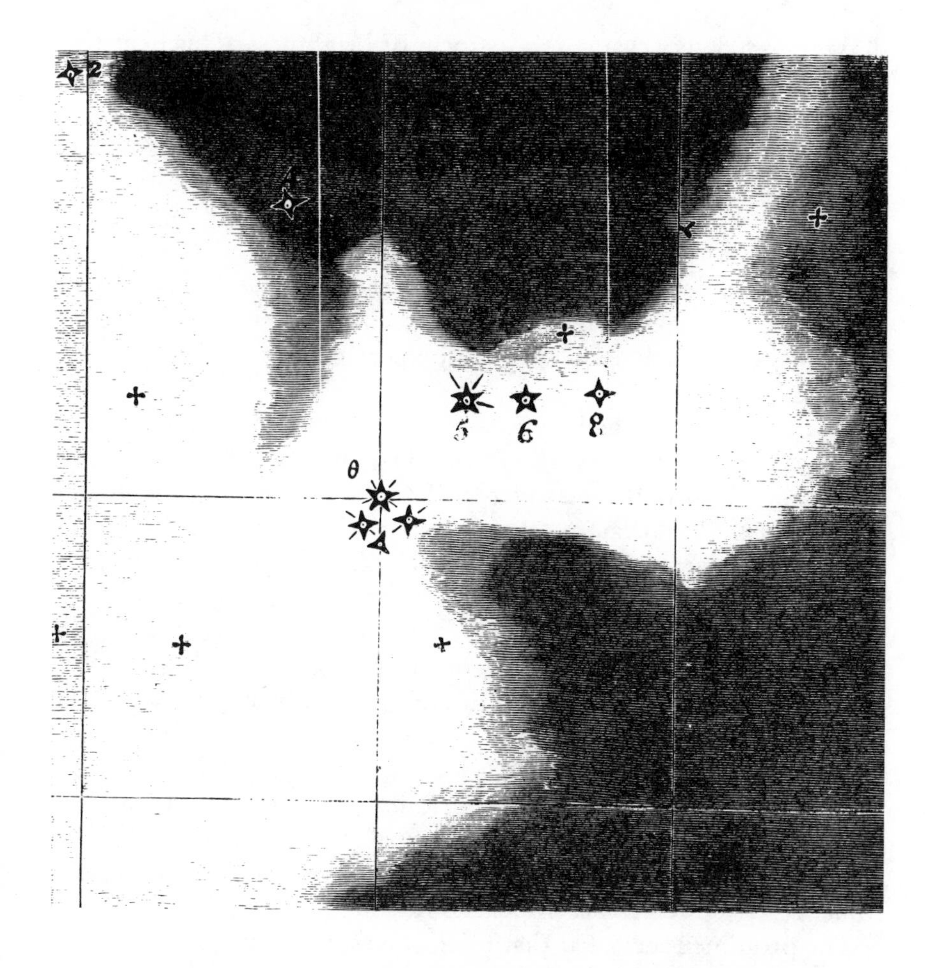

Among the better 18th-century pictures of the Orion nebula is this one published by the French comet hunter Charles Messier in 1771. Nevertheless, the star positions are very rough, and therefore the shape of the nebula as a whole is somewhat distorted. Messier used a 3.3-inch refractor at 68 power. Reproduced from E. S. Holden, *Monograph of the Central Parts of the Nebula of Orion,* Washington, 1882.

state, but gradually subsiding by the mutual gravitation of its molecules into stars and sidereal systems." To learn more about the nature of the nebulae would require detection of evolutionary changes, he said. Hence, he set for himself the task of making a very careful drawing of the Orion nebula that could be compared with later drawings to detect changes in its form or brightness. This depiction was prepared from sketches and notes taken on several favorable nights and was afterward compared with the sky and corrected.

Amateurs today may still find use for the largely forgotten set of names that Herschel introduced in 1826 for the main features of the Orion nebula. This nomenclature is shown in the chart on the next page. Certain of the names are derived from "a rude resemblance which the whole nebula presents to the head, snout and jaws of some monstrous animal." Other regions are named after various early observers of the nebula. Here are the main features of this terminology.

Regio Huygeniana, the Huygenian region, is the bright central part. It was named after Huygens (1629–1695), whom Herschel mistakenly believed was the discoverer of the Orion nebula. This area surrounds the Trapezium, the multiple star Theta^1 Orionis. The uneven brightness of this region was vividly compared by Herschel to "a curdling liquid, or a surface strewed over with flocks of wool, or to the breaking up of a mackerel sky." This texture is well seen in Leopold Trouvelot's drawing on page 386. As the chart shows, Regio Huygeniana corresponds to the upper part of the head of the "monstrous animal." From this association derive the names *Frons* (forehead), *Occiput* (back of the head), and *Rostrum* (beak).

Extending southeastward from *Rostrum* is *Proboscis Major* (the greater trunk), a milky lane that was first observed in 1773 by Messier with a 3.3–inch refractor. The darkness between it and *Proboscis Minor* is *Regio Messieriana* (Messier's region).

The prominent dark bar that reaches westward almost to the Trapezium, and which corresponds to the open mouth of the monster, was named by John Herschel *Sinus Magnus* (the Great Gulf). In Trouvelot's drawing and that by G. P. Bond this gulf is seen to be spanned from north to south by one or more faint streaks. The most prominent of these, not named by Herschel, later became known as *Pons Schroeteri* (Schröter's Bridge), after the well-known German amateur who described it in 1797.

On the key chart, south of Regio Huygeniana lies another conspicuous dark gulf, *Sinus Gentilii.* This designation commemorates the French astronomer G. H. Le Gentil, who detected the feature in 1758 with a Gregorian reflector of six-foot focus.

To round out Herschel's nomenclature, we note the series of six

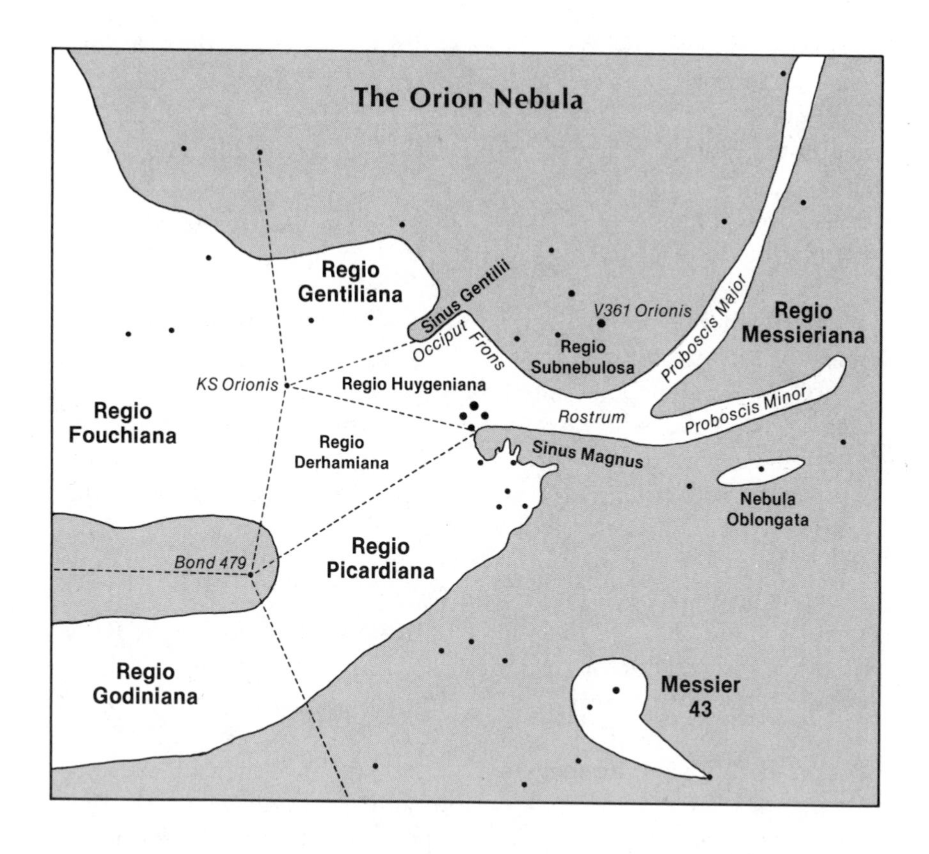

South is up, east to the right in this sketch map of the inner parts of the Orion nebula, showing some of the names introduced by John Herschel in 1826. His original chart is here turned left for right and relettered, since he observed with a front-view reflector. The terminology is described in the text; also, *Nebula Oblongata* and *Regio Subnebulosa* mean Elongated Nebula and Slightly Nebulous Region, respectively. KS Orionis and V361 Orionis are modern variable star designations; their visual magnitude ranges are 9.9–10.9 and 7.8–9.6. The star Bond 479, so numbered in G. P. Bond's catalogue, is of magnitude 10.5. Adapted from *Memoirs* of the Royal Astronomical Society, *2*, 1826.

regions or provinces in the left half of the chart, named after pioneer observers of the Orion nebula. Proceeding counterclockwise from the top these men are Le Gentil, J. P. Grandjean de Fouchy, L. Godin, J. Picard, W. Derham, and Huygens.

Herschel, we recall, had made his detailed 1826 drawing of the Orion nebula in hopes of detecting eventual changes. He made a second study of this object in 1837 during his expedition to South Africa. As a preliminary, he determined the positions of 150 stars in the nebula with a 5-inch refractor to insure accurate placement of nebulous features as drawn with his 18¼-inch reflector. His conclusion, from a comparison of his work at these two epochs, was that the Orion nebula was virtually unchanged, any differences being mainly due to inaccuracies in his earlier drawing.

Although John Herschel lived until 1871, he did not observe the nebula further. But other astronomers were quick to follow him in charting ever-growing detail.

William Cranch Bond, the first director of Harvard Observatory, was 57 years old when its 15-inch refractor was mounted in June, 1847. This instrument surpassed Herschel's reflector in effective light-gathering power and was far more convenient to use. The following winter Bond paid close attention to the Orion nebula, which he believed could be resolved into stars on the best nights, when its brighter portions seemed full of points of light.

Meanwhile, M. V. Lyapunov at Kazan Observatory in Russia used a 10-inch refractor for a minute scrutiny of special areas in the nebula. Otto Struve was also investigating it with the 15-inch Pulkovo refractor, twin of the Harvard instrument. Because Struve criticized William Bond's work, the latter's son and successor George P. Bond undertook a new and elaborate visual study to vindicate Harvard Observatory's reputation. This work continued, with some interruptions, from 1857 to 1865. With the Harvard 15-inch, the nebulosities of Messier 42 and 43 could be traced over an area of 2.3 square degrees, whereas Lyapunov with his smaller refractor found an area of only 0.12 square degree for Messier 42. The faint extensions gave George Bond a strong impression of a spiral structure, a finding which he announced in 1861.

Bond determined the relative positions of about 1,100 stars in the nebula. One of his assistants in this work was Asaph Hall, later famous as discoverer of the satellites of Mars. Hall recalled years later: "I have a very distinct recollection of how cold my feet were when he was making his winter observations on Orion. I sat in the small alcove of the great dome behind a black curtain, and noted on the chronometer the transits of stars when Professor Bond called them out, and wrote down also the readings for declination. . . . Sometimes I was called to the telescope to examine a very faint star,

G. P. Bond's engraving of the Orion nebula is perhaps the most faithful of all visual depictions. Slightly reduced in scale from the original, the area shown here is about 22 minutes of arc from east to west, and 19 from north to south (top). This engraving, as told in the text, is based on observations in 1857–64 with a 15-inch telescope. From *Astronomical Engravings . . . of Harvard College Observatory,* 1876.

or some configuration of the nebula. Professor Bond had one of the keenest eyes I have ever met with."

Bond did not live to publish his work on the Orion nebula. During the winter of 1864–1865, as his tuberculosis progressed, he strove anxiously to complete the manuscript. After he became too feeble to hold a pen he continued to dictate until February 16, 1865, the day before his death. This work was edited by T. H. Safford for publication in 1867 as Volume 5 of the *Annals* of Harvard College Observatory.

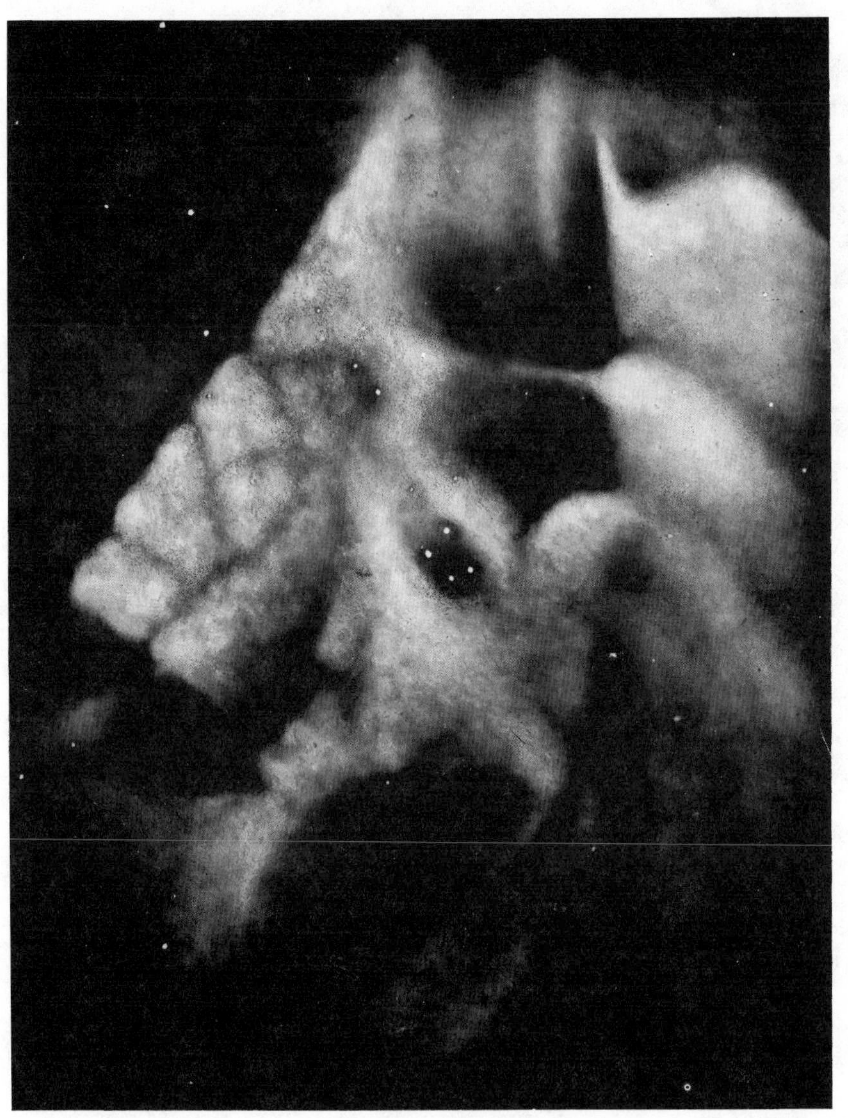

The inner parts of the Orion nebula as drawn by Leopold Trouvelot in 1874, from his observations with the 15-inch Harvard Observatory refractor. South is up as viewed with page sideways, and the area shown is about seven minutes of arc wide. The four bright stars of the Trapezium are in an illusory dark hole that was known as Sinus Lamontii (Lamont's Gulf). Reproduced from *Astronomical Engravings . . . of Harvard College Observatory*, 1876.

Fortunately, Bond's famous steel engraving of the Orion nebula that forms the frontispiece of that book was completed before his death. It is characterized as follows by Edward S. Holden, who was both Bond's cousin and a later authority on the nebula. "I am acquainted with but one drawing of the nebula which is entirely above criticism, that of the late G. P. Bond. He was himself a skilled artist, and he had been familiar with the nebula for fifteen or twenty years. He made scores of drawings, in white on black, and the reverse, in colors, etc. Each of these was revised and re-revised many times. The final drawing in water-color was copied by Mr. Watts, a skillful engraver, who himself was extremely familiar with the nebula, from repeated views and studies of it through the Harvard refractor. The revisions of the original plate lasted many months, and I have myself examined from fifteen to twenty *final* revises of the plate. Color, form, and relative brilliancy were all successively and exhaustively criticized, and Professor Bond expressed himself as fully satisfied with the plate in every essential feature."

That quotation is from Holden's monumental *Monograph of the Central Parts of the Nebula of Orion*, published in 1882 as an appendix to the Washington *Astronomical and Meteorological Observations* for 1878. At the U. S. Naval Observatory, Holden had used the 26-inch refractor between 1873 and 1880 for the last of the great visual studies of the Orion nebula. Holden was particularly interested in the possibility of changes in small structural details. He made many comparisons of the relative brightness of various nebulous patches, sometimes with a crude visual photometer. These observations were supplemented by micrometer measurements of positions and orientations. He was also aware that the nebula was rich in variable stars, for the Bonds had noted light changes in the stars now known as T, AF, AP, AQ, and NV Orionis. Holden made many estimates of these and other stars suspected to vary.

His 230-page monograph is also a remarkably complete history of the Orion nebula, for he went to great pains to collect all earlier accounts and pictures of it. Here you can read, for example, what Messier said in his own words. Holden managed to obtain photographic copies of some important unpublished drawings, such as J. F. J. Schmidt's remarkably detailed view with the Athens Observatory 6-inch refractor in 1861. Unfortunately, with few exceptions, Holden has reproduced all these pictures as woodcuts.

Holden's study of the Orion nebula ended when he left the Naval Observatory to become director of Washburn Observatory in Wisconsin. And, practically simultaneously, photography intervened so powerfully that no one since has followed in the footsteps of Herschel, the Bonds, and Holden.

74. E. P. Mason and the nebulae

FEW PERSONALITIES in American astronomy are as haunting as Ebenezer Porter Mason, who died in 1840 when he was only 21 and at the start of a brilliant career. As a Yale undergraduate, he and two classmates built the largest telescope in the United States. He used this instrument so effectively as to win Sir John Herschel's warm praise: "Mason, a young and ardent astronomer, [was] a native of the United States of America, whose premature death is the more to be regretted, as he was (so far as I am aware) the only other recent observer who has given himself, with the assiduity which the subject requires, to the exact delineation of nebulae, and whose figures I find at all satisfactory . . ." Here, too, is the judgment of Prof. Denison Olmsted (1791–1859) of Yale on his young protégé and collaborator: "[Mason] in early youth developed talents for Practical Astronomy so extraordinary, as to leave no doubt, that, had his life been spared, he would have risen to a rank among the first astronomers of the age. . . . Science will long mourn the loss of a youth signally endowed with that rare assemblage of qualities essential to the structure of a great astronomer in which are united the refined artist and the profound mathematician."

Mason was born on December 7, 1819, in the village of Washington, Litchfield County, Connecticut, where his father was pastor of the Congregational church. When eight years old, he was sent to live with an aunt in Richmond, Virginia. His interest in astronomy began about this time, as he learned the constellations with the aid of a celestial globe. In 1830 he rejoined his father, who had moved to the island of Nantucket, Massachusetts. As a young teenager,

Ebenezer was already beginning to attract attention for his artistic skill and his flair for mathematics.

When in 1835 the 16-year-old Mason entered Yale College his talents flowered quickly. The professor of astronomy was Olmsted, an outstanding teacher who wrote clear and thorough textbooks and encouraged his students to do practical observing. Yale in 1830 had acquired a 5-inch Dollond refractor of good optical quality. But it had a crude altazimuth mount on casters, so that the instrument could be pushed from one window of the octagonal observing tower to another. It was with this instrument that Olmsted and Elias Loomis (1811–1889) recovered Halley's Comet on August 31, 1835, before word of its detection at Rome on the 5th had crossed the Atlantic.

As a freshman, Mason was not eligible to take Olmsted's astronomy course, but the two became close friends. Borrowing a 6-inch Herschelian reflector from a classmate, he made observations of Jupiter's satellites, the rings of Saturn, occultations, and other phenomena. But his interest turned more and more to the nebulae.

Soon after, Mason and several friends made a number of small Newtonian and Herschelian telescopes with speculum-metal mirrors. We are told: "One of Mason's letters gives an account of an evening in his room devoted to figuring a mirror with two classmates, each one taking his turn in the tedious task until a late hour, with frequent trips outdoors to make a hasty test sight of the stars, for the Foucault test (see Chapter 29) was then unknown."

At that time the largest telescope in America was perhaps the 8½-inch reflector built by Amasa Holcomb of Southwick, Massachusetts, and exhibited by him in 1835 at the Franklin Institute in Philadelphia. The Yale students surpassed this with a 12-inch Herschelian of 14 feet focal length. Mason gives these details: "The instrument was first planned and begun in the summer of 1838, by my friend and classmate, Mr. H. L. Smith. A tolerably good metal was cast, after several failures, and the speculum was finally polished near the close of the summer. Mr. Smith and Mr. Bradley shared the expenses attending the formation of the mirror and erection of the telescope, and divided the long labor of grinding the speculum, and I united with them in the less tedious task of giving the mirror its final polish and figure . . ."

"The mode of mounting the telescope was similar to Ramage's, but ruder. The base consisted of three beams, forming a triangle, which revolved on a circular ledge of plank, by means of rollers at the angles, and which was guided truly in its circuit by a crosspiece, through which rose a central bolt, firmly driven into the ground. From the angles of this base rose three beams, meeting at a height of sixteen or seventeen feet from the ground, and a rope passed through a pulley fixed at this height, and sustained the weight of the upper part of the telescope. The lower end, containing the speculum, rested on a small platform at one of the solid angles of the base, and revolved with the frame. The quick motion in altitude was by means of the rope just mentioned, which passed down to a windlass at the base, while a slow motion was gained by a . . . combination of ropes within the immediate command of the observer. In azimuth, the whole frame could be wheeled about by a single person, and a slower motion was obtained by simply swinging the telescope by hand, which could be done by the observer, in following a star, with perfect steadiness."

Although Smith was chiefly responsible for building this big telescope, Mason led the way in putting it to scientific use. In the summer of 1839, he and Smith made a detailed study on many nights of three important objects: the Trifid and Omega nebulae in Sagittarius and the Veil nebula in Cygnus. Their purpose was to obtain accurate descriptions and delineations for use in detecting possible future changes. This work was reported in a masterful 49-page paper in the *Transactions* of the American Philosophical Society. For each nebula, Mason began by carefully charting all the faint stars in its vicinity to furnish reference points. Many of these stars were measured with the 5-inch refractor and filar micrometer. This operation was difficult with an altazimuth-mounted instrument since the micrometer had to be reoriented for each measure. Other stars were plotted by eye estimation.

In addition to making meticulous pencil drawings of nebulae, Mason was the first astronomer (as far as I know) to prepare brightness-contour maps. Reproduced here is his contour map of the Trifid nebula in which the "isophotes" depend on visual estimates rather than photometric measurement. His simple method is worth trying by present-day amateurs. He defined the innermost contour line as that one "imagined to surround all those portions of the

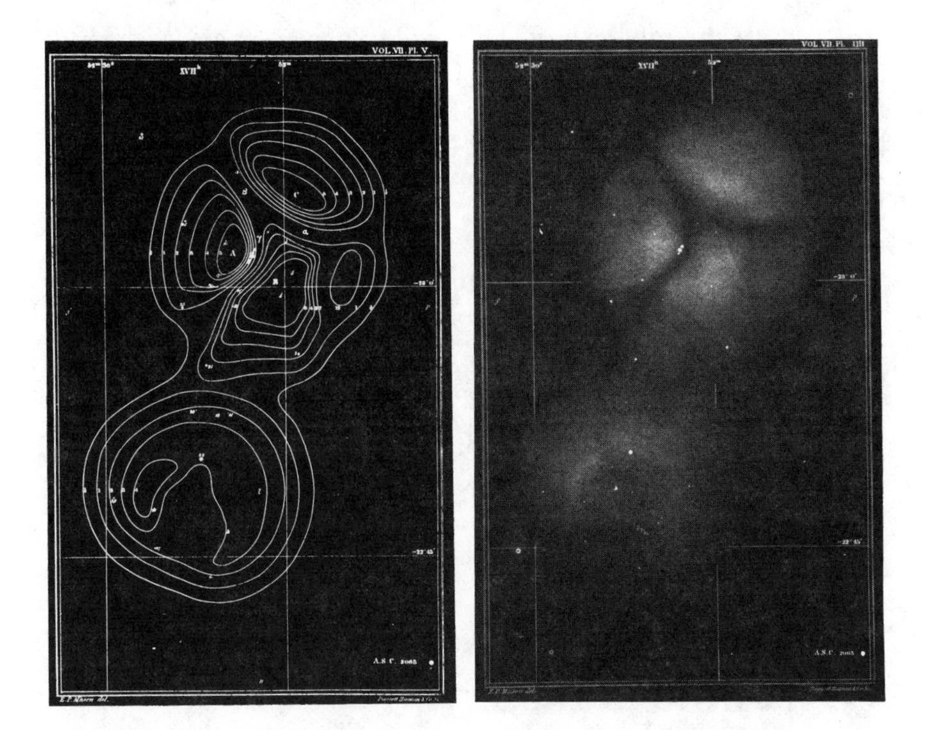

Messier 20, the Trifid nebula in Sagittarius, as depicted in two ways by E. P. Mason in 1839. Because a Herschelian (front-view) reflector was used, south is up and west to the right, so the view is a mirror image of that in a refractor or Newtonian. Both the drawing and the isophotal map are at about half the scale of the originals in the *Transactions* of the American Philosophical Society.

nebula which are of uniform brightness, and brighter than any other part." The next outer line represents "the first perceptible gradation or diminution of light," and so on, until the outermost contour, "which represents the utmost bounds of the visible nebula." In practice, Mason visualized each line as winding its way among the stars of the field and then drew it on a star chart. He then corrected the contours by frequent comparisons with the telescopic view. Another feature of this memoir was a catalogue of the stars in the field of each nebula, containing 29, 37, and 196 entries for the Trifid, Omega, and Veil regions, respectively.

This investigation was submitted by its 20-year-old author to the American Philosophical Society in April, 1840, when he was already seriously ill with tuberculosis. During his senior year at Yale he had helped Professor Olmsted in preparing a new edition of the latter's *Introduction to Astronomy*. He also began work on a supplement to this book, entitled *Practical Astronomy,* which explained the use of astronomical instruments and the methods of reducing observations. It had chapters on the prediction of eclipses and on finding geographical longitude from observed times of stellar occultations.

Meanwhile, Mason's health continued to deteriorate. Hoping that an outdoor life might be beneficial, in the summer of 1840 he took a job with a field party that was surveying the disputed boundary line between Maine and Canada. This strenuous work in the wilderness was unsuitable, and he returned in October. With his tuberculosis now far advanced, he managed to complete the last pages of his *Practical Astronomy*. Finally, he boarded ship for Richmond to stay with his uncle and aunt, but he died 11 days after his arrival. It was December 26th, three weeks past his 21st birthday.

75. A hole in the sky

IN THE SUMMER of 1833 Sir John Herschel was making the final preparations for his famous expedition to the Cape of Good Hope. There he would extend to the southern skies the survey of northern stars and nebulae conducted by his father, Sir William Herschel (see Chapter 9). His 83-year-old aunt Caroline Herschel took a lively interest in his plans, for she too was an enthusiastic astronomer and had been the faithful assistant of her brother for many years. On August 1st Caroline wrote from her home in Hanover, Germany: "Dear Nephew: As soon as your instrument is erected I wish you would see if there is not something remarkable in the lower part of the Scorpion to be found, for I remember your Father returned several nights and years to the same spot, but could not satisfy himself about the uncommon appearance of that part of the heavens. It was something more than a total absence of stars (I believe)."

After John Herschel arrived in South Africa and began work with his 18¾-inch reflector, he wrote to his aunt on June 6, 1834: "I have not been unmindful of your hint about *Scorpio*. I am now *rummaging*

Through persistent correspondence, Caroline Herschel (1750–1848) encouraged her nephew John to make careful star counts in the region around Rho Ophiuchi, the location of her brother William's "hole in the heavens." She was nearly 80 when she sat for this portrait by Tielemann.

the recesses of that constellation and find it full of beautiful globular clusters. . . ."

Replying on September 11th, Caroline was not satisfied: "It is not *Clusters of Stars* I want you to discover in the body of the Scorpion (or thereabouts) for that does not answer to my expectation, remembering having once heard your Father, after a long awful silence, exclaim, 'Hier ist wahrhaftig ein Loch im Himmel! [Here is truly a hole in the heavens!]' "

What was it that so impressed and tantalized William Herschel? As usually told in the popular literature, the story ends here, with perhaps some allusion to the dark nebulae in the Scorpius Milky Way. Actually, Sir William's own writings tell a good deal more.

With the newly completed 18¾-inch reflector of 20-foot focus he began in 1783 an extensive series of sweeps to discover new clusters, nebulae, and double stars. During the course of these sweeps, made with a 157x eyepiece, he frequently stopped to count the number of stars visible in the field of view, which was 15 minutes of arc in diameter. Such a *star gauge* gave the number of stars per unit area, as seen in a particular direction, brighter than the limiting magnitude of the telescope. (A rough comparison with modern star counts suggests that this limit was about magnitude 15.) For greater accuracy, Herschel often averaged the counts from as many as 10 neighboring fields. "Where the stars happened to be uncommonly crowded," he wrote, "no more than half a field was counted, and even sometimes only a quadrant; but then it was always done with the precaution of fixing on some row of stars that would point out the division of the field, so as to prevent any considerable mistake."

Observations of this kind, made in many parts of the sky and especially the Milky Way, were the basis of William Herschel's famous paper, "On the Construction of the Heavens," which he presented to the Royal Society of London in February, 1785. This classic in stellar statistics was a brilliant attempt to determine the form and approximate dimensions of the sidereal system. In this paper Herschel also advanced some theoretical ideas on the tendency of clusters to form in an initially nearly uniform scattering of numberless stars. He speculated that both globular and open clusters might be formed in that way, adding, "as a natural consequence . . . there will be formed great cavities or vacancies by the retreat of the stars towards the various centers which attract them. . . ."

Later in the same monograph Herschel presents observational evidence for such vacancies, in a section headed *An Opening in the Heavens*. This passage reads: "Some parts of our system indeed seem already to have sustained greater ravages of time than others, if this way of expressing myself may be allowed: for instance, in the body of the Scorpion is an opening, or hole, which is probably owing to

---

The region of Rho Ophiuchi, site of Herschel's "hole in the heavens." E. E. Barnard wrote: "One very striking thing about all the nebulosity [photographed] in this region is the fact that it is so faint that it cannot be seen with the eye even in a powerful telescope." Photo by Alan McClure.

this cause. I found it while I was gaging in the parallel from 112 to 114 degrees of north polar distance [declination −22° to −24°]. As I approached the milky way, the gages had been gradually running up from 9.7 to 17.1; when, all of a sudden, they fell down to nothing, a very few pretty large [bright] stars excepted, which made them shew 0.5, 0.7, 1.1, 1.4, 1.8; after which they again rose to 4.7, 13.5, 20,3, and soon after to 41.1.

"This opening is at least 4 degrees broad, but its height [extent in declination] I have not yet ascertained. It is remarkable that [Messier 80], which is one of the richest and most compressed clusters of small stars I remember to have seen, is situated just on the western border of it, and would almost authorise a suspicion that the stars, of which it is composed, were collected from that place, and had left the vacancy. What adds not a little to this surmise is, that the same phaenomenon is once more repeated with [Messier 4]: which is also on the western border of another vacancy, and has moreover a small, miniature cluster, or easily resolvable nebula of about 2½ minutes in diameter, north following it, at no very great distance."

Clearly, Caroline Herschel in her letters to her nephew half a century later was recalling this fascination of William's for starless voids in close association with rich clusters. The catalogue of star gauges in that paper indicates an extent from about $16^h\ 00^m$ to $16^h\ 17^m$ in right ascension and at least −22° to −25° in declination. This is the vicinity of Rho Ophiuchi, and Herschel's "Loch im Himmel" is unquestionably the Rho Ophiuchi dark nebula, familiar in Milky Way photographs ever since E. E. Barnard's time.

During his stay in South Africa from 1834 to 1838, John Herschel did pay special attention to starless fields encountered in the course of his sweeps. He published a list of about 50 such positions, nearly half of them in the general neighborhood of Rho Ophiuchi. Thus, on May 24, 1835, he described the region at right ascension $16^h\ 14^m$ between declinations −23° 03′ and −24° 11′ as containing "not the smallest star. Sky perfectly pure and superb."

These observations by the Herschels were conducted with large but somewhat inefficient reflectors. A modern amateur using a good-sized rich-field telescope at a very favorable site might gain a rather different impression of this remarkable region of the Milky Way.

Lord Rosse (1800–1867). From A. v. Schweiger-Lerchenfeld's *Atlas der Himmelskünde,* 1898.

76. The discovery of spiral structure in galaxies

IN ALL OF William and John Herschel's many observational descriptions of the forms of nebulae, the word spiral does not seem to occur. The term *spiral nebula* dates back only to 1845, when the Earl of Rosse introduced it to describe his discovery of this characteristic. By the start of the 20th century, photographic surveys showed that the faint nebulae existed by the hundred thousands and that a large fraction of them were spirals. Today, explaining why so much of the material in the universe has assumed such a shape is an important problem in astronomy.

William Parsons (1800–1867) was one of the very few amateur astronomers who dreamed of building the largest telescope in the world and then did so. He enjoyed the advantages of being heir to a wealthy Irish peer and of being a highly talented engineer. At Birr

Castle he began making small speculum-metal mirrors in 1827 and completed a 36-inch altazimuth reflector by 1839 (the same year William Herschel's 48-inch was finally dismantled). Parsons was well advanced with the construction of a 72-inch reflector of novel design when, upon his father's death, he became the third Earl of Rosse.

This instrument was sufficiently complete in February, 1845, for test observations, which demonstrated that its light grasp and definition were excellent for its appointed task, a survey of the Herschel nebulae. With a focal length of 53 feet, the 72-inch leviathan was used as a Newtonian. Slung from chains between two stonework walls 56 feet high, the tube was supported at its lower end by an enormous universal joint. The telescope was restricted to viewing a strip of sky along the meridian roughly one hour wide.

"The spiral arrangement of 51 Messier was detected in the spring of 1845," Lord Rosse told the Royal Society in 1850. Presumably he made the discovery personally, for although in later years most of the observing with the 72-inch was done by others, the first regular assistant (Johnstone Stoney) was not appointed until 1848. One of the earliest to see the spirality must have been J. P. Nichol of Glasgow, for he made the drawing of M51 that Lord Rosse displayed at the 1845 meeting of the British Association for the Advancement of Science.

A second spiral, M99 in Coma Berenices, was recognized during the spring of 1846. Lord Rosse wrote in June, 1850: "The other spiral nebulae discovered up to the present time are comparatively difficult to be seen, and the full power of the instrument is required, at least in our climate, to bring out the details. It should be observed that we are in the habit of calling all objects spirals in which we have detected a curvilinear arrangement not consisting of regular re-entering curves; it is convenient to class them under a common name, though we have not the means of proving that they are similar systems. They at present amount to fourteen, four of which have been discovered this spring." This list included M33 in Triangulum, NGC 2903 in Leo, and NGC 7479 in Pegasus.

Lord Rosse's announcement concerning M51, the Whirlpool nebula in Canes Venatici, raised much excitement at the time because of its bearing on current cosmological thinking. M51 had been care-

This photograph of Lord Rosse's 72-inch telescope, near the present town of Birr, County Offaly, Ireland, may have been taken around 1880, to judge from the growth of ivy. The instrument had two primary mirrors of speculum metal, each six inches thick, weighing 3½ and four tons, respectively. While one mirror was being repolished, the other was used in the telescope. Above the western (left) wall, note the movable gallery upon which the observer stood to reach the Newtonian focus, when the telescope pointed near the zenith. The dark horizontal bar below the mouth of the tube is the screw that drove the telescope in hour angle. This was cranked by an assistant, giving a motion smooth enough to permit rough micrometer measurements at the eyepiece. Science Museum photograph:

This drawing of the Whirlpool nebula as seen in the large Birr Castle reflector was published in 1850. Lord Rosse later concluded that the lesser core (NGC 5195) was also a spiral; photographs, however, show it as an irregular galaxy.

Sir John Herschel's drawing of M51, made more than a decade before Lord Rosse's discovery, shows a split ring surrounding the central condensation. Such a depiction closely resembled Herschel's idea of the Milky Way's structure. From *Philosophical Transactions* of the Royal Society of London, *123,* Part 2, 1833.

fully studied by John Herschel with an 18¾-inch reflector. He described it in 1833 as consisting of a bright round nebula and a companion, the former surrounded at some distance by a ring which was double in its southwestern part. "Supposing it [the ring] to consist of stars, the appearance it would present to a spectator placed on a planet attendant on one of them excentrically situated towards the north preceding quarter of the central mass, would be exactly similar to that of our Milky Way. . . . Can it be, then, that we have here a brother-system bearing a real physical resemblance and strong analogy of structure to our own?"

The spiral arrangement in M51 and similar objects immediately suggested internal motions, governed by dynamical laws that Lord Rosse in 1850 thought "almost within our grasp." This lead was followed up by an American scientist, Stephen Alexander (1806–1883), in a remarkable series of papers in the *Astronomical Journal* for 1852. Alexander argued that "the Milky Way and the stars within it constitute a spiral and several (it may be *four*) branches, and a central (probably spheroidal) cluster. . . ." He even attempted to trace the course of these spiral arms across the sky, on the basis of John Herschel's maps and description of the naked-eye Milky Way. Bravely, he sketched in qualitative terms the evolutionary history of a typical galaxy. The initial state was a huge rotating mass in dynamical equilibrium. As it cooled it contracted and, becoming unstable as the rotation accelerated, began to shed its matter in the form of spiral arms.

As far as I know, Alexander was the first to suggest seriously that our galaxy has a spiral form. This man, who deserves to be better remembered today, was professor of mathematics and astronomy at the College of New Jersey, now Princeton University.

The number of known and suspected spiral systems was increased by the publication of another major memoir by Lord Rosse in 1861. But after his death in 1867 the difficult art of polishing and refiguring the 72-inch mirrors was gradually lost. (With the copper-rich speculum metal, this chore was necessary every two or three years.) The large telescope at Birr Castle was in active service for several decades more, but it was little used for nebulae.

Curiously, the Great Andromeda galaxy M31 was not recognized to be spiral until very late. However, the dark "canals" discovered in 1847 by G. P. Bond with the 15-inch Harvard refractor are the interarm spaces. The true form was not revealed until long-exposure photographs were taken in 1888 by Isaac Roberts with a 20-inch reflector and in 1890 by E. E. Barnard with a 6-inch f/5 refractor. Strangely enough, neither Roberts nor Barnard used the word "spiral" to describe their pictures, but spoke of "rings." The correct interpretation of these photographs must have followed very soon, however.

Interwoven with all this history is the practical question of how large a telescope is needed to recognize spiral structure visually. The veteran amateur Walter Scott Houston devoted much care to this point, with telescopes up to the 36-inch Steward Observatory reflector. He informs me that he has never seen spiral form clearly enough to have justified a *discovery* situation. Seeing what is already known to exist requires less optical power than discovery. Lord Rosse tells us that the first Otto Struve saw the spirality of M51 distinctly with the 15-inch Pulkovo refractor. In the Sagot-Texereau *Revue des Constellations* (1963), we read that G. Gauthier could view distinctly the whorls of that galaxy with a 13-inch reflector at 80x. On the other hand, according to Houston, such experienced observers as Edgar Everhart and Leslie Peltier could not see a spiral pattern in M33 or M51 with a 12-inch, even under very favorable conditions. These facts help explain why the discovery of spiral nebulae remained the

Bond's "canals," dark dust lanes in the Andromeda galaxy, are evident in this 1874 drawing by Leopold Trouvelot. Also depicted are the satellite galaxies Messier 32 (above) and NGC 205. When compared to a modern long-exposure photograph, the fidelity of Trouvelot's rendition, made from observations with Harvard Observatory's 15-inch refractor, is amazing. South is up. From *Astronomical Engravings . . . of Harvard College Observatory*, 1876.

exclusive province of Lord Rosse's leviathan until astronomical photography came of age.

77. The nucleus of the Andromeda galaxy

THE LONG-EXPOSURE PHOTOGRAPHS of Messier 31 inserted in most astronomy books are intended to display that galaxy's spiral arms. Hence the bright central regions are strongly overexposed, concealing a very small, almost starlike core at the precise center.

This innermost feature was concisely described in 1929 by Edwin P. Hubble from his observations with the 100-inch reflector: "The nucleus itself, on the shortest exposures, is sensibly round, with a diameter of about 3″ . . . and with an apparent photographic magnitude about 14.0. . . ." This is a good description still, but photoelectric data imply that the nucleus is about a magnitude brighter than this. A large, long-focus refractor like the 30-inch, f/18.5 Thaw telescope at Allegheny Observatory is also effective for recording the tiny heart of the Andromeda system, as the accompanying photograph by Walter Feibelman demonstrates. He writes, "The plate scale of the Thaw is 14.6 seconds of arc per millimeter. Measured on an enlarged print of the original negative, the nucleus comes out to be 4.5 seconds in diameter."

This feature should be carefully distinguished from the central condensation, or general brightening of the Andromeda nebula toward its midpoint. The condensation is a gross detail visible in small telescopes. In fact, the total light from an area 42 seconds of arc in diameter is equivalent to a star of photographic magnitude 9.5, according to J. Stebbins and A. Whitford's photoelectric measurements.

Thus, the inner part of the Andromeda nebula can be described in the same language that we use for a comet head: a coma containing a faint starlike nucleus. The problems of specifying the magnitude

The nucleus of the Adromeda galaxy photographed with the 30-inch Thaw refractor by W. A. Feibelman. This is a 9½-times enlargement from an Eastman 103a-O plate exposed 10 minutes. On the same scale, the three-degree length of M31 that is ordinarily seen would stretch 23 feet! North is up, and Lamont's star is near the right edge, about 125 seconds of arc from the nucleus. In 1885 the supernova S Andromedae appeared 0.4 inch to the right of the nucleus and 0.1 inch below. Spectroscopic observations reveal that the nucleus is rotating rapidly, in a period of about 520,000 years. Physically, it resembles an abnormally massive and luminous globular cluster. The four faint starlike objects below the nucleus are globulars of about 16th magnitude.

of a comet are closely paralleled here. A comet may look five magnitudes brighter in a small telescope than on a photograph taken with only enough exposure to record its nucleus; similarly, the central condensation of M31 may appear approximately 8th magnitude in an amateur's 3-inch refractor and 13th magnitude on a short-exposure plate with the 100-inch telescope. In fact, by comparing visual descriptions of M31 as seen with telescopes of different size, we may gain some insight into the problems of comparing comet magnitude observations.

The published visual reports of the M31 nucleus are extensive, though scattered in the literature, and go back many years. On the very first nights (July 15 and 20, 1847) that the 15-inch refractor at Harvard Observatory was used, William C. Bond noted that the great nebula in Andromeda had a nucleus resembling a star. "I do not recollect to have seen any notice of this appearance," he stated in a letter to President Edward Everett of Harvard that was published in the *Astronomische Nachrichten*. But Bond was not the discoverer of the nucleus, for on October 13 and 14, 1836, F. Lamont had seen it with the 10½-inch refractor of Munich Observatory. With a filar micrometer, Lamont determined its diameter as 6.9 seconds of arc and measured its coordinates relative to a nearby 12th-magnitude star. Conspicuous on the Feibelman photograph, this neighbor may for convenience be called *Lamont's star*.

Especially important among the visual observations of the nucleus are E. E. Barnard's with the 40-inch Yerkes refractor, which he reported in the *Astronomical Journal* in 1917. He had long before thought to answer the question of whether the Andromeda nebula was another galaxy or lay inside our Milky Way by determining if the nebula had an appreciable proper motion. For this purpose, Barnard measured micrometrically the position of the nucleus relative to three faint stars on many nights from 1898 to 1916. One of these three reference objects was Lamont's star, which Barnard judged to be magnitude 12.3; its position angle and distance from the nucleus were determined as 261°.29 and 124″.65, the mean of 29 nights. "The nucleus of the nebula," wrote Barnard, "is about 2″ to 3″ in diameter, but it is so strongly condensed that under good conditions it can be bisected with almost the same accuracy as the comparison stars. From its nature, the brightness of the nucleus varies greatly with the size of the telescope and the magnifying power used: With the 40-inch telescope it is of about 13 to 14 magnitude, or about the same brightness as star *c*." The latter, rated by Barnard as magnitude 13.8, lies 158 seconds of arc south of the nucleus.

In the April, 1960, *Publications* of the Astronomical Society of the Pacific, A. Lallemand, M. Duchesne, and Merle F. Walker reported how the nucleus of the Andromeda galaxy looked at the coudé focus of the 120-inch Lick reflector, where the scale is 1.9 seconds of arc per millimeter. "Even with such a large scale," they noted, "the visual appearance of the nucleus on the slit was nearly stellar, with only a slight haze visible around it; the inner parts of the nebula, so bright on ordinary photographs, were completely invisible." The photographic diameter of the nucleus is 4.4 seconds, according to these investigators.

At its brightest, the 1885 supernova in the Andromeda nebula dramatically changed the appearance of the galaxy's core. From A. v. Schweiger-Lerchenfeld's *Atlas der Himmelskünde,* 1898.

The central part of Messier 31 received searching attention from a host of observers upon the appearance of the supernova S Andromedae in August, 1885, at a point only 16 seconds of arc from the nucleus, in the direction of Lamont's star. At maximum light, the supernova was approximately 6th magnitude, and its gradual fading could be followed in large telescopes until early 1890. At first, when S Andromedae was still very bright, its glare obliterated its immediate surroundings, and for a short while it was believed that an outburst of the nucleus itself had occurred. But as the supernova began to fade, the "old nucleus" became increasingly visible, and many observers measured the relative coordinates of the two with micrometers.

E. Hartwig, a young German astronomer then at Dorpat Observatory in Russia, suggested that the nucleus was variable in brightness. He noted that E. Schönfeld at Mannheim Observatory (presumably using a 6-inch refractor) had reported it "about 10th magnitude, easily seen even through thin clouds," and that H. L. d'Arrest had assigned a magnitude of 9 or 10. On the other hand, Hartwig himself in late 1885 found "even on a clear moonless night, only a very faint star certainly fainter than 12." With a power of 500 on the 9.6-inch Dorpat refractor, Hartwig could see that the image was not quite stellar. The correct interpretation of these statements, I suspect, is not that the nucleus is variable, but simply that Schönfeld and d'Arrest were describing the central condensation and Hartwig the true nucleus.

STAR ATLASES AND OTHER PUBLICATIONS

78. Johann Bayer and his star nomenclature

PARADOXICALLY, JOHANN BAYER (1572–1625) is both famous and obscure today. Every amateur astronomer has heard that Bayer introduced the Greek-letter designations that are still used for the brighter naked-eye stars. Also, the beautiful constellation maps from his sky atlas of 1603 are frequently reproduced in astronomical books and magazines. Nevertheless, much of what the popular literature says about him and his star letters is misleading or inaccurate.

Bayer was a German lawyer who lived most of his life at Augsburg in Bavaria, but he was born at the village of Rain, about 80 miles northeast. Some of the confusion about him arose because Bayer was a very common name in southern Germany. But the right man has been pinpointed by Basil Brown, an English historian of astronomical maps, who consulted the Augsburg city archives. Brown also found Bayer's grave in the Dominican cemetery; on it is an epitaph declaring him "Publicly known for his most excellent work, the URANOMETRIA".

Evidently our lawyer was a person of some consequence, since he served as adviser to the Augsburg city council at a salary of 500 silver guldens a year. About his activities as an astronomer hardly anything is known, except what can be gleaned from his *Uranometria* itself. This popular guide to the sidereal heavens was dedicated to two prominent citizens and to the city council, which acknowledged the compliment with an award of 150 guldens.

The *Uranometria* contains 51 star maps beautifully engraved on copper by Alexander Mair. In addition to two hemispheres, there are charts for each of Ptolemy's 48 classical constellations plus a map showing the far southern sky with a dozen newer constellations – Apus, Chamaeleon, Dorado, Grus, Hydrus, Indus, Musca, Pavo, Phoenix, Triangulum Australe, Tucana, and Volans. Altogether, about 2,000 stars are depicted. The foundation is the catalogue of 1,005 stars observed by Tycho Brahe and published in 1592. Ordinarily Bayer copied Tycho without change, but he did revise the magnitudes of some of the fainter stars, and he added a number of stars omitted by Tycho. The part of the sky too far south to be observed from Europe was copied from the celestial globe made in

ΟΥΔΕΙΣ ΕΙΣΙΤΩ
ΑΓΕΩΜΕΤΡΗΤΟΣ
AETERNITATI.
IOANNIS BAYERI
RHAINANI I.C.
VRANO-
METRIA,
OMNIVM ASTERISMORVM
CONTINENS SCHEMATA,
NOVA METHODO
DELINEATA,
AEREIS LAMINIS EXPRESSA.
ATLANTI
VETVSTISS.
ASTRONOM.
MAGISTRO.
HERCVLI
VETVSTISS.
ASTRONOM.
DISCIPVLO.

1601 by Iodocus Hondius of Amsterdam. This in turn was based upon the catalogue of southern stars by the Dutch navigator P. D. Keyzer from observations made in Madagascar and Sumatra during voyages in the 1590's. Altogether, Bayer borrowed 129 stars from Hondius' globe, together with the 12 constellations listed above. For some reason he did not assign letters to any stars in these constellations.

The great popularity of the *Uranometria* is evident from new editions in 1639, 1648, 1655, and 1661. Not only was this handsome atlas much more extensive than any previous one, but also the use of letters as star designations was a great convenience to practical astronomers. Previously, astronomers had generally followed the usage of Ptolemy, who identified stars by their locations in the mythological constellation figures. "At the tip of the tail of the Lesser Bear" is an awkward if unambiguous way of specifying Polaris. But "the preceding of two in the right rear foot" is a decidedly clumsy way of referring to ν Ursae Majoris. There was considerable scope for error since no two drawings of a constellation figure quite matched. Moreover, *right* and *left* as used in these descriptions referred to the constellations as seen on a globe (*i.e.* viewed from outside the celestial sphere) rather than from inside as we see the sky.

Bayer was not the first to try to alleviate these inconveniences. For example, numbers were assigned to stars in the two sky maps that Albrecht Dürer engraved about 1515 to illustrate a geographical treatise by Johann Stabius of Vienna, court astronomer of Emperor Maximilian I. In each constellation the stars were numbered in the order in which Ptolemy mentioned them. The idea of using letters to designate stars was also not original with Bayer, but goes back to the Italian scientist Alessandro Piccolomini (1508–1578), who was Archbishop of Patras. His book *De le Stelle Fisse,* which went

The title page of Bayer's star atlas, published at Augsburg in 1603. It reads: *Uranometria, containing charts of all the constellations, drawn up by a new method and engraved on copper plates.* Beneath the title are an emblem of Capricornus and a view of the city of Augsburg. The panel at lower left is inscribed *To Atlas, the earliest teacher of astronomy,* while the panel at lower right reads *To Hercules, the earliest student of astronomy.*

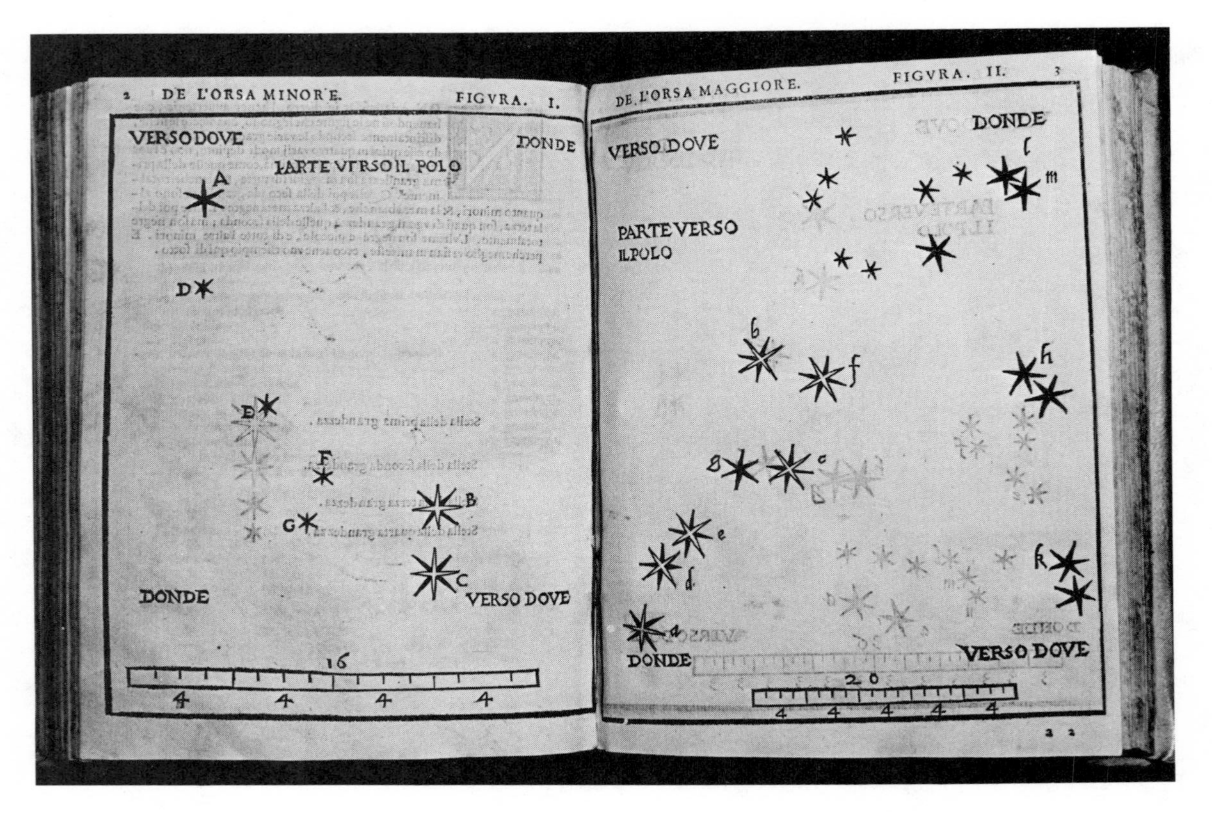

Ursa Minor and Ursa Major as depicted in the first (1540) edition of Alessandro Piccolomini's *De le Stelle Fisse,* the first star atlas to label stars with letters. The Italian words "verso dove," meaning whither, and "donde," meaning whence, were meant to indicate the constellation's sense of rotation as it revolves around the sky. Courtesy Houghton Library, Harvard University.

through several editions between 1540 and 1579, contains 48 woodcut constellation maps in which the stars are labeled with letters.

Piccolomini's first map, reproduced here, shows Ursa Minor, with seven stars labeled *A, B, C, D, E, F, G*. Their order in Ptolemy's catalogue is 1, 6, 7, 2, 3, 4, 5, so evidently the Italian did not blindly follow the catalogue order. But curiously these same stars are named α, β, γ, δ, ϵ, ζ, and η Ursae Minoris, respectively, in Bayer's *Uranometria* of 1603. This parallelism between Bayer's sequence of Greek letters and Piccolomini's Roman ones suggests that the German cartographer sometimes borrowed from his Italian predecessor. Brown tells us: "The method of distinguishing stars by Latin letters was adopted by French constructors of celestial globes in the later sixteenth and early seventeenth centuries, apparently derived from Piccolomini's maps."

The Greek Alphabet

α	Alpha	ι	Iota	ρ	Rho
β	Beta	κ	Kappa	σ	Sigma
γ	Gamma	λ	Lambda	τ	Tau
δ	Delta	μ	Mu	υ	Upsilon
ϵ	Epsilon	ν	Nu	φ	Phi
ζ	Zeta	ξ	Xi	χ	Chi
η	Eta	ο	Omicron	ψ	Psi
θ	Theta	π	Pi	ω	Omega

In his *Uranometria* Bayer assigned each star in a constellation one of the 24 letters of the Greek alphabet; when these were exhausted he continued with Latin letters. There has long prevailed an erroneous impression that he intended to affix letters strictly in the order of brightness. Actually, Bayer nowhere stated just what he did, but his procedure can be inferred in part from the star lists in the *Uranometria* where each constellation's stars are divided into magnitude classes.

Consider, for example, the constellation Ursa Major. Tycho lists seven 2nd-magnitude stars, and to these Bayer gives the letters α to η, running west to east along the Big Dipper. The three stars of 3rd magnitude get the next letters: θ, ι, and κ. The 4th-magnitude stars follow with λ to ψ; finally, the 5th-magnitude objects continue with ω, *A*, *b*, and so on to *h*. (Bayer always used *A* instead of *a*.) Clearly, Bayer was not particularly interested in the precise brightnesses of the stars but only in placing stars into the proper magnitude classes. Within each class he usually arranged the stars along the constellation pattern from head to foot, but sometimes apparently at random.

Loose-jointed though this system is, some working rules can be recognized. If there is one star much brighter than the others in a constellation, it was usually called α by Bayer, irrespective of location. Draco is an apparent exception, since α Draconis is actually as faint as magnitude 3.6. However, Bayer erroneously rated it as 2 and thus the brightest star in the constellation. Tycho had made the same mistake, and in this case the Augsburg lawyer seems to have followed him blindly without checking the stars themselves.

In Gemini, Castor (α) is actually about 0.4 magnitude fainter than Pollux (β). According to W. T. Lynn of Greenwich Observatory, "Bayer seems to have thought that α and β were equally conspicuous stars, each in the head of one of the twins, and he probably lettered them in their order of right ascension." Tycho called both stars magnitude 2.

Evidently we cannot use Bayer's nomenclature as a safe guide to the relative brilliance of the stars in 1603. The German astronomer F. W. Argelander carefully weighed the reliability of the *Uranometria's* magnitudes in a memoir published in 1842. He concluded: "I hope that I have demonstrated to my readers . . . that in any question of a star's variability or disappearance they should not use Bayer as an authority."

Not all the Greek and Latin letters we see on modern naked-eye sky maps go back to Bayer. Later astronomers made extensive additions, some of which are still in use. John Flamsteed, the first Astronomer Royal, had died in 1719 before the publication of his catalogue of about 3,000 stars in 1725 and the accompanying *Atlas Coelestis* in 1729. In his scheme the stars belonging to a constellation were arranged in order of increasing right ascension.

Later, sequential numbers were attached to these listings, giving the star designations that are familiar today: 61 Cygni, 32 Orionis, and so on. Although these numbers were not Flamsteed's, he did attach letters to some stars. Examples include *c*, *d*, *e*, *f*, and *g* Cas-

In Alexander Mair's engraving of Corvus for Bayer's atlas, a handsome crow is perched on a parallel of declination. Bayer's magnitudes are very inaccurate in this southern constellation. Apparently he followed Ptolemy's values for stars α through δ, and Tycho's for the other three, without making new observations. The unlabeled bright star just below the Crow's rear foot is Gamma Hydrae.

siopeiae, and *e, f, g* in Pegasus. Francis Baily (1774–1844), who prepared a new edition of Flamsteed's catalogue, offered this conjecture: "In all these [cases], I consider that Flamsteed had inserted the letters in his MS. maps, for a temporary purpose only . . . and that such letters have been inadvertently and improperly retained by his editors."

Much more extensive were the additions introduced by the French astronomer N. L. de Lacaille (1713–1762), whose atlas of the southern skies was based on observations he made at the Cape of Good Hope in 1751–1752. In this work, letters were assigned to stars even as faint as o Octantis (magnitude 7.2) and A Octantis (7.8). Baily criticized this exuberance: "In order to show the confusion caused by such a profusion of letters . . . I would remark that Lacaille has, in the constellation *Argo* alone, used (besides the Greek alphabet) the whole of the English alphabet, both in *small* and in *capital* letters, *each* of them more than *three* times: in fact, he has used nearly 180 letters in that constellation alone; and upwards of 80 in *Centaurus*. Thus we have in *Argo* 3 stars marked *a,* and 7 marked *A;* 6 marked *d,* and 5 marked *D;* and so on with several others: and these stars are not always such as follow each other in regular sequence . . . but are frequently situate in distant parts of the heavens. It is high time that this state of confusion and perplexity should be wholly abolished . . ." Similarly, the northern and zodiacal constellations received a rain of new letter designations in J. E. Bode's large star atlas published at Berlin in 1801. To him we owe the label λ Ursae Minoris for a 6.6-magnitude star just one degree from the north celestial pole. Star numbers of the form 80 B. Piscium and 162 B. Geminorum are also used by Bode and are occasionally encountered today, especially among observers of occultations.

The Bayer nomenclature has suffered still another form of misuse. Bayer himself had occasionally adopted names like k^1 and k^2 Tauri for two stars forming a naked-eye pair. Many more such examples were introduced in 19th-century atlases and catalogues. But some authorities used the superscripts in order of brightness of the two stars, other authorities in order of right ascension. Still others discarded the superscript for the brighter star. Such a tangle is offered by 30, 31, and 32 Cygni, which in various modern reference works compete for the designations o, o^1, and o^2 Cygni.

Fortunately, today we are spared most of the complications that grew up around Bayer's nomenclature during more than three centuries. Modern atlases of the naked-eye stars use essentially the pure Bayer Greek letters and "Flamsteed numbers" in the northern and zodiacal constellations and a simplified version of Lacaille's letters for the far-southern ones. This reform was largely the work of three astronomers between 1830 and 1925: Francis Baily in England, Benjamin Apthorp Gould at Cordoba Observatory in Argentina, and Frank Schlesinger at Yale University Observatory. Thanks to them we have a simple and easy way of referring to the bright naked-eye stars – just as Bayer intended.

79. Hevelius: his star catalogue and atlas

After three centuries of narrow escapes from destruction, an astronomical relic of exceptional historical importance found a safe haven at Provo, Utah, in the Clark Library of Brigham Young University. This leather-bound 183-page manuscript is the original of the famous star catalogue of Johannes Hevelius (1611–1687), in the handwriting of the Danzig astronomer himself.

Its first escape was from the fire of September 26, 1679, that gutted Hevelius' observatory, then the finest in Europe. During the siege of Danzig in 1734 by Saxon and Russian armies, the family home was severely damaged by artillery bombardment, and many of the remaining instruments and papers perished, but not the catalogue manuscript. Finally, in the 1780's a descendant willed the volume to the Danzig Society of Natural History, where it remained until World War II. Early in 1945, as Soviet troops were about to capture Danzig, the society's library was removed to a nearby village for safety, but instead it suffered almost total destruction. The manuscript was lost until the late 1960's, when it mysteriously reappeared in the hands of a dealer in rare books at Altadena, California.

Although Hevelius did not live to see his catalogue of 1,564 stars in print, it was published in 1690 by his widow Elisabeth, herself an able astronomer and her husband's collaborator in many of his scientific endeavors. Titled *Prodromus Astronomiae,* the volume contained not only the catalogue but also the handsome star atlas of Hevelius.

The catalogue was not so much a major scientific advance as the last flowering of a dying tradition. Hevelius was profoundly influenced by the example of Tycho Brahe, even to the extent of naming his observatory *Stellaburgum* in imitation of the Danish astronomer's *Uraniborg*. The large and highly perfected quadrants and sextants that Tycho described in his book of 1598 served as models for the instruments that Hevelius erected at Danzig. And when the latter

Johannes Hevelius (1611–1687). This portrait appeared in his *Selenographia,* 1647.

began observing star positions in earnest in 1658, it was with the aim of compiling a star catalogue more extensive than Tycho's.

The attitude of Hevelius toward the recently invented telescope was ambivalent. He built telescopes and used them effectively for charting the Moon and for observing eclipses and a transit of Mercury, but he retained naked-eye sights like Tycho's for his positional instruments. Meanwhile, about 1640 William Gascoigne in England began to use a refractor with cross hairs in its focal plane for making accurate pointings, and within a few decades this arrangement became standard among European astronomers, except for Hevelius.

Altogether, about a dozen different instruments were used by the Danzig astronomer in his catalogue work, but his two favorites were a large azimuthal quadrant and a brass sextant. With the former he could measure the altitude and azimuth of a star, noting the time with a clock, from which the star's celestial longitude and latitude could be calculated. (In the 17th century, astronomers generally employed these coordinates instead of the right ascension and declination that we commonly use today.) The sextant, which had a radius of six feet, was mounted in such a way that it could measure the angular distance between any two sky objects. This was a complicated operation that required two observers, each sighting at one object. The longitude and latitude of a star could be determined by finding its sextant distances from two other stars with known coordinates.

The naked-eye pointing devices used by Hevelius were adapted from those of Tycho, who had discarded as unsatisfactory the ancient method of lining up a star with a pair of holes. In one Tycho arrangement, each hole was replaced by a metal plate bearing four slits arranged in a square. The observer's line of vision was directly toward a star if he could see the star when he looked through each pair of slits in turn. In another arrangement, used by Hevelius on his brass sextant, the front plate was replaced by a short cylinder whose diameter was the same as the spacing between the slits in the plate next to the observer's eye. The pointing was exact if the eye, when brought to each slit in turn, saw the star at the edge of the cylinder. Through long practice, Hevelius gained great skill in this mode of observing.

In 1674 he was drawn into a controversy with the English astronomer Robert Hooke, who denied the reliability of naked-eye sights and strongly advocated the superiority of telescopic instruments. Much misunderstanding arose from each party's unfamiliarity with the other's observing procedures. Therefore in May, 1679, Edmond Halley (then aged 22) traveled under the auspices of the Royal Society to Danzig with his own instruments for a two-month visit as the guest of Hevelius. On the very night of Halley's arrival the two astronomers commenced their comparative observations. Within days Halley became convinced that his host could indeed measure angular distances to within half a minute of arc with plain sights. Halley found that he himself could not do better than about one minute. Conversely, the Danzig astronomer was not impressed with the accuracy he could achieve with his young guest's unfamiliar telescopic appliances.

The two astronomers parted on the best of terms. On leaving Danzig, Halley wrote a testimonial letter thanking Hevelius for his kindness and candor and offering to bear witness "to the scarce credible accuracy of his host's instruments, against all who may in future call his observations in question, having himself with his own eyes seen, not one or two, but many observations of stars made with

Fig. M.

the great brass sextant agree most accurately and almost incredibly with one another. . . ."

Today a more objective test is possible by a comparison of old observers' measured distances with modern data. Here – for eight pairs of stars observed by Tycho around 1585 and Hevelius about 1670, using plain sights, and by John Flamsteed around 1680, using telescopic sights – are the differences in seconds of arc between the measured and the actual distances.

Star Distance	*Tycho*	*Hevelius*	*Flamsteed*
α Arietis to α Tauri	−11″	+1″	−9″
α Tauri to β Geminorum	0″	−28″	−27″
β Gem to α Leonis	−49″	−50″	+1″
α Leonis to α Virginis	−14″	−37″	−24″
α Virginis to δ Ophiuchi	+34″	+7″	−3″
δ Ophiuchi to α Aquilae	−56″	−14″	−20″
α Aquilae to α Pegasi	+30″	−14″	−19″
α Pegasi to α Arietis	+2″	+1″	+3″
Standard deviation	34″	27″	18″

Flamsteed's observations are clearly superior. No successors used Hevelius' outdated methods, and the rapid advance of astronomical tools and techniques made his star catalogue obsolete within a few decades. It retains, however, permanent importance as a monument to its author's untiring enthusiasm for astronomy and his skill as an observer.

Hevelius' atlas was published in 1687 and reissued in 1690 as a supplement to his catalogue. It covers the entire sky in 54 maps, plus two large folding sheets for the northern and southern hemispheres. Its title, *Firmamentum Sobiescianum,* honors the great soldier and un-

Hevelius and his wife, Elisabeth Margarethe, measured star positions with this large brass sextant made by Guenter in 1658. From an engraving in Hevelius' *Machinae Coelestis,* 1673.

successful statesman John Sobieski, King of Poland from 1674 to 1696, who raised the siege of Vienna by the Turks in 1683.

This historic atlas, though well known by name, has become very rare. Fortunately, in 1968 a facsimile edition was issued by a distinguished Soviet historian of astronomy, V. P. Sheglov. Most of the explanatory text is given in English as well as Russian. Sheglov devotes some of his pages to a distinguished forerunner of Hevelius, Ulugh Beg (1394–1449). This Uzbek astronomer and prince, grandson of Tamerlane, compiled a catalogue of 1,018 stars from observations at Samarkand. The site of Ulugh Beg's observatory was discovered in 1908–1909 but was not fully excavated until 1948. Like the Hevelius catalogue, Ulugh Beg's was also republished by Francis Baily in 1843.

Anyone interested in constellation lore will enjoy leafing through the star maps of Hevelius. Here and there strange obsolete constellations catch the eye, as in Plate 25, where between Triangulum and Aires we find Triangulum Minor, the Lesser Triangle, and Musca, the Fly (the latter not to be confused with the present southern constellation of the same name). Elsewhere is one of the earliest representations of Camelopardalis. Though sometimes said to have been introduced by Hevelius, the Giraffe had been placed in the sky by Jakob Bartsch a half-century before. Eight constellations appear for the first time in *Firmamentum Sobiescianum,* including such familiar ones as Canes Venatici, Lacerta, Leo Minor, and Lynx. Vulpecula cum Ansere, the Fox and Goose, is today simply called Vulpecula. To honor his royal patron, Hevelius formed from the stray stars southwest of Aquila the new constellation Scutum Sobiescianum, Sobieski's Shield, that we know as Scutum. Just above it on Plate 16 there is an ornamental scroll dedicating the asterism "To the Most Serene, Most Powerful, and Most Invincible John III, King of Poland. . . ." Two of the Hevelius constellations have long been abandoned. One of these was Mons Maenalus, in the feet of Bootes; the other was Cerberus, a three-headed serpent (not a dog, as one might expect) grasped in the left hand of Hercules.

All of these innovations appear together in the elaborate frontispiece of the atlas, an allegorical engraving by C. de la Haye. Seated on a cloud is the muse Urania, flanked by the standing figures of the 10 greatest astronomers, who include Ptolemy, Ulugh Beg, and Copernicus. Hevelius approaches them with bowed head, carrying

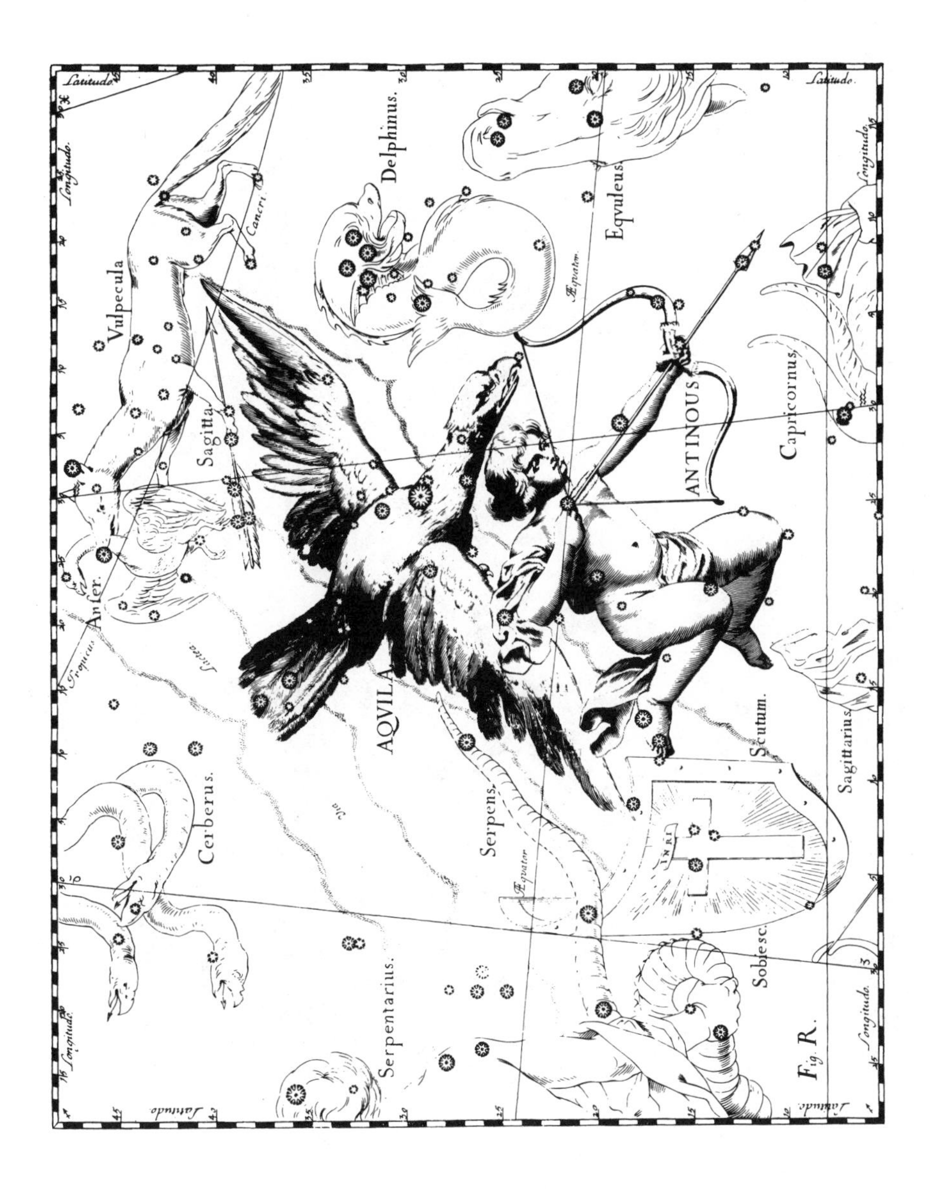

Aquila, the Eagle, with the now-defunct constellation Antinous, a youth said to have been installed in the sky by Emperor Hadrian. As on celestial globes, Hevelius' maps depict the star patterns left for right – mirror images of the constellations as they appear in the sky. From V. P. Sheglov's reprint of *Prodromus Astronomiae*.

Hevelius introduces his new constellations to the Muse of Astronomy in this portion of his atlas frontispiece.

his sextant in one hand and the shield of Sobieski in his other. Behind him troop a lizard, a fox carrying a limp goose in its mouth, a lion cub, a lynx, a pair of hunting dogs, and a three-headed snake. Overhead fly cherubs carrying scrolls to complete this curious period piece.

One reason why Hevelius' atlas never enjoyed the popularity of Bayer's or Flamsteed's is that the stars are unlabeled. The Danzig astronomer clung to the old-fashioned system of identifying the stars by their places in the constellation figures. Thus, in the star catalogue of Hevelius, the four stars we call Alpha, Beta, Gamma, and Delta Andromedae are, respectively, the Chained Woman's head, the bright star in her belt, the bright star in her foot, and her left shoulder. It is very odd that Hevelius stuck by so awkward a system, so quite possibly he was an astronomical reactionary in more ways than one.

80. How the BD was made

HOW DOES AN UP-TO-DATE ASTRONOMER go about compiling a new star atlas complete to, say, magnitude 8? This is a straightforward task for a good computer programmer. The positions and magnitudes of all such stars have been recorded on magnetic tape and can be computer plotted to yield star maps. As recently as the 1970's, our atlas maker would have laboriously plotted the stars by hand from printed catalogues. But how would he have proceeded a century or more earlier still, when no comprehensive star catalogue existed?

This was the problem faced in 1852 by the German astronomer F. W. A. Argelander, the director of Bonn Observatory from 1837 to 1875. Argelander possessed an unusual ability to organize large observing projects and bring them to a successful conclusion. In this case the result was the famous *Bonner Durchmusterung,* a set of 37 large charts and a three-volume catalogue giving the approximate positions of 324,198 stars, down to magnitude 9 and fainter, lying between declinations +90° and −2°.

The BD, as it is generally called, remained for a century a constantly used reference work among astronomers. Even today stars

F. W. A. Argelander, born on March 23, 1799, became director of Bonn Observatory and was the leading German astronomer when he died in 1875. He gave this photograph to G. P. Bond of Harvard Observatory.

are sometimes designated by their BD numbers. For example, BD +19°2777 means the 2,777th star in the declination zone +19° to +20°; it is better known as Arcturus. The catalogue gives the stars' magnitudes to 0.1 and their right ascensions and declinations to 0.1 second of time and 0.1 minute of arc, respectively. The great usefulness of this work in star identification was enhanced by its relative freedom from errors.

The extraordinary success of the BD, produced as it was with very modest equipment, makes this project a good case study in the planning of astronomical research.

Argelander, who was a native of Memel in East Prussia, had entered the University of Königsberg in 1817 at the age of 18 to study finance. But soon he was turned to astronomy by the lectures of Friedrich Wilhelm Bessel, who later was to measure the distance of 61 Cygni. The young student quickly demonstrated his exceptional astronomical talents in varied calculations that he made for Bessel, and in 1820 Argelander became his assistant at Königsberg Observatory.

Stellar astronomy was Bessel's great interest, and he developed with unparalleled skill improved methods of finding star positions and proper motions that remained standard for many years. Bessel also felt the need for a comprehensive survey of telescopic stars, and as a beginning he undertook extensive star cataloguing with the Königsberg meridian circle. Argelander was deeply influenced by Bessel's ideas, which foreshadowed the BD. When he came to Bonn, his intention was to extend Bessel's meridian observations of stars, but it was several years before suitable instruments became available. This interim period Argelander devoted to making a new atlas of naked-eye stars, based on careful magnitude estimates. He also monitored the changing brightness of variable stars by his newly invented step method.

The need for good celestial maps had become acute by around 1850. A dozen asteroids had been discovered already, and their number was growing yearly. Observing these faint bodies and searching for new ones had become popular activities, while work on variable stars added to the demand. As Argelander was winding up a large meridian-circle program, it was natural for him to consider the cartographic problem as his next task.

Prior to the BD, the most ambitious sky-mapping project had been the Berlin Academy star charts, of which Bessel was the moving spirit. The first installment had been published in 1830, but the work was still incomplete. In this enterprise, intended to cover the sky between declinations $+15°$ and $-15°$, each hour of right ascension was assigned to an individual astronomer. The general plan of the Berlin charts was to plot all stars whose positions were known from catalogues of precision; then, at the telescope, all other stars down to about magnitude 9 were to be inserted, either by eye estimation or by micrometer comparison with known stars.

As a young astronomer, Argelander had taken part in the Berlin project and published the map for Hour 22 in 1832. He was very dissatisfied with the procedures, though his section was one of the best done of the 24. The observer at the telescope had to look back and forth between the illuminated chart and the dark sky, dulling the eye and making the magnitude estimates uncertain. The micrometer comparisons were very time consuming, and errors in the star positions taken from catalogues easily went undetected, especially in star-poor regions.

Argelander's intention was that the BD should contain, for the

Seen here in the old south dome of Bonn Observatory is the tiny 78-mm (3.1-inch) Fraunhofer comet seeker with which F. W. A. Argelander's associates observed the positions of more than 300,000 stars for the *Bonner Durchmusterung*, using the eyepiece reticle shown on page 432. Note the sturdy equatorial mount and the large finely divided circles, read with verniers.

northern sky and part of the southern, all stars down to magnitude 9 and as many of magnitude 9–10 as possible, but no objects that did not actually exist. To assure this, he decided to measure the right ascension and declination of every star to about one minute of arc. The utmost economy in time and labor would be needed to complete the enormous observing task within a few years. This meant that star positions had to be observed wholesale in zones rather than individually.

After some experimenting, he adopted this procedure: The star observations were made with a 78-mm (3.1-inch) comet seeker, equatorially mounted on a rigid iron pier in the observatory's south dome. This small Fraunhofer refractor had a 10-power orthoscopic eyepiece giving a 6° field. In the focal plane of the telescope was mounted a thin glass semicircular plate whose straight edge, oriented north-south, was visible as a fine black line against the night sky. This was called the *hour line*. Perpendicular to it was painted a *middle line* extending to the edge of the field. Above and below the middle line were 10 *declination lines* spaced seven minutes of arc apart and with every third line slightly longer than the others. These marks were done in heavy oil paint, so that they could be seen without artificial illumination. Although the painted lines were about 1½ minutes of arc broad and not quite even, observational accuracy was scarcely impaired.

To observe a zone of stars took two men: *A* sitting or reclining at the eyepiece in the darkened dome, while his recorder *B* sat in a lighted room underneath, facing a sidereal clock. Before they started work, the telescope was set to the zone's declination and clamped in both coordinates. Ordinarily the stars were observed in a zone 2° wide. As each one drifted through the field of the stationary telescope, *A* would estimate its magnitude and call it out at the moment the star passed behind the hour line. *B* would then record the clock time and magnitude in his observing book, which was ruled in columns, with a line for each star. *A* also noted the declination line and estimated the tenth of division. Without removing his eye from the eyepiece, *A* wrote down the number on a pad faced with cardboard strips to guide his hand. (If only Argelander had had access to tape recorders!) At the end of each sheet *A* signaled to *B*, who drew a line in his book, so that their records would stay in step.

This kind of observing was tiring, and zones were never contin-

This diagram of the BD reticle is from Argelander's description. Its use is explained in the text.

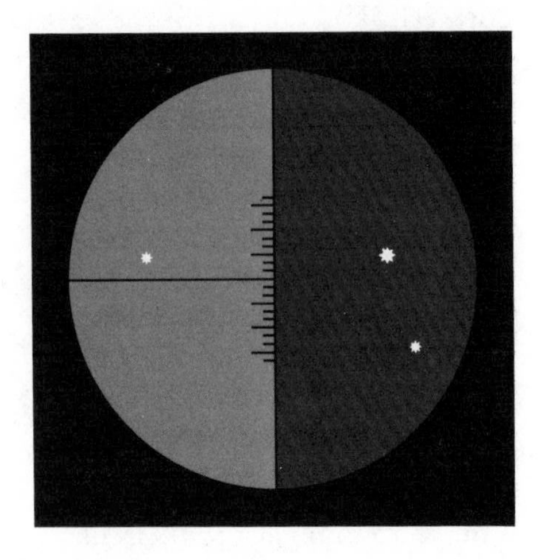

ued beyond an hour and a quarter. The moment a zone ended, *B* would ring a bell to summon a fresh pair of observers, while *A* read the circles of the telescope and labeled his record sheets. The two observers would immediately confer, while their memories were still fresh, about any inconsistencies between their records. As soon as possible, usually the next day, *A*'s data would be transcribed into *B*'s observing book, with any uncertainty in interpretation duly noted.

The reduction of the observations to get approximate star coordinates for 1855 was very simple because the zones were so small. Some 16 to 20 catalogue stars were selected in each zone and compared with the observations. This comparison provided one constant which, when added to the clock time for each star, gave its right ascension, and a second constant which, when added to the scale reading (converted to arc), gave its declination.

Many careful precautions were taken by Argelander to ensure that the BD should be virtually error-free. Each part of the sky was surveyed twice, and the Milky Way regions three times. This procedure was essential, for in the richest regions 30 stars might come per minute, so inevitably some would be passed by. No star was admitted to the BD unless there were two accordant positions; a star observed in only one zone was not accepted unless it had been previously catalogued. If for any reason one of the two zones covering some area was unsatisfactory, a *revision zone* was observed with a larger telescope. Finally, ambiguities or inconsistencies about individual stars were checked out with the Bonn meridian circle. All original records and calculations for the project were preserved at the observatory, so that any seeming error found in the BD at some future time could be investigated. The systematic procedures and orderliness of the papers make it easy even today to verify a BD star's credentials if necessary.

This very effective final plan was adopted only after much experiment and a slow start. In February, 1852, Argelander's 20-year-old assistant Friedrich Thormann began observing zones with the comet seeker, aided by J. F. J. Schmidt (see Chapter 50). The director himself took little part at this time, being preoccupied with finishing his big program of meridian-circle observing. The original idea was to cover the sky to as far south as declination $-23°$ or beyond. But the work went very slowly, and in the first year only about 100 zones were observed. In the early spring of 1853 the discouraged Thormann resigned, giving up astronomy to become a mining engineer in France. Schmidt, who had been ill, left the observatory to take another astronomical position where he had more opportunities for his favorite lunar studies.

At this time Wilhelm Struve, the famous double star expert who was director of Pulkovo Observatory in Russia, visited Bonn. On his advice, Argelander cut back the BD plan to declinations $+90°$ to $-2°$, although this meant the abandonment of a third of the work already done. He then left on a four-month tour of observatories with Struve. In September Argelander returned to Bonn Observatory in the best of spirits and full of vigor. Meanwhile, two very capable new assistants had been enlisted and had begun observing zones. One was Eduard Schönfeld (1828–1891), who later succeeded Argelander as director at Bonn; the other was Adalbert Krüger (1832–1896), who married Argelander's daughter and became director successively at Helsinki, Gotha, and Kiel observatories.

Now the enterprise began to advance rapidly, as procedural problems became settled. Argelander soon found that his sharper-eyed young associates could observe much fainter stars than he with the comet seeker. Hence, to avoid unevenness in limiting magnitude, he left the actual work with that telescope to Schönfeld and Krüger, while he undertook the revision zones and the big task of checking out discrepancies with the meridian circle. As Schönfeld later recalled: "The winter and spring of 1854 were very favorable for observing, but since the reductions were kept up to date, Argelander reconciled himself to the idea that the BD would contain 50,000 to 100,000 more stars than he had assumed. Always concerned with preserving the unity of the whole, he personally organized much of the material, and he was reluctant to let a single zone pass without taking some part in it. The extensive routine work was spiced with minor discoveries; new variable stars and stars with large proper motions were found, and the uncovering of erroneous positions in older star catalogues often called for interesting detective work."

By the start of 1857 the first four charts of the 37 were lithographed and issued, with more on the way, while the number of zones exceeded 1,100. Argelander relieved Schönfeld and Krüger of as much of the routine as possible, to speed the project. Finally, in March, 1859, the zone observations had been completed to their northern limit, declination +81°. No zones were needed for the remaining polar cap because positions of all the necessary stars were already available in the star catalogues by R. C. Carrington and F. M. Schwerd (see Chapter 65), but the magnitudes of these stars were carefully determined with the Bonn comet seeker. Elsewhere in the sky some checking observations were made as late as 1861. The publication of the BD catalogue and charts was finished in 1863.

It was a splendid achievement. The completion of this enormous task with such limited means was a tribute not only to Argelander's organizing powers and enthusiasm but also to his warm personality, which evoked full and willing cooperation from his staff. As a reliable and convenient record of the sky, the BD had a usefulness to 19th-century astronomers that can be compared only to the role of the Palomar Sky Survey in the 20th century.

Appropriately, Schönfeld undertook a southward extension of the BD after he became director at Bonn following Argelander's death in 1875. For this purpose a 159-mm (6-inch) refractor with powers of 26 and 81 served. An analogous eyepiece reticle was used but with field illumination. The result, known as the SBD, was published in 1886 in chart and catalogue form. It gives the places for 1855 of 133,659 stars between declinations −2° and −23°.

In the introduction to that work Schönfeld hinted that future work of this sort belonged to photography. The great *Cape Photo-*

+19°

m	1941–2000 8ᵘ ′ ″	+19° ′	
9.0	1 38.0	28.0	K
9.5	2 10.1	14.1	
9.5	23.9	11.0	
9.3	25.2	27.9	K
9.5	43.8	50.6	
9.4	48.1	58.4	
9.2	50.0	33.3	K
9.2	54.3	41.7	K
9.5	3 38.5	14.4	
9.5	38.8	32.8	
9.5	44.7	22.2	
9.5	4 10.1	22.7	
9.5	19.5	28.9	
9.5	34.0	18.0	
9.5	5 36.2	25.5	
9.5	6 3.8	46.4	
9.5	17.4	57.1	
9.4	19.9	32.2	B
9.5	7 12.1	13.7	
9.5	14.1	37.9	
9.5	19.4	25.0	
9.2	32.7	0.1	B
7.6	37.4	7.5	K
9.1	42.1	8.2	K
9.5	8 26.0	3.4	

12′ 8′ 4′ VIII 56′

19° 18° 17° 16°

var

At left is a part of a BD catalogue page for the +19° declination zone, at right a portion of the corresponding BD chart. The first star in the catalogue excerpt is BD +19°1941, which is described as of magnitude 9.0, at right ascension 8h 01m 38.0s, declination +19° 28′.0. Both catalogue and chart are for the epoch 1855. (The BD follows an obsolete usage in having ′ and ″ stand for minutes and seconds of time as well as minutes and seconds of arc.) In the last column, the letter *K* means that the star occurs in the Königsberg zone catalogues, *B* that the star has been observed with the Bonn meridian circle. On the chart, the star marked "var" is the long-period variable V Cancri, discovered in 1870 after the first edition of the BD was published; this chart reproduction is from the second edition. An easy way to identify a given BD star on the chart is as follows. Suppose we want BD +19°1955 (9.5), at 8h 05m 36.2s, +19° 25′.5. Inspection of the catalogue shows that this is the fourth star following 8h 04m of right ascension in the +19° zone. Going to the chart above, we count leftward from the 8h 04m line to the fourth star image, and verify that it matches the appropriate magnitude and declination.

graphic Durchmusterung (CPD) by David Gill and J. C. Kapteyn covers the sky from −19° declination to the south pole. It contains the positions of nearly half a million stars measured on photographs taken in 1885–1889 at the Royal Cape Observatory.

Simultaneously, a visual survey of the southern sky was underway at the Argentine National Observatory. This was the *Cordoba Durchmusterung* (CoD), begun by J. Thome and continued by C. D. Perrine (see Chapter 60). This huge enterprise, a compilation involving 2,500,000 observations of 600,000 stars, was not completed until about 1930.

Now nearly forgotten is another visual durchmusterung, begun at Cincinnati Observatory about 1880 by Ormond Stone. His observations were made with a 4-inch Clark refractor, eyepiece reticle, and chronograph. Only a small part of this work was ever published, after Stone went to the University of Virginia. Known as the *Virginia Durchmusterung,* this fragment is a catalogue of 6,671 stars between declinations $-23°$ and $-24°$. Collectors of astronomical curiosa can find it in Vol. 1, Part 5, of the *Publications* of Leander McCormick Observatory (1905).

81. The brothers Henry and the *Carte du Ciel*

In the southeastern part of the Moon lies a close pair of twin craters bearing a single name, Frères Henry. This unusual joint nomenclature commemorates the French astronomers Paul and Prosper Henry, brothers who collaborated so closely in their scientific work as to appear a single person. They were important pioneers in astronomical photography, but, paradoxically, their success was so great that it crippled astronomy in France for a generation.

The brothers were born at Nancy – Paul in 1848 and Prosper in 1849 – and each at the age of 16 entered the meteorological department of the Paris Observatory. But primarily their interests were astronomical, and at their home they built a 12-inch reflecting telescope and began the construction of star charts.

In 1870–1871 the Franco-Prussian war altered their lives. German armies swept across France and besieged Paris, where food became so scarce the inhabitants ate animals auctioned off by the zoological garden. Paul and Prosper Henry manned one of the 12 permanent observation posts around the city that maintained a round-the-clock watch on the German batteries and troop movements.

With the return of peace, the brothers were transferred to the observatory's astronomical division, which gave them the use of a large refractor to search for new minor planets. Of these they discovered 14, beginning with 175 Liberatrix on September 11, 1872. With strict impartiality, these discoveries were announced as having been made alternately by Prosper and by Paul.

A necessary first step in this planet hunting was the laborious construction of maps of faint stars along the ecliptic (the plane of the Earth's orbit projected on the sky), by means of old-fashioned visual methods. But when the brothers arrived at the parts of the sky where the ecliptic crosses the Milky Way, the number of stars became so excessive that the task of plotting them one by one seemed beyond human endurance and skill. Fortunately, at this time photographic dry plates were beginning to come into use, and the Henrys decided to exploit them. At their Montrouge home they erected an optical shop and constructed a 6-inch objective especially designed for photography. It gave striking results for that time; a 44-minute exposure of the Perseus Double Cluster showed stars as faint as magnitude 12, with perfectly round images.

The importance of these experiments deeply impressed Admiral E. Mouchez, who had become director of Paris Observatory in 1878 as U. J. J. Le Verrier's successor and was famous for his defense of Le Havre during the war of 1870–1871. He encouraged the brothers to build a 13½-inch photographic refractor of 11-foot focus, which was completed in May, 1885.

The success of this instrument was described by a French scientist 20 years later: "Once celestial photography was realized, the French astronomers divided the universe among themselves: Janssen took the sun, Loewy and Puiseux the moon, and the brothers Henry the stars."

Admiral Mouchez now proposed an extensive international program for the cataloguing and mapping of faint stars by photography. In April, 1887, 57 astronomers from 19 nations met at Paris to draw up plans. The entire sky was apportioned among 18 observatories, each to use a 13½-inch telescope of the Henry design. It was decided to obtain precise positions for all stars brighter than magnitude 11,

The brothers Henry at Paris Observatory with the 13-inch astrograph they designed, which was adapted for the *Carte du Ciel* celestial mapping program. The boxlike tube contains both the photographic telescope and a visual guiding instrument of the same focal length. From *l'Astronomie,* No. 5, 1886.

numbering about one million, and to make charts showing the 25 million stars brighter than magnitude 14.

The 13½-inch astrographic telescope of the Henrys had a field of good definition only about two degrees square, hence some 20,000 photographs would be needed to cover the sky twice. The Paris congress turned down the plea of the American astronomer E. C. Pickering for the adoption of wide-field doublet lenses, on the ground that these novelties were untried. The assembled astronomers were much more impressed by the excellent performance of the prototype Henry instrument. Besides contributing the adopted design, the Henry brothers made with their own hands about half of the 13½-inch objectives for the program.

Although this huge *Carte du Ciel* project has never been completed, the portions that were finished have been of very great value to astronomy. The star positions in the catalogues have served astronomers in many ways, and mastering the project's technical problems stimulated many advances in the power and versatility of photographic methods of observation. There is a very interesting account of these matters in H. H. Turner's book, *The Great Star Map,* New York, 1912. We should not overlook the significance of the *Carte du Ciel* as an early example of large-scale international cooperation in science.

But when the work was started, astronomical photography was primitive. Wide-angle lenses for the observations, and computing machines for the reductions, were unknown when the rigid plans were adopted. In fact the venture was premature; the techniques of later years would have made more manageable the enormous labor of the enterprise. Some idea of the scale of the work is afforded by the experience at Oxford, one of the observatories that completed its share of the catalogue: 20 years and £34,000 were spent in observing, measuring, reducing, and publishing the places of almost 200,000 stars in its zone of declination, +25° to +31°. At Potsdam, Germany, if its original plan of publication had been carried to completion, that observatory's share of the catalogue would have consisted of 387 large volumes filling 45 feet of shelf space and weighing a ton!

It is hardly to be wondered that the cooperating French observatories, Paris, Bordeaux, Toulouse, and Algiers, had their energies committed for decades, while the new science of astrophysics was

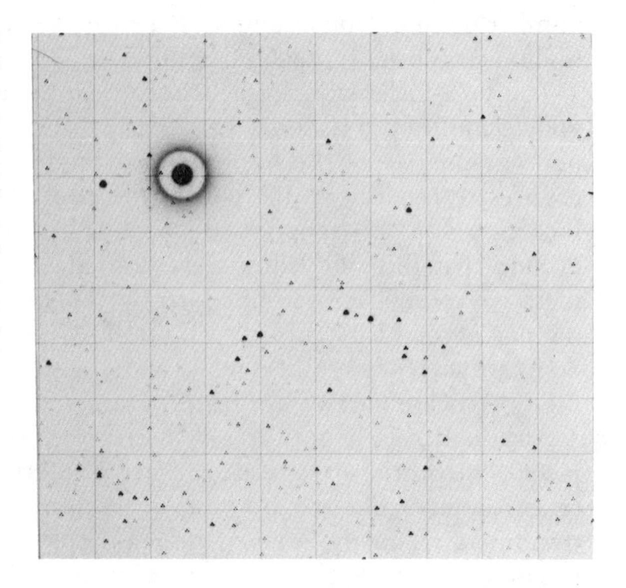

A small part of a *Carte du Ciel* atlas sheet, prepared at Uccle Observatory, showing the region around Epsilon Cassiopeiae (which exhibits a photographic halo). Images of the fainter stars appear triangular because three exposures were taken of every field, with the telescope being moved slightly between each. The grid was exposed directly on the plate. North is up.

rising in America, England, and Germany. The whole outlook of French astronomy was confined by the *Carte du Ciel* project. Perhaps astrophysical research in France would have begun to flower earlier if the brothers Henry had not made such good lenses!

82. A. G. Pingré and his annals

PARIS in the last decades before the French Revolution was the center of intense astronomical activity. Visitors to the meetings of the Royal Academy of Sciences listened to Laplace, Lagrange, and Lalande report epoch-making discoveries about lunar and planetary motions. Charles Messier was searching the heavens for comets with a success that has seldom been equaled. Any important celestial event of the 1770's – an eclipse or the occultation of a 1st-magnitude star – would be watched from half a dozen observatories within the city, by astronomers in knee breeches and powdered hair. Their telescopes were usually small reflectors with metal mirrors, or else the long and cumbersome nonachromatic refractors just going out of fashion. Up-to-date observers were praising the newly introduced 3-inch achromatic refractors made by John Dolland.

In this thriving astronomical community of prerevolutionary Paris, one of the best-known figures was Alexandre Guy Pingré (1711–1796), canon at the abbey of Ste. Geneviève. As a young man he had been active in theological controversy, and he gained a reputation for unorthodoxy that seems to have hindered his ecclesiastical preferment and led him into a scientific career.

He first attracted the attention of French astronomers by the accurate prediction of a lunar eclipse in 1749, and a few years later he took charge of the little observatory at Ste. Geneviève. There Pingré worked for the rest of his life, except for astronomical expeditions. One of these took him to the remote island of Rodrigues in the Indian Ocean to observe the 1761 transit of Venus; another voyage was to Santo Domingo in the Caribbean for the 1769 transit of the same planet.

Of Pingré's many writings, the one that added most to his fame was his *Cométographie* (1783–1784). Even now, these two fat quarto volumes remain a mine of information about early comets. Their author ransacked an enormous mass of classical writings, medieval chronicles, and recent publications in order to collect all useful data about individual comets. The excellent library of Ste. Geneviève – important today as one of the largest in France – helped him greatly in this task of many years. Pingré was no mere compiler; he carefully weighed and analyzed his comet material, making many orbit computations by the laborious methods of his day.

Soon after he had come to Paris, Pingré formed the plan of another enormous undertaking, which he explained in detail to the Academy of Sciences in 1756. This was the systematic collection and publication of all observational data obtained by astronomers during the 17th century. The Academy officially praised the project as very valuable to astronomy and urged its speedy completion. Actually, the gigantic undertaking took Pingré three decades. When he presented his finished manuscript to the Academy it was received with enthusiasm, and the National Assembly in 1791 granted 3,000 livres to pay for its printing. But now a strange chain of events intervened.

Barrois, the printer, proceeded so slowly that when Pingré died on May 1, 1796, only 364 pages – about two thirds of the book – had gone through the press. At that point, the growing wartime inflation that was choking the French economy caused Barrois to suspend work. In the confusion that followed, both the printed portion of the book and the original manuscript disappeared.

For a century, astronomers regretted the loss of this great survey of their science during the eventful period from Tycho Brahe to Edmond Halley. Finally, about 1898 the Parisian book collector Victor Advielle found a battered astronomical treatise on sale as waste paper in a village in southern France. This he took to G. Bigourdan, of Paris Observatory, who recognized it as a copy of the 364 printed pages of Pingré's lost *Celestial Annals of the 17th Century*. It bore an inscription in Lalande's handwriting: "Lalande, 1796. Barrois gave me these pages in November 1796. The rest will be done after peace comes." Soon afterward, Bigourdan managed to find the remainder of the original manuscript in the archives of Paris Observatory. It had been mislabeled as *Manuscrit Erasme Bartholin. Copie des Observations de Tycho-Brahé*. As a comparison with known autographs proved, the large sheets were in the trembling handwriting of the elderly Pingré.

Thanks to these fortunate discoveries, the Academy of Sciences was finally able in 1901 to publish the book under its original title of *Annales Célestes du Dix-Septième Siècle*. Bigourdan was the editor.

As its name suggests, the work is a compilation of the main astronomical events of each year from 1601 to 1700. For example, under 1601 we read of Johannes Kepler's observations of the solar eclipse on December 24th at Prague, using Tycho's instruments; there are also corresponding records from Norway, Holland, and Italy. We have lunar eclipse observations from places as far afield as Peking, and positional measurements of planets. The nova P Cygni that had appeared in 1600 was still bright, and Pingré lists valuable records of its irregularly changing light. Finally, 1601 was the date of Tycho's death, and we find a long account of his career.

This same plan is followed for later years, becoming more varied as telescopes came into use and were improved. The first occultation listed is one of Aldebaran on March 4, 1607, watched with the naked eye by David Fabricius in Holland. By about 1630 timings of occultations start to be accurate enough to be of value today in studies of the Moon's motion. Timings of eclipses of Jupiter's satellites begin in 1652, when Giovanni Hodierna in Sicily recorded a disappearance of Io on June 27th.

An interesting aspect of this panorama of 17th-century astronomy is the worldwide distribution of observers. Even in that age missionaries and traders carried to the far corners of the world telescopes, quadrants, and clocks. Pingré mentions many observations made at Peking, and one of 1613 in Japan. A particular curiosity is the series of eclipse timings of Jupiter's satellites in late 1689 made at Malacca in Malaya by the French missionaries Comille and de Bèze. Pingré writes: "The observers were in prison; it was therefore difficult for them to be certain of their clock corrections."

Among the events of 1612 we find the following: "Simon Marius on December 25th discovered the nebula in the belt of Andromeda. . . . The discovery is also attributed to him by Pierre Petit, Hevelius, and Mairan. And in fact, Marius gave so exact a description that no misunderstanding is possible. Hevelius regarded this object as variable in brightness, believing it had disappeared in 1657, 1658, and 1659. Bullialdus, in a letter to Hevelius, agreed that the nebula varied, but denied that it vanishes; he maintained it had been seen 150 years before. Later it was found to have been listed in a 10th-century catalogue. Marius, however, did not hesitate in pronouncing it a new object." This passage, of course, refers to Messier 31, the great Andromeda galaxy. We know today that its brightness does not vary.

I have looked through Pingré's book for early records of astronomical activity in the New World. The oldest observation listed as made in North America refers to the lunar eclipse of November 8, 1631. This event was seen at Charlton Island, in James Bay, Canada, by the explorer Thomas James, during a voyage in search of the Northwest Passage, when he discovered the southern part of Hudson Bay now named for him.

These timings of lunar eclipses in remote parts of the world were very valuable to 17th-century mapmakers, because geographical longitudes could be deduced from them. One instance described by Pingré is the total eclipse of April 14, 1642, when the Moon became too faint to be seen with the naked eye, according to some but not all observers. This eclipse was timed at Quebec by Joseph Bressan, and by an anonymous Canadian "at Sainte Marie, in the country of the Huron Indians." Another record of the same event came from Fort Ceulen, Brazil.

The first allusion to an astronomical observation made within the present United States comes from Rhode Island, dated June 25, 1638. It pertains to an eclipse of the Moon, "At a place in New England, called Aquedniek or Aquednuick, in latitude $+40° 50'$, there elapsed $1^h 05^m$ between emersion and sunrise."

These few examples will give some idea of the curious and sometimes useful data in Pingré's *Annales*. Whether your taste runs to early Americana, or you need old observations of variable stars or Jupiter's satellites, or you have some historical problems, it is still worthwhile to consult the book written by the canon of Ste. Geneviève and rescued by Bigourdan.

83. An enigmatic astronomical woodcut

THE PICTURE on the opposite page may be more or less familiar to you. In an antique spirit, it shows an extensive landscape beneath the Sun and Moon enclosed by the sphere of the fixed stars. At lower left, where Earth and sky meet, a crawling figure with the cloak and staff of a pilgrim has thrust his head through the sphere and is gazing at the wonders outside.

This illustration has been reproduced in many popular books. There are two curious features about all these reproductions, though. The original source of the woodcut is seldom if ever mentioned, and the captions are oddly various and seem to be based on the writers' imaginations.

As one example, Donald H. Menzel's coffee-table *Astronomy* (1971) prints this picture with the description: "The medieval concept of the sky as a star-studded globe, through which a fortunate traveller might poke his head and view the glories of Heaven beyond. The interlocked wheels were those described by Ezekiel,

Once thought to date from the 15th or 16th century, this woodcut appears to have been drawn in the 19th century by the famous French popularizer of astronomy Camille Flammarion.

which in reality are parhelic phenomena caused by ice crystals in the earth's atmosphere." Contrast that description with the caption used for the same illustration in Louis MacNeice's *Astrology* (1964): "Some astrologers regard the planet Uranus as the patron of the heavens . . . others connect it with mechanical invention. A 16th-century German woodcut shows that these two ideas were linked long before the discovery of the planet Uranus: A curious human peers through the vault of the universe and sees the mechanism that moves the stars."

In the spring of 1976, at the University of Erlangen in West Germany, there was a fine exhibition of 15th- and 16th-century astronomical instruments, books, and prints, to honor the quincentennial of the death of the astronomer Regiomontanus in 1476. One of the displays was our same woodcut, with no indication of origin but labeled as the cosmic model of Cardinal Nicolas Cusanus (1401–1464).

Where did this work of art originate? In 1957 the German historian of astronomy Ernst Zinner wrote an article, "A Notable Old German Woodcut," expressing his bewilderment. He was convinced that it had probably been engraved sometime between the years 1530 and 1550, but he could find no trace of it earlier than as an illustration in a 1906–1907 popular scientific book. There the author W. Förster *did* give his source for the picture as *Astronomie Populaire* (1880) by Camille Flammarion. But when Zinner turned to this famous French book, he found that it contained no such woodcut! Evidently Foerster had made a mistake, and so the trail was lost despite Zinner's remarkable familiarity with early astronomical works.

The man who picked up the trail was Arthur Beer, a German-born astrophysicist and science historian at Cambridge University. He discovered that Flammarion *had* published the mysterious woodcut, but in another book, *L'Atmosphère, Météorologie Populaire* (1888). Beer went on in 1958 to propose the solution to the problem.

Independently, the same solution was advanced in a 1973 article by Bruno Weber, who was in charge of rare books at the Central Library in Zurich, Switzerland. He, incidentally, seems to have been unaware of Beer's priority, for his article does not mention the latter's name.

The first step in clearing up the problem was the realization that the woodcut was not nearly as old as the 16th century. To the trained eyes of art historians its composition contains incongruities, as if a modern artist had copied features from several originals of different ages. Also, as Weber tells us, there are telltale signs of modern technique. Note for example the dotted hillsides in the landscape. Such stippling is not to be expected in 16th-century printing blocks in which the wood was cut away with a knife. Yet it is characteristic of wood engravings done with a burin, which is a technique introduced about the year 1800 by Thomas Bewick in England.

If, then, the picture dates from the 19th century, the artist was probably Camille Flammarion himself, according to both Beer and Weber. Support for this idea comes from the varied career of this outstanding French popularizer of astronomy who lived from 1842 to 1925.

Camille Flammarion (1842–1925), from A. v. Schweiger-Lerchenfeld's *Atlas der Himmelskünde,* 1898.

Flammarion's long career was dazzling. His many eloquent books and articles on scientific and philosophical subjects made him one of the most widely read authors in France. At the age of 23, he started an annual review of astronomy, which he continued to publish for 60 years. He fathered the amateur movement in France by founding the journal *l'Astronomie* in 1882 and the Société Astronomique de France in 1887. Both of these endeavors continue to flourish today. His scientific works include a famous two-volume compendium of all telescopic observations of Mars since the mid-17th century.

At the observatory that Flammarion established in 1883 at Juvisy, south of Paris, he carried on not only astronomical work but many kinds of meteorological and geophysical studies. For example, he systematically photographed clouds and halo phenomena, obtaining 1,500 exposures over three decades. Flammarion made a dozen scientific balloon ascensions, and in 1897 he pioneered in the astronomical use of motion-picture photography. Each spring from 1871 until

1925 he made drawings of the budding trees along the Avenue de l'Observatoire in Paris.

Flammarion was a gifted artist. His biographer E. Touchet tells us: "His planetary drawings were very reliable. His representations were sober and truthful, showing fewer details than he actually saw, but these details were trustworthy." Furthermore, when as a 12-year-old boy he moved to Paris, he was apprenticed to an engraver. It is highly likely, says Weber, that many of the illustrations in Flammarion's books were engraved from his own drawings and probably under his supervision. There is no obstacle, then, to believing that Flammarion was capable of drawing the picture reproduced here. But to clinch the demonstration that he did draw it, Weber has been able to trace with much plausibility how Flammarion got the idea for the picture.

The versatile French astronomer was an omnivorous reader from the age of four, and at his death owned 12,000 books. This collection was rich in printed works of the 15th and 16th centuries, for as a true bibliophile Flammarion wished to make himself as independent as possible of public libraries. He tells the story of how as a youngster he found himself at a book auction when there came up for sale a copy of the 1553 Venice edition of Ptolemy's *Almagest,* bound in gold-ornamented parchment. "Somebody else bid five francs. Having counted all the money in my pocket, I cried out '17 francs 50 centimes!' My competitor, taken aback by this bizarre sum, must have thought I was placing a bid ordered by some library. At any rate, the book went to me."

Hence Weber believes that Flammarion must have been familiar with Sebastian Münster's *Cosmographia* of 1550, an astronomical work very popular in the 16th century. The first illustration in that book is also a hilly landscape beneath the Sun and Moon and enclosed in a star-dabbled sphere. The placement of the features, however, is quite different. Only the pilgrim poking his head through the sphere is missing, and there are angels and demons outside the sphere instead of wheels and the rest.

So far so good. But where did Flammarion get the idea of his crawling pilgrim? Weber points out that the clue is supplied by Flammarion's own caption for the picture in his 1888 book: "A missionary of the middle ages has reached the place where the sky and the earth touch . . ." Weber goes on to cite a legend familiar in medieval times that tells how St. Macarius Romanus was one of the few mortals who had traveled to beyond where the sky meets the Earth. This very same legend is told at length by Flammarion in his 1865 book *Les Mondes Imaginaires.*

The conclusion reached by Beer and by Weber that Flammarion himself drew the picture appears inescapable. Almost certainly it is a composite of the ideas of the *Cosmographia* illustration and the Macarius legend, as Weber proposes in considerable detail. We should not call the picture a fake, though, but rather an imitation with a purpose. And to this day the French astronomer's artistic creation continues to stir our imaginations.

Notes and references

Immediately following each chapter number is a citation to the issue of *Sky & Telescope* magazine containing Joseph Ashbrook's original "Astronomical Scrapbook" article. It is from these that this anthology has been prepared. The bibliographies do not include all of the references consulted by Ashbrook; they are intended only to suggest sources for further reading. Where possible, citations are given to works that appeared subsequent to the original magazine articles.

CHAPTER 1
S&T: June, 1965, page 353. Heinrich Matiegka's "Report on the Investigation of the Skeleton of Tycho Brahe" was presented to the Royal Bohemian Academy of Sciences on October 11, 1901, and was printed in the academy's publications.

CHAPTER 2
S&T: February, 1967, page 92. Burton, "a freebooting scholar", is quoted from Sir B. Ifor Evans, *A Short History of English Literature,* Baltimore, page 198, 1962. All Burton quotations are from Floyd Dell and Paul Jordan-Smith's edition of the *Anatomy of Melancholy,* New York, 1927, in which "Digression of Air" is on pages 407 to 438. In this edition, which contains useful notes, the spelling has been modernized and Burton's innumerable Latin quotations have been translated into English.

CHAPTER 3.
S&T: December, 1958, page 85.

CHAPTER 4
S&T: June, 1979, page 545. For more about Sewall see *The Diary of Samuel Sewall 1674–1729,* edited by M. Halsey Thomas, New York, 1973, and Samuel Eliot Morrison, *Three Centuries of Harvard 1636–1936,* Cambridge, Mass., 1936.

CHAPTER 5
S&T: March, 1976, page 163. Most of the information about Kew Observatory and the Demainbrays is from *The Observatory 5,* 279–285, 1882; *13,* 182–183, 1890; and the *Dictionary of National Biography*. The latter has a good article on James Ferguson, which I have used together with F. Henderson, *Life of James Ferguson,* Edinburgh, 1867. The facts on William Herschel are mainly from C. A. Lubbock, *The Herschel Chronicle,* Cambridge, England, 1933, and E. S. Holden, *Sir William Herschel,* London, 1881. For the Kew instruments, see *Irish Astronomical Journal, 9,* 57–68, 1969, and Patrick Moore, *Armagh Observatory 1790–1967,* 17, 1967.

CHAPTER 6

S&T: June, 1963, page 326. The little book of family history on which this chapter is based is *Wilhelm Struve,* by Otto Struve, Karlsruhe, 1895. Copies of this work are rare; mine was a present from its author's grandson and namesake, who died in April, 1963. By a curious coincidence, when I consulted Harvard Observatory's copy of Vol. 4 of the *Astronomische Nachrichten* to read Struve's account of the Dorpat telescope, I found that its title page bears the autograph of the same Wilhelm Wrangell who escorted the telescope from Munich.

CHAPTER 7

S&T: October, 1969, page 224. In addition to Agnes Clerke's articles on Pond and Airy in the *Dictionary of National Biography,* and John Pond's own papers in the *Philosophical Transactions* and the *Memoirs* of the Royal Astronomical Society, the following sources were particularly useful: Pond's obituary in *Monthly Notices* of the Royal Astronomical Society, 4, 31, 1837. T. Lewis, "Notes on some Historical Instruments at the Royal Observatory, Greenwich," *The Observatory, 13,* 200, 1890. G. B. Airy, *Autobiography,* London, 1896, pages 127–129. H. P. Hollis, "The Greenwich Assistants During 250 Years," *The Observatory, 48,* 388, 1925; H. C. King, *The History of the Telescope,* London, 1955. P. S. Laurie, "The Board of Visitors of the Royal Observatory," *Quarterly Journal* of the Royal Astronomical Society, *7,* 169, 334, 1966. Laurie also sent me more information by letter.

CHAPTER 8

S&T: April, 1979, page 342.

CHAPTER 9

S&T: June, 1969, page 358. My principal source was *Herschel at the Cape: Diaries and Correspondence of Sir John Herschel, 1834–1838,* edited by David S. Evans, Terence J. Deeming, Betty Hall Evans, and Stephen Goldfarb, Austin, Texas, 1969.

CHAPTER 10

S&T: June, 1964, page 342. Among the literature consulted were: Simon Newcomb, *Compendium of Spherical Astronomy,* New York, 1906; G. B. Airy, *Autobiography of Sir George Airy,* London, 1896; E. W. Maunder, *The Royal Observatory, Greenwich, A Glance at its History and Work,* London, 1902.

CHAPTER 11

S&T: November, 1974, page 283. August Sonntag's career at Altona Observatory has been reconstructed from his papers in the *Astronomische Nachrichten, 27–35,* and from Axel V. Nielsen's "H. C. Schumacher and the Observatory at Altona During the War of 1848–50," in *Meddelelser* No. 22 of Ole Römer Observatory, Aarhus, 1951. Information about Sonntag's Arctic travels is mostly from George W. Corner's excellent *Doctor Kane of the Arctic Seas,* Philadelphia, 1972, which contains full references to original sources. Another account of the Kane and Hayes expeditions can be found in G. Hartwig's *The Polar and Tropical Worlds,* Springfield, Massachusetts, 1873; some of its details differ from those presented in this chapter. Benjamin Boss, *History of the Dudley Observatory, 1852–1956*, Albany, 1968, has a little about Sonntag on pages 30 and 79, and there are some leads in J. C. Poggendorff's *Biographisch-Literarisches Handwörterbuch* (2nd ed.).

CHAPTER 12

S&T: April, 1955, page 236.

CHAPTER 13

S&T: February, 1973, page 90 and March, 1973, page 152. The main facts of C. H. F. Peters life are from the *Dictionary of American Biography* and obituaries in *Astronomische Nachrichten, 125,* 127, 1890; *The Observatory, 13,* 311, 1890; and *Monthly Notices* of the Royal Astronomical Society, *51,* 199, 1891. Minor details about Peters appeared in *Sky & Telescope, 45,* 271 and 339, 1973. Details concerning the Dudley Observatory controversy can be found in B. Boss, *History of Dudley Observatory,* Albany, 10–27, 1968. See also Trudy E. Bell's excellent article in *Griffith Observer, 38,* 2–9, 1974. For information about intramercurial planets see *Astronomische Nachrichten, 94,* 321, 1879; *95,* 101, 319, 1879. O. J. Eggen also writes about these alleged planets in *Leaflet* No. 287 of the Astronomical Society of the Pacific, 1953, as does R. T. Gould in *Oddities,* New York, 1965. The latter is very readable but has some errors. The great star-catalogue case is described by S. Newcomb, *Reminiscences of an Astronomer,* Boston, 1903; this work is basic but biased against Peters. Lawyers will find the case reported in 9 N.Y.S. 789 (Sup. Ct. Special Term, Oneida Co., 1889); 20 N.Y.S. 189 (Sup. Ct. General Term. 4th Dept., 1892); and 36 N.E. 814 (Ct. App. N.Y. 1894).

CHAPTER 14

S&T: October, 1972, page 236. In addition to reading many of John Tebbutt's papers in the *Monthly Notices* of the Royal Astronomical Society and his annual *Report* of the Windsor Observatory, I have consulted his *History and Description of Mr. Tebbutt's Observatory, Windsor, New South Wales,* Sydney, 1887. Much information came from the obituary notice by H. H. Turner in *Monthly Notices, 78,* 252, 1918. Turner mentions an autobiography of Tebbutt that I have not seen; it was published in *Astronomical Memoirs,* Sydney, 1908. A brief description of Tebbutt's observatory as it appeared in the early 1970's was published in *Sky & Telescope, 45,* 219, 1973.

CHAPTER 15

S&T: June, 1972, page 364. For Lewis Swift's life I have consulted chiefly: *Dictionary of American Biography;* Ralph Bates and Blake McKelvey, "Lewis Swift: The Rochester Astronomer," *Rochester History, 9,* 1, 1947; Ralph Bates, "The Story of Lewis Swift," *The Sky, 4,* 7, 1940; *Astronomische Nachrichten, 194,* 133, 1913; *Monthly Notices* of the Royal Astronomical Society, *73,* 217, 1913. James L. Richards of Arcadia, California, lent me his valuable collection of newspaper and magazine articles about Thaddeus Lowe and his observatory. Most valuable were the

Pasadena *Post* for September 16, 1936; Pasadena *Star News* for June 18, 1956, and December 20, 1959; Donald Duke, "Mount Lowe Railway," *Pacific Railway Journal, 1*, No. 4, 1954; and John Dewar, "Railway to the Clouds," *Quarterly* of the Los Angeles County Museum, *1*, No. 2, 1962.

CHAPTER 16

S&T: January, 1975, page 17. Clyde Fisher's obituary of Garrett P. Serviss appeared in *Popular Astronomy, 37*, 364, August-September, 1929. An extensive chronology and bibliography of Serviss was prepared in 1978 by Paul W. Luther, Bernardston, Mass., as Vol. 1, No. 1 of *Books and Astronomers Monthly*. The American Astronomical Society of Brooklyn has also been described by Trudy E. Bell, *Griffith Observer*, April, 1978.

CHAPTER 17

S&T: April, 1970, page 225 and June, 1970, page 363. In preparing this chapter the most useful sources of information have been: W. Abney, *Monthly Notices* of the Royal Astronomical Society, *54*, 277–283, 1894; E. E. Barnard, *Popular Astronomy, 29*, 309–324, 1921; S. W. Burnham, *Publications* of Yerkes Observatory, *1*, vii-xi, 1900; O. J. Eggen, *Leaflet* No. 295 of the Astronomical Society of the Pacific, 1953; E. B. Frost, *Astrophysical Journal, 54*, 1–8, 1921; F. J. Neubauer, *Popular Astronomy, 58*, 201–222, 318–334, 369–388, 1950.

CHAPTER 18

S&T: April, 1963, page 198. For more information about W. R. Brooks, see *Monthly Notices* of the Royal Astronomical Society, *82*, 246, 1922; *Dictionary of American Biography, 3*, 91, 1929; *Dictionary of Scientific Biography, 2*, 502, 1970.

CHAPTER 19

S&T: October, 1964, page 218. The following sources were used in writing this chapter: *Die Sterne, 6*, 210, 1926; *Popular Astronomy, 6*, 253, 1898; *Astronomische Nachrichten, 145*, 333, 1898, and *146*, 295, 1898; *The Observatory, 21*, 141 and 359, 1898.

CHAPTER 20

S&T: August, 1957, page 477. Federico Rutllant, who was director at Santiago in 1957, provided useful information, but he should not be held responsible for my interpretation. I have drawn freely from the accounts by R. Prager, *Vierteljahrsschrift der Astronomische Gesellschaft, 49*, 14, 1914, and by R. Grandón, *Anuario* for 1952 of the National Astronomical Observatory of the University of Chile.

CHAPTER 21

S&T: December, 1978, page 515. Michael Heim's book is *Spiridion Gopcevic: Leben und Werk*, Wiesbaden, 1966. There is a complete set of Leo Brenner's *Astronomische Rundschau* at Harvard Observatory. This chapter could not have been written without the generous help of Jürgen Blunck, H. von Socher, Richard Baum, and Estelle Karlin.

CHAPTER 22

S&T: October, 1962, page 193. For the dark story of T. J. J. See's difficulties at Lowell Observatory, see William Graves Hoyt, *Lowell and Mars,* Tucson, 1976, especially pages 119–123.

CHAPTER 23

S&T: October, 1966, page 203.

CHAPTER 24

S&T: August, 1959, page 559. This article is based on the very complete account of Huygens' telescopes in Vol. 15 of *Oeuvres Complètes de Christiaan Huygens,* 9–26, 1925, published at The Hague under the auspices of the Netherlands Academy of Sciences.

CHAPTER 25

S&T: September, 1977, page 174. Much of this chapter is based on J. A. Bennett's full account of William Herschel's "large 20-foot" telescope in *Journal for the History of Astronomy,* 7, 75–108, 1976.

CHAPTER 26

S&T: June, 1967, page 346. More information about William Kitchiner can be found in *Dictionary of National Biography* and J. C. Poggendorff, *Handwörterbuch,* Vol. 1, 1266, 1863. In the latter the entry is spelled Kitchener.

CHAPTER 27

S&T: October, 1959, page 664.

CHAPTER 28

S&T: December, 1969, page 380. For more about James Nasmyth see: *Dictionary of National Biography;* Agnes M. Clerke, *A Popular History of Astronomy during the 19th Century* (2nd ed.), Edinburgh, 1887; H. C. King, *The History of the Telescope,* New York, 1979; Colin A. Ronan, *Astronomers Royal,* New York, 1969; Richard Learner, *Astronomy Through the Telescope,* New York, 1981.

CHAPTER 29

S&T: July, 1975, page 26. Leon Foucault's pendulum demonstration is described by J. F. Cox in "A Propos du Centenaire de l' Expérience du Pendule de Foucault," *l'Astronomie,* July-August, 1951. Concerning Foucault's telescope making, see his "La Construction des Télescopes en Verre Argenté," *Annales de l'Observatoire Impérial de Paris,* 5, 197, 1859; also his "Description of an Improvement in the making of large Reflecting Telescopes with Silvered Glass Specula," *Monthly Notices* of the Royal Astronomical Society, *19,* 186, 1859; and "Description of Various Processes made use of for Finding out the Configuration of Optical Surfaces," *Monthly Notices* of the Royal Astronomical Society, *19,* 284, 1859. The last article was reprinted in its entirety in *Sky & Telescope,* 5, 18, 1945.

CHAPTER 30

S&T: February, 1969, page 74. Alvan Clark's autobiography was published, among other places, in *Sidereal Messenger, 8,* 109–117, 1889, but this, like most of its reprints, incorrectly gives his father's name as Abram, rather than Alvan. A correct version with an introductory letter by William A. Richardson is in *Contributions* of the Old Residents' Historical Association, Lowell, Mass., *55,* 164–178, 1891. See also "The Alvan Clark Establishment," *Scientific American,* September 24, 1887, page 198; "Possibilities of the Telescope," by A. G. Clark, *North American Review, 156,* 48–53, 1893; and "What Shooters Owe to the Genius of Alvan Clark," *The American Rifleman,* July, 1976, page 34–36.

CHAPTER 31

S&T: February, 1979, page 141. I am particularly indebted to Deborah Jean Warner's excellent article in *Technology and Culture, 12,* 190, 1971. She gives much information about Lewis Rutherfurd's scientific activities lying outside the scope of this article. Additional material came from Rutherfurd's article "Astronomical Photography," *American Journal of Science, 39,* 304, 1865, and from the two volumes of Rutherfurd results published in 1898 and 1910 by the Observatory of Columbia University.

CHAPTER 32

S&T: December, 1961, page 331.

CHAPTER 33

S&T: January, 1978, page 20. For information about A. A. Common's career, see *Monthly Notices* of the Royal Astronomical Society, *64,* 274, 1904, and the *Dictionary of National Biography*. The history of the Crossley reflector is detailed in *Publications* of Lick Observatory, *8,* 11, 1908. Common's own detailed account of his 5-foot reflector is in *Memoirs* of the Royal Astronomical Society, *50,* 113, 1892. Some odds and ends about it were also gleaned from *The Observatory, 8,* 181, 1885; *11,* 415, 1888; *12,* 56, 1889; *14,* 314, 1891; *15,* 44, 1892; and *18,* 183, 1895. The history of the 60-inch after Common's death is mainly from the *Annual Reports* of Harvard Observatory for the years given and *Harvard Circular* 83, 1904. Ernest Cherrington's remarks appeared in *Sky & Telescope, 55,* 224, 1978. George Field of the Harvard–Smithsonian Center for Astrophysics told me of the sale of the Common mirror to Varo Electronics.

CHAPTER 34

S&T: April, 1962, page 210.

CHAPTER 35

S&T: August, 1958, page 509.

CHAPTER 36

S&T: August, 1961, page 85. The quotation is from page 250 of J. G. F. Bohnenberger's *Anleitung zur Geographischen Ortsbestimmung,* Göttingen, 1851 (2nd ed.).

CHAPTER 37

S&T: October, 1960, page 207.

CHAPTER 38

S&T: April, 1964, page 219. Most of the facts about the 1780 dark day have been taken from Sidney Perley, *Historic Storms of New England,* 105–114; John Stetson Barry, *History of Massachusetts, 3,* 183; D. Hamilton Hurd, editor, *History of Middlesex County, 1,* 300, 352, 753, and *2,* 288; Samuel Adams Drake, *History of Middlesex County, Massachusetts, 2,* 176; H. P. De Forest and E. C. Bates, *The History of Westborough,* 182–184. Thanks are due to Ruth Rusden, Cicely Botley, and the editors of *Yankee* for suggestions. Helen Sawyer Hogg in *Journal* of the Royal Astronomical Society of Canada, *59,* 183–191, 1965, gives information about many other cases of dark days observed since 1706 in the United States and Canada. All these events are explained by forest fires. A dark day on September 18, 1938, was widely observed in northwestern Siberia. It is described in the Russian journal *Priroda* by V. N. Andreyev (No. 2, 62, 1940) and Yu. M. Yemilianov (No. 6, 87, 1968).

CHAPTER 39

S&T: August, 1967, page 92. The quotations from Galileo are from pages 41–45 of Stillman Drake's translation in *Discoveries and Opinions of Galileo,* New York, 1957. For modern results, see Chapter 15 (by A. Danjon) in *The Earth as a Planet,* edited by G. P. Kuiper, Chicago, 1954.

CHAPTER 40

S&T: August, 1971, page 78 and February, 1972, page 95. Sources used to prepare this chapter include: *Circular* of the Lunar Section, British Astronomical Association, April, 1971. *Journal* of the British Astronomical Association, *21,* 355, 1911; 27, 36, 1916; *42,* 219, 1932. *The Observatory, 34,* 162, 203, 305, 344, 374, 1911; *44,* 308, 1921; *103,* 26, 1983. *Monthly Notices* of the Royal Astronomical Society, *70,* 527, 1910. *Astronomische Nachrichten, 71,* 201, 1868. *Kepler's Somnium,* Madison, 1967, pages 136–7 and 144. *Keplers Traum vom Mond,* Leipzig, 1898, pages 21 and 145–152. *Cycle of Celestial Objects,* London, 1844, Vol. 1, pages 123–124. *l'Astronomie, 46,* 57, 1932. *Bulletin de la Société Astronomique de France, 50,* 57, 1936; *62,* 347, 1948. *Astronomical Journal, 72,* 808, 1967. *Sky & Telescope, 43,* 295, 1972.

CHAPTER 41

S&T: December, 1956, page 68. Recent accurate calculations by Steven C. Albers show that mutual occultations of planets are much more frequent than had been supposed. Between the years 1557 and 2230 he finds that 21 occur, but through bad luck we are living in a period of drought. On January 3, 1818, Venus occulted Jupiter, but the event seems to have gone unnoticed. The next mutual occultation is due on November 22, 2065, when Venus again passes in front of Jupiter. Albers gives full details in *Sky & Telescope, 57,* 220, 1979.

CHAPTER 42

S&T: August, 1970, page 86. The primary sources for this article have been E. F. MacPike's *Hevelius, Flamsteed and Halley,* London, 1937, and Colin A. Ronan's *Edmond Halley: Genius in Eclipse,* Garden City, New York, 1969. Some information has been gleaned from Francis Baily's sketch of Halley's life in *Memoirs* of the Royal Astronomical Society, *13,* 35–40, 1843, and the article by Agnes C. Clerke in the *Dictionary of National Biography.* John Flamsteed's letter to Richard Townely is quoted from O. J. Eggen, *Occasional Notes* of the Royal Astronomical Society, *3,* 211–221, 1958. The sources for the site of Halley's observatory are Mrs. Gill, *Six Months in Ascension,* London, 1880, pages 28–33, and G. C. Kitchins' *A Handbook and Gazetteer of the Island of St. Helena,* 1937, pages 31 and 50. The latter work is unpublished, but Arthur Mawson sent me excerpts from the typewritten copy preserved at Plantation House, St. Helena.

CHAPTER 43

S&T: April, 1961, page 213. The principal source for this chapter was C. L. Littrow, *P. Hell's Reise nach Wardoe . . . ,* Vienna, 1835. See also G. Sarton, "Vindication of Father Hell," *Isis, 35,* 97, 1944.

CHAPTER 44

S&T: June, 1966, page 340. Much information about Rev. S. J. Perry can be found in *Dictionary of National Biography.* Details of the expeditions appeared in Perry's "Notes of a Voyage to Kerguelen Island . . . ," *Month and Catholic Review,* reprinted by Henry S. King & Co., London, 1876; see also Ladislaus Weinek's *Die Reise . . . ,* Prague, 1911.

CHAPTER 45

S&T: October, 1979, page 324. Jean Meeus has also calculated circumstances for transits of Mercury between 1920 and 2080, *Journal* of the British Astronomical Association, *67,* 30, 1956.

CHAPTER 46

S&T: June, 1962, page 322. An English translation of Plutarch's book about the Moon was prepared by H. Cherniss. It was published in 1957 by the Harvard University Press as Vol. 12 of Plutarch's *Moralia,* in the Loeb Classical Library. Further information about pretelescopic observations of the Moon can be found in N. T. Bobrovnikoff, *Modern Astrophysics,* Paris, 1967.

CHAPTER 47

S&T: February, 1965, page 92. An extensive summary of lunar cartography from ancient naked-eye studies to 1968 is Antonio Paluzíe Borrell, *Historia de la Cartografia Lunar,* Tarragona, Spain, 1967.

CHAPTER 48

S&T: June, 1957, page 378. I am uncertain about the nationality of Roger Boscovich. The *Dictionary of Scientific Biography* calls him a Croatian by birth, but a well-informed article by Georges Nikolitch *(l'Astronomie, 51,* 91–98, 1937) says that he was born of Slovene parents and was a citizen successively of Ragusa, the Italian republic of Lucca, and France.

CHAPTER 49

S&T: December, 1955, page 62. A facsimile reproduction of the 1878 edition of Lohrmann's atlas was published in 1963 as Wilhelm Gotthelf Lohrmann, *Mondkarte in 25 Sektionen* by the Leipzig firm of Johann Ambrosius Barth. Lohrmann's original drawings had been deposited for safekeeping at Leipzig Observatory but were burned in World War II during a bombing raid on December 3, 1943. The same air raid also destroyed the printing plates and the publisher's stock of the 1878 edition.

CHAPTER 50

S&T: April, 1958, page 290. The title of J. F. J. Schmidt's manuscript in the Harvard Observatory library is *Beobachtungen der Mondfinsternisse 1842–1879.* The Charles Pritchard quotation is from *The Observatory, 6,* 171, 1883. I am obliged to P. van de Kamp for the translation of A. Nijland, taken from the obituary of the latter in *Hemel en Dampkring* for September, 1936. In 1958, a reduced-scale reproduction of Schmidt's map was being offered by Scientific Book Service, Columbus, Ohio. Many of Schmidt's records, such as the major part of his variable star journals, were deposited at Potsdam Observatory; this material may include the unpublished lunar observations, as has been suggested by D. Eginitis in his history of Athens Observatory.

CHAPTER 51

S&T: February, 1961, page 92.

CHAPTER 52

S&T: October, 1965, page 202. Descriptions of Hörbiger's *Cosmic Ice Doctrine* and its role in Hitler's Germany are given in Dusty Sklar's *Gods and Beasts: The Nazis and the Occult,* New York, 1977, 73–80; and Martin Gardner's *Fads and Fallacies in the Name of Science,* New York, 1957, 37f., 170f. The definitive English presentation of Hörbiger's doctrines is *Moons, Myth, and Man* by Hans Schindler Bellamy, New York, 1936.

CHAPTER 53

S&T: August, 1960, page 87. A full bibliography of the lunar crater Linné would contain several hundred titles; only a few of the more significant ones can be listed here. W. G. Lohrmann, *Topographie der sichtbaren Mondoberflaeche,* Dresden, 1824. W. Beer and J. H. Mädler, *Der Mond,* Berlin, 1837. J. F. J. Schmidt, *Charte der Gebirge des Mondes, Erläuterungsband,* Berlin, 1878. W. Prinz, "Have There Been Changes in the Lunar Craters Messier and Linné?" *Ciel et Terre, 14,* 32, 49, 121, 1894. P. Fauth, "The Lunar Objects Alpetragius d and Linné," *Astronomische Rundschau, 3,* 172, 1901. J. Ashbrook, "An Analysis of Schmidt's Observations of Linné," *Strolling Astronomer, 17,* 85, 1963. R. J. Pike, "The Lunar Crater Linné," *Sky & Telescope, 46,* 364, 1973.

CHAPTER 54

S&T: August, 1954, page 333. Paul Stroobant's memoir is "Etude sur le satellite énigmatique de Vénus," forming part of Vol. 49 of the *Mémoires couronnés* of the Académie Royale des Sciences de Belgique, Brussels, 1887.

CHAPTER 55

S&T: October, 1957, page 588. Joseph Meurers' book, *Astronomische Experimente,* is in German; it was published by Akademie-Verlag, Berlin in 1956. Walter Villiger's own detailed report of his Venus experiments is in *Neue Annalen* of Munich Observatory, *3,* 1898. His paper includes a valuable but rather overlooked critique of all the principal Venus observations made up to that time. W. Spangenberg's paper was in *Astronomische Nachrichten, 281,* 6, 1952. Dollfus' experiments with different magnifications are described in *Sky & Telescope, 15,* 397, 1956.

CHAPTER 56

S&T: February, 1955, page 140.

CHAPTER 57

S&T: July, 1977, page 20. In writing this chapter, I received much information and encouragement from Margaret Hall, granddaughter of Asaph Hall. Very helpful were Owen Gingerich's articles on Hall in the *Dictionary of Scientific Biography* and in the *Journal for the History of Astronomy, 1,* 109, 1970, and G. W. Hill's "Biographical Memoir of Asaph Hall," *Biographical Memoirs* of the National Academy of Sciences, Vols. 5–6, 1909. For more information see *Sky & Telescope, 15,* 494 and 495, 1956.

CHAPTER 58

S&T: December, 1957, page 74.

CHAPTER 59

S&T: December, 1970, page 361 and August, 1978, page 99. The information about Johann Palisa, Pola Observatory, and the early observations of Scylla is mainly from *Vierteljahrsschrift der Astronomische Gesellschaft, 16,* 227, 1881; *Astronomische Nachrichten, 86,* 381, 1875; *87,* 355, 1876; *88,* 41, 1876. For more information about Joel Metcalf and his innovative observing techniques see *Astrophysical Journal, 23,*

306, 1906; *Astronomische Nachrichten, 178,* 254, 1908; *Popular Astronomy, 33,* 496, 1925. A. Kopff's and Palisa's observations in 1907 are given in *Astronomische Nachrichten, 176,* 194, 236, 263, 1907. The later history of Scylla is based on *Minor Planet Circulars* 3085 and 3086, and on material kindly furnished by Paul Herget and C. M. Bardwell of Cincinnati Observatory, University of Cincinnati.

References to the discovery, observations, and orbit studies of Albert are given in *Publications* of the Lick Observatory, *19,* 418, 1935. There are useful summaries in *The Observatory, 34,* 398, 1911; *35,* 265, 1912; *Publications* of the Astronomical Society of the Pacific, *35,* 133, 1913; and *Monthly Notices* of the Royal Astronomical Society, *72,* 296, 1912. Again, Herget gave me much information by letter.

CHAPTER 60

S&T: September, 1976, page 164. C. D. Perrine's discovery announcements of Jupiter's sixth and seventh satellites appeared, respectively, in *Publications* of the Astronomical Society of the Pacific *17,* 22, 1905, and *17,* 62, 1905. His life and work are summarized in *Dictionary of Scientific Biography.* An unrecognized predis-covery observation of Jupiter **VI** is described in Harvard Observatory *Annals, 60,* No. 2, 1908.

CHAPTER 61

S&T: May, 1978, page 380 and June, 1978, page 474.

CHAPTER 62

S&T: October, 1955, page 501, April, 1959, page 312, and April, 1969, page 229. Among the sources consulted are the following: *Memoirs* of the British Astronomical Association, *39,* No. 3, 1961; *40,* No. 2, 1966. *l'Annuaire du Bureau des Longitudes* for 1950 and 1962. *Sky & Telescope, 14,* 501, 1955; *18,* 312, 1959. *Correspondance Astronomique, 4,* 456, 1820. *Astronomische Nachrichten, 170,* 99, 131, 147, 1905–1906; 189, 87, 1911; *198,* 449, 1914. *Berliner Astronomisches Jahrbuch* for 1805, page 128; 1806, page 266. *Nature, 14,* 311, 1876.

CHAPTER 63

S&T: April, 1971, page 223. For the early history of meteoritic astronomy I used Fritz Heide, *Meteorites,* Chicago, 1964, and A. von Humboldt's *Cosmos,* New York, 1850. See also *Sky & Telescope, 12,* 210, 1953; *15,* 292, 1956. Thomas Jefferson's opinions are from the *Proceedings* of the American Philosophical Society, *87,* 234, 1943. Nathaniel Bowditch's work has been taken secondhand from Bernhard von Lindenau's *Zeitschrift für Astronomie, 1,* 137, 1816. Another account of the Weston meteorite, and also of the great display of Leonid meteors seen in 1833, appeared in *The People's Magazine, 1,* 172, 1834. For other information on the early belief in meteoritic satellites of the Earth, see *Sky & Telescope, 14,* 334, 1955.

CHAPTER 64

S&T: June, 1960, page 463.

CHAPTER 65

S&T: December, 1960, page 324. Menzel's interpretation of Carrington's flare appeared in *Publications* of the Astronomical Society of the Pacific, *73,* 194, 1961.

CHAPTER 66

S&T: August, 1968, page 87. The observing technique employed by Jules Janssen should not be confused with the familiar "open-slit" method first used in 1869 by Sir William Huggins. The latter employs a spectroscope of considerable dispersion with its slit tangent to the solar limb. By cautiously widening the slit, a monochromatic image of the prominence can be seen in hydrogen-alpha light.

CHAPTER 67

S&T: June, 1961, page 320.

CHAPTER 68

S&T: May, 1974, page 296. Bernhard von Lindenau's article was in *Zeitschrift für Astronomie, 2,* 199, 1816. For the life of Stephen Groombridge, I used mainly George B. Airy's introduction to his 1838 edition of Groombridge's *A Catalogue of Circumpolar Stars;* the obituary by R. Sheepshanks in *Monthly Notices* of the Royal Astronomical Society, *2,* 145, 1833; and Agnes Clerke's article in the *Dictionary of National Biography.* F. W. Argelander's articles on the proper motion of Groombridge 1830 appeared in *Astronomische Nachrichten, 19,* 393, 1842; *20,* 163, 312, 1843. Peter van de Kamp and Wulff Heintz furnished information by letter about this star's motion and duplicity.

CHAPTER 69

S&T: July, 1973, page 27. Information about William Dawes' life and work is found in *Monthly Notices* of the Royal Astronomical Society; *29,* 116, 1869; *The Observatory, 36,* 419, 1913; and the *Dictionary of National Biography,* which also summarizes his planetary and solar work. Dawes' letters to Knott were printed in *The Observatory, 33,* 343, 383, 419, 473, 1910. For his double star observations, see *Memoirs* of the Royal Astronomical Society, *5, 8, 19, 35.*

CHAPTER 70

S&T: April, 1957, page 265. A. V. Nielsen's 1942 paper in *Meddelelser* No. 16 of the Ole Römer Observatory gives full details about the supposed binary nature of Mizar B and J. H. Mädler's observation. The facts on Sidus Ludovicianum are from the 1894 edition of Wilhelm Olbers' collected works, Vol. 1, pages 523, 649, and 654.

CHAPTER 71

S&T: April, 1960, page 342. Data for the (updated) table of very red stars were taken from Alan Hirshfeld and Roger W. Sinnott's *Sky Catalogue 2000.0,* Cambridge, Mass., 1982.

CHAPTER 72

S&T: October, 1963, page 204. Schmidt's exquisite pencil drawings of the Milky Way were found in the Potsdam archives and published in 1923 as Vol. 14, Part 2, of the *Annalen* of Leiden Observatory. Pannekoek's career is described by Bart J. Bok in *Sky & Telescope, 20,* 74, 1960.

CHAPTER 73

S&T: November, 1975, page 299. John Herschel's account of the Orion nebula was published in *Memoirs* of the Royal Astronomical Society, *2,* 459, 1826. George Bond's "Observations upon the Great Nebula of Orion" appeared in Harvard Observatory *Annals, 5,* 1867.

CHAPTER 74

S&T: December, 1972, page 366. Most of the material for this chapter is from T. R. Treadwell, "Notes on a Forgotten Episode," *Popular Astronomy, 51,* 497, 1943, and Mason's "Observations on Nebulae with a Fourteen Feet Reflector, made by H. L. Smith and E. P. Mason, during the year 1839." *Transactions* of the American Philosophical Society, 7, 165, 1841. The eventual fate of the 12-inch telescope is not known to me. After Smith graduated from Yale in 1839, he took the instrument to Ohio City, Ohio, where he remounted it. He later taught at Hobart College.

CHAPTER 75

S&T: July, 1974, page 9. The letters between Caroline and John Herschel have been taken from *Herschel at the Cape,* edited by David S. Evans and others, Austin, Texas, 1969, pages 71–73 and 143, and from *The Herschel Chronicle,* edited by Constance A. Lubbock, Cambridge, England, 1933, pages 372-373. William Herschel's "On the Construction of the Heavens" can be found in *The Scientific Papers of Sir William Herschel,* London, 1912, Vol. 1, 223–259. See page 253 for *An Opening in the Heavens,* and Vol. 2, pages 712–713, for his observations in 1785–1787. John Herschel's list of starless fields is in his *Results of Astronomical Observations made . . . at the Cape of Good Hope,* London, 1847, pages 381–382. Interesting correspondence about the "hole in the heavens" is found in *Sidereal Messenger, 1,* 98, 1882, and *2,* 125, 1883; also *Journal* of the British Astronomical Association, *54,* 19, 74, 105, 1944.

CHAPTER 76

S&T: June, 1968, page 366. In preparing this chapter the following references were consulted: M. A. Ellison, "The Third Earl of Rosse and his Great Telescopes," *Journal* of the British Astronomical Association, *52,* 267, 1942. *Philosophical Transactions,* 1833, 496; 1850, 505; 1861, 681. *Astronomical Journal, 2,* 101, 1852. *Monthly Notices* of the Royal Astronomical Society, *49,* 65, 1888; *49,* 121, 1889; *50,* 310, 1890.

CHAPTER 77

S&T: February, 1968, page 97. A very important source of information concerning the nucleus of M31 is H. M. Johnson, *Astrophysical Journal, 133,* 309, 1960. See also T. D. Kinman, *Astrophysical Journal, 142,* 1376, 1965.

CHAPTER 78

S&T: May, 1973, page 292. Facts about Johann Bayer's life and *Uranometria* have been gleaned from the *Dictionary of Scientific Biography,* edited by C. C. Gillispie, New York, 1970, Vol. 1, 530–531; Basil Brown, *Astronomical Atlases, Maps & Charts,* London, 1932, 19–25; and E. Zinner in *Veröffentlichungen der Remeis-Sternwarte,* Vol. 2, 28–30, 1926. Tycho's star catalogue, edited by Francis Baily, appeared in *Memoirs* of the Royal Astronomical Society, *13,* 1843. Much historical information about star nomenclature is found in Francis Baily, *The Catalogue of Stars of the British Association for the Advancement of Science,* London, 1845, 63–72; B. A. Gould, *Resultados del Observatorio Nacional Argentino en Cordoba,* Vol. 1, 48–67, 1879; and L. Boss, *Preliminary General Catalogue of 6188 Stars,* Washington, 1910, xvii-xix.

CHAPTER 79

S&T: December, 1968, page 370 and April, 1972, page 223. For a biography of Hevelius, as well as a description of his equipment and observations, see A. J. Szanser, "Johannes Hevelius (1611–1687) – Astronomer of Polish Kings," *Quarterly Journal* of the Royal Astronomical Society, *16,* 488, 1976. See also *Johannes Hevelius and His Catalogue of Stars,* Brigham Young University, 1971; and *Jan Hevelius, The Star Atlas* (V. P. Sheglov, ed.), Tashkent, 1981 (4th ed.).

CHAPTER 80

S&T: April, 1980, page 300. F. W. A. Argelander's *Bonner Durchmusterung* was published as *Atlas des nördlichen gestirnten Himmels für den Anfang des Jahres 1855,* Bonn, 1863. His obituary by colleague E. Schönfeld appeared in *Vierteljahrsschrift der Astronomische Gesellschaft, 10,* 150, 1875. See also *Sky & Telescope, 29,* 276, 1965.

CHAPTER 81

S&T: June, 1958, page 394.

CHAPTER 82

S&T: December, 1962, page 321.

CHAPTER 83

S&T: May, 1977, page 356. This chapter is heavily indebted to Bruno Weber's very thorough account in *Gutenberg Jahrbuch,* 1973, 381–407, from which Flammarion's woodcut has been copied. E. Zinner's article was in *Börsenblatt für den Deutschen Buchhandel* (Frankfurt edition), March 18, 1957. Both were made available by Owen Gingerich, who also told me about Arthur Beer's work.

Index

Index

Index